Nanofluid Dynamics and Transport Phenomenon

The text offers a detailed presentation of mathematical, numerical, and experimental techniques for nanofluids. It further covers the synthesis, characterization, stability, and heat transport. The book comprehensively discusses topics such as the comparison of heat transfer models, flow features of ternary hybrid nanofluids, thermodynamics and mass diffusion, and natural convection in triangular cavities.

This book:

- Emphasizes the enhancement of heat transfer processes through nanoparticles, extending beyond heat transfer to applications in renewable energy.
- Explores the applications of nanofluids in enhancing food processing and agricultural practices.
- Covers thermal instability of couple-stress on viscous-elastic nanofluid flow and natural convection in a triangular cavity.
- Explains concepts including nanofluid-based energy storage, mass diffusion, thermodynamics, and nanofluid synthetic techniques.
- Presents topics such as numerical methods, fluid dynamics simulation, magnetohydrodynamics, heat and mass transfer, and radiation.

It is primarily written for senior undergraduates, graduate students, and academic researchers in the fields of mechanical engineering, aerospace engineering, automotive engineering, industrial and production engineering, energy engineering, fluid dynamics, and tribology.

Engineering Tribology, Manufacturing and Applied Energy

Series Editors

Yashvir Singh and Nishant Kumar Singh

About the Series:

Tribology is the study of parts when they are in contact with each other and have a considerable impact during their sliding or rotating motion. The series will publish the books highlighting the latest developments and research in tribology with its application in the manufacturing and energy sector. The series will include books covering established areas of tribology and emerging fields such as tribo-chemistry, surface engineering, nano-tribology, and bio-tribology. The series will focus on the recent developments and advancements in green tribology for example environmentally friendly composites. Aimed at senior undergraduate, graduate students, academic researchers and professionals, the proposed series will cover the latest techniques in the field of tribology with applications in diverse fields such as manufacturing, energy, and medical.

Modeling and Simulation of Fluid Flow and Heat Transfer
Edited by Reshu Gupta, Mukesh Kumar Awasthi

Nanofluid Dynamics and Transport Phenomenon
Edited by Reshu Gupta, Mukesh Kumar Awasthi, Dhananjay Yadav, and Yashvir Singh

For more information on series page, please visit https://www.routledge.com/Engineering-Tribology-Manufacturing-and-Applied-Energy/book-series/CRCETMAE?srsltid=AfmBOoqxt3CH9TkUJO7QNmpmV_eemj3e18Ac8XYolO9jGbSneK8YysHq

Nanofluid Dynamics and Transport Phenomenon

Edited by
Reshu Gupta, Mukesh Kumar Awasthi,
Dhananjay Yadav, and Yashvir Singh

CRC Press
Taylor & Francis Group
Boca Raton London New York

CRC Press is an imprint of the
Taylor & Francis Group, an **informa** business

Designed cover image: shutterstock

First edition published 2025
by CRC Press
2385 NW Executive Center Drive, Suite 320, Boca Raton FL 33431

and by CRC Press
4 Park Square, Milton Park, Abingdon, Oxon, OX14 4RN

CRC Press is an imprint of Taylor & Francis Group, LLC

ISBN: 9781032863757 (hbk)
ISBN: 9781032863733 (pbk)
ISBN: 9781003527251 (ebk)

DOI: 10.1201/9781003527251

Typeset in Sabon
by KnowledgeWorks Global Ltd.

Contents

Aim and Scope

The book "**Nanofluid Dynamics and Transport Phenomenon**" offers a comprehensive exploration of the transformative impact of nanofluid dynamics on various scientific and technological domains. One of the central themes of this book is the profound implication of nanofluid dynamics in enhancing heat transfer processes. Heat transfer is a fundamental aspect of industrial and technological applications, encompassing areas as diverse as electronics cooling and energy production. The integration of nanoparticles into fluids represents a groundbreaking approach to increase thermal conductivity significantly, thereby making heat dissipation more efficient. This property holds the potential to revolutionize the design of cooling systems and heat exchangers, ultimately leading to the development of more energy-efficient devices. The book thoroughly examines the principles, methodologies, and applications of nanofluid dynamics in the context of heat transfer enhancement.

The book begins by offering readers a solid foundation in the fundamental concepts of nanofluids, including their composition, properties, and synthesis methods. It then moves on to discuss the diverse applications of nanofluids in thermal solutions, ranging from electronics cooling to renewable energy systems. The authors highlight the superior thermal conductivity and heat transfer properties of nanofluids, showcasing their potential to enhance the efficiency of heat exchange processes in various engineering applications.

In addition, nanofluid dynamics holds significant promise in the realm of renewable energy. The book explores how advanced nanofluid-based solar collectors and photovoltaic cells have the potential to enhance the efficiency of solar energy harvesting. By improving energy absorption and utilization, these technologies can play a pivotal role in sustainable energy production. Similarly, the book addresses the development of nanofluid-based battery and super capacitor technologies in the field of energy storage. These innovations aim to enhance energy density and charging rates, addressing the pressing challenges in energy storage and distribution.

The book is organized into thirteen chapters, each delving into distinct aspects of nanofluid dynamics and transport phenomena. Chapter 1 provides an overview of the fundamental concepts, history, and characteristics of nanofluid. Chapter 2 refers to the methods and procedures used to characterize and validate the properties of nanofluid. Chapter 3 explores the application of nanofluids in various thermal management systems including cooling of electronics, heat exchangers, and solar collectors. It discusses the mechanisms underlying the enhanced thermal conductivity of nanofluids, the methods of nanofluid preparation, and their key characteristics. Subsequent chapters delve into topics such as carbon nanotube (CNT) application in solar systems, enhancement of tri-hybrid nanofluid heat transfer over a spinning disk, natural convection in a triangular cavity, and application of nanofluid in food technology and agriculture, among others.

Furthermore, the book underscores the interdisciplinary significance of nanofluid dynamics. The insights gained from studying the behavior of nanoparticles in fluids have reverberated throughout other scientific disciplines, including colloidal science, interfacial phenomena, and even quantum mechanics at the nanoscale. These connections reveal the multifaceted nature of nanofluid dynamics and its pivotal role in shaping modern science and engineering. The book presents a cohesive narrative that bridges the gap between fundamental principles and diverse applications, offering readers a comprehensive understanding of the far-reaching implications of nanofluid dynamics in a wide range of fields.

Editors

Dr. Reshu Gupta
Dr. Mukesh Kumar Awasthi
Dr. Dhananjay Yadav
Dr. Yashvir Singh

Preface

The book "**Nanofluid Dynamics and Transport Phenomenon**" provides a comprehensive exploration of the transformative impact of nanofluid dynamics on various scientific and technological domains. A central theme of the book is the significant role of nanofluid dynamics in enhancing heat transfer processes. Heat transfer is fundamental to numerous industrial and technological applications, spanning areas such as electronics cooling and energy production. The integration of nanoparticles into fluids represents a groundbreaking approach to significantly increase thermal conductivity, thereby improving heat dissipation efficiency. This property has the potential to revolutionize the design of cooling systems and heat exchangers, ultimately leading to the development of more energy-efficient devices. The book thoroughly examines the principles, methodologies, and applications of nanofluid dynamics in the context of heat transfer enhancement.

Moreover, the book delves into the expanding realm of nanofluid dynamics, demonstrating its applications in the fields of food technology and agriculture. Nanofluid can enhance the efficiency of thermal processes like pasteurization and sterilization in food processing. To improve the protection qualities of food packaging materials against microbiological contaminants, moisture, and oxygen pollutants, nanofluids can be incorporated. This is helpful in increasing the packaged food products' quality and shelf life. The rheological of food products can be changed by adding nanoparticles dispersed in the liquid, which enhances the texture, stability, and sensory aspects of the product. The viscosity of the food formulations, for instance, can be modified by using nano-cellulose or nano-emulsions. To enhance food safety and quality control, nanofluid-based sensors and biosensors can be created for the quick identification of pollutants, spilling indicators, and food-borne pathogens. Nanofluid plays a vital role in the agricultural field. Nanofluid-based sensors can be utilized for real-time monitoring of soil conditions, nutrient levels, and plant health parameters.

The book is organized into thirteen chapters. **Chapter 1** provides an introductory overview of nanofluid dynamics, covering basic concepts and principles essential for understanding the behavior of nanofluids. The experimental techniques used for testing and validating nanofluid properties are discussed in **Chapter 2**, providing insights into practical methodologies and challenges. **Chapter 3** explores the application of nanofluids in enhancing heat transfer efficiency and optimizing thermal performance. A computational approach for optimizing Solar Air Heater (SAH) performance through the integration of nanofluids is presented in **Chapter 4**, offering insights into computational modeling techniques. The potential of nanomaterials in solar energy applications is explored in **Chapter 5**. **Chapter 6** presents optimization algorithms for enhancing the application of carbon nanotubes in solar energy systems, focusing on computational optimization techniques. **Chapter 7** investigates the use of carbon nanotubes and graphene in improving heat transfer over spinning disks.

The comprehensive overview of nanofluid factors and their impact on the effective deployment of robotic technology in healthcare are explored in **Chapter 8**. **Chapter 9** analyzes the impact of Hall current and chemical reactions on flow behavior. **Chapter 10** examines natural convection in a triangular cavity filled with MoS2-water nanofluid and microorganisms, exploring heat transfer phenomena. **Chapter 11** investigates the convective transportation of Casson nanofluid under the influence of activation energy and inclined magnetization, considering the impact of thermal radiation. The numerical computation techniques in non-Darcy porous media with Forchheimer correction was applied in **Chapter 12**. **Chapter 13** explores the applications of nanofluids in enhancing food processing and agricultural practices.

The book presents a cohesive narrative that bridges the gap between fundamental principles and diverse applications, offering readers a comprehensive understanding of the far-reaching implications of nanofluid dynamics in a wide range of fields. This book primarily targets undergraduate and graduate students pursuing degrees in various engineering disciplines. Additionally, it caters to researchers actively engaged in the study of nanofluid dynamics and heat transfer, as well as professionals seeking a comprehensive reference guide within this domain.

In conclusion, we hope that this comprehensive and meticulously designed book will contribute significantly to the understanding and advancement of nanofluid flow and transport studies. We extend our gratitude to all the contributors and reviewers who have been instrumental in shaping this endeavor. We are optimistic that this book will inspire readers to unravel the complexities of nanofluid dynamics, leading to innovative solutions and transformative breakthroughs in diverse fields of engineering and scientific research.

Dr. Reshu Gupta
Dr. Mukesh Kumar Awasthi
Dr. Dhananjay Yadav
Dr. Yashvir Singh

About Editors

Reshu Gupta has done her Ph.D. on the topic "Study of Some Problems of Heat Transfer in the Flows of Some Fluids" from Chaudhary Charan Singh University, Meerut. She is an Assistant Professor in the Applied Science Cluster (Mathematics) at the University of Petroleum and Energy Studies, Dehradun, India. She has 20 years' experience of teaching. She is a dedicated, resourceful, and goal-driven professional educator with a solid commitment to the social and academic growth and development of every student. She is a caring faculty member who wants all students to be successful learners and works to create a classroom atmosphere that is stimulating, encouraging, and adaptive to the varied needs of students. Her research area includes fluid dynamics, differential equations, heat, and mass transfer, nanofluids, entropy, and artificial neural networks. She published several papers in SCI/Scopus journals and conference proceedings. Dr. Gupta introduces concepts into the curriculum related to life and social skills. Her teaching philosophy is grounded in the understanding that it is her responsibility to create an environment that supports and inspires student learning. She believes that the most significant learning occurs in situations that are both meaningful and realistic. Her general teaching focus at university is to promote learning by engaging students in their learning environment, promoting higher-order thinking skills, and respecting their opinions and backgrounds. She has taught Linear Algebra, Numerical Methods, Differential Equations, Calculus, Abstract Algebra, Ring Theory, Discrete Mathematics, Engineering Mathematics, Fourier Series, and Laplace Transform.

Mukesh Kumar Awasthi has done his Ph.D. on the topic "Viscous Correction for the Potential Flow Analysis of Capillary and Kelvin-Helmholtz instability." He is working as an Assistant Professor in the Department of Mathematics at Babasaheb Bhimrao Ambedkar University, Lucknow. Dr. Awasthi is specialized in the mathematical modeling of flow problems. He has taught courses of Fluid Mechanics, Discrete Mathematics, Partial Differential Equations, Abstract Algebra, Mathematical Methods, and Measure Theory to postgraduate students. He has acquired excellent knowledge in the mathematical modeling of flow problems and he can solve these problems analytically as well as numerically. He has a good grasp of the subjects like viscous potential flow, electro-hydrodynamics, magneto-hydrodynamics, heat, and mass transfer. He has excellent communication skills and leadership qualities. He is self-motivated and responds to suggestions in a more convincing manner. Dr. Awasthi has qualified National Eligibility Test (NET) conducted on all India level in the year 2008 by the Council of Scientific and Industrial Research (CSIR) and got Junior Research Fellowship (JRF) and Senior Research Fellowship (SRF) for doing research. He has published 115 plus research publications (journal articles/books/book chapters/conference articles) in Elsevier, Taylor & Francis, Springer, Emerald, World Scientific, and many other national and international journals and conferences. Also, he has published eight books. He has attended many symposia, workshops, and conferences in mathematics as well as fluid mechanics. He has got the "Research Awards" consecutively four times from 2013 to 2016 by the University of Petroleum and Energy Studies, Dehradun, India. He has also received the start-up research fund for his project "Nonlinear study of the interface in multilayer fluid system" from UGC, New Delhi. He is also listed in the top 2% influential researchers in the world prepared by Stanford University based on Scopus data in 2022. His Orcid is 0000-0002-6706-5226, Google Scholar web link is https://scholar.google.co.in/citations?user=Dj3ktGAAAAAJ, and research gate web link ishttps://www.researchgate.net/profile/Mukesh-Awasthi-2.

Dhananjay Yadav received his Ph.D. degree from Department of Mathematics, Indian Institute of Technology (IIT) Roorkee, India, in 2013 and postgraduation (M.Sc.) in Mathematics from Deen Dayal Upadhyaya University Gorakhpur, India in 2007. Currently, he is working as an Associate Professor in Department of Mathematics at University of Nizwa, Oman. Prior to his appointment to University of Nizwa, Oman, he had worked as Principal Research Scientist at Athabasca University, Canada; Yonsei University, South Korea; and Jeju National University, South Korea. He is a leading expert in CO_2 capture, storage and oil recovery, computational sustainability and environmental analytics, fluid mechanics, numerical analysis, hydrodynamic and hydromagnetic stability, nanofluids, and fluid flow in porous media. He has published more than 100 research articles (high impact factor) in various reputed international journals. He is also listed in the top 2% influential researchers in the world prepared by Stanford University based on Scopus data.

Yashvir Singh is presently working as an Associate Professor at Mechanical Engineering Department of Harcourt Butler Technical University, Kanpur, Uttar Pradesh, India. He has completed his Ph.D. in the area of tribology from the University of Petroleum and Energy Studies, Dehradun, Uttarakhand, India, and his Master of Technology from Dr. APJ Abdul Kalam Technical University, Lucknow, Uttar Pradesh, India. He has a rich experience of more than fourteen years as an academician and researcher in various Universities and Colleges. He has published more than 100 research papers in SCI/Scopus/Web of Science Journals with high impact factors. He has attended 37 Conferences in India and four International Conferences. He has written three books with International Publishers – CRC Press and Nova Publishers. He is a reviewer and editorial board member of various reputed International Journals in Elsevier, Springer, Bentham Science, and Taylor and Francis. He is an active member in organizing many international seminars, workshops, and conferences in India, Taiwan, and Malaysia. Dr. Singh has guided various students and presently working in joint collaboration with various countries' professors and scientists. He is also listed in the top 2% influential researchers in the world prepared by Stanford University based on Scopus data in 2022.

Contributors

Ali Ahadi
Department of Mechanical Engineering, Babol Noshirvani University of Technology, Babol, Iran
Renewable Energy Systems and Nanofluid Applications in Heat Transfer Laboratory, Babol Noshirvani University of Technology, Babol, Iran

Sahin Ahmed
Department of Mathematics, Rajiv Gandhi University, Papum Pare, Arunachal Pradesh, India

Mohd Anas
Mechanical Engineering Department, Integral University, Lucknow, Uttar Pradesh, India

Mukesh Kumar Awasthi
Department of Mathematics, Babasaheb Bhimrao Ambedkar University, Lucknow, Uttar Pradesh, India

Kanchan Bhardwaj
Botany Department, Shri Guru Ram Rai University, Dehradun, Uttarakhand, India

Rashi Bhargava
Mathematics Department, Shri Guru Ram Rai University, Dehradun, Uttarakhand, India

Bikash Das
Department of Mathematics, Rajiv Gandhi University, Papum Pare, Arunachal Pradesh, India

Paramananda Deka
Department of Mathematics, Dibrugarh University, Dibrugarh, Assam, India

Devender
Department of Mathematics, National Institute of Technology Kurukshetra, Haryana, India

Nitesh Dutt
Department of Mechanical Engineering, COER University Roorkee, Uttarakhand, India

Arun Kumar Singh Gangwar
Mechanical Engineering Department, Integral University, Lucknow, Uttar Pradesh, India

Velu Kaliyannan Gobinath
Department of Mechatronics Engineering, Kongu Engineering College, Erode, Tamil Nadu, India

Mayzul Alom Hussain
Department of Mathematics, Rajiv Gandhi University, Papum Pare, Arunachal Pradesh, India

Shaik Javeed Hussain
Global College of Engineering and Technology, Muscat, Oman

Hossein Khademi
Renewable Energy Systems and Nanofluid Applications in Heat Transfer Laboratory, Babol Noshirvani University of Technology, Babol, Iran

Ashwani Kumar
Department of Mechanical Engineering, Technical Education Department, Kanpur, Uttar Pradesh, India

Vivek Kumar
Department of Mathematics, National Institute of Technology Kurukshetra, Haryana, India

Seyed-Mostafa Moussavi
Department of Mechanical Engineering, Babol Noshirvani University of Technology, Babol, Iran
Renewable Energy Systems and Nanofluid Applications in Heat Transfer Laboratory, Babol Noshirvani University of Technology, Babol, Iran

Anil Nangkar
Department of Mathematics, Rajiv Gandhi University, Papum Pare, Arunachal Pradesh, India

Nagarajan Nithyavathy
Department of Mechanical Engineering, Vellalar College of Engineering and Technology, Erode, Tamil Nadu, India

Rathinasamy Parameshwaran
Department of Mechatronics Engineering, Kongu Engineering College, Erode, Tamil Nadu, India

Bhagyashri Patgiri
Department of Mathematics, Cotton University, Guwahati, Assam, India

Ashish Paul
Department of Mathematics, Cotton University, Guwahati, Assam, India

Gunasekaran Raja
Department of Mechatronics Engineering, Kongu Engineering College, Erode, Tamil Nadu, India

Paras Ram
Department of Mathematics, National Institute of Technology, Kurukshetra, Haryana, India

Vishal Saikia
Department of Mathematics, Dibrugarh University, Dibrugarh, Assam, India

Neelav Sarma
Department of Mathematics, Cotton University, Guwahati, Assam, India

Kushal Sharma
Department of Mathematics, Malaviya National Institute of Technology, Jaipur, Rajasthan, India

Mohsen Sheikholeslami
Department of Mechanical Engineering, Babol Noshirvani University of Technology, Babol, Iran

Renewable Energy Systems and Nanofluid Applications in Heat Transfer Laboratory, Babol Noshirvani University of Technology, Babol, Iran

Kandasamy Suganeswaran
Department of Mechatronics Engineering, Kongu Engineering College, Erode, Tamil Nadu, India

Sivasakthivel Thangavel
Global College of Engineering and Technology, Muscat, Oman

Dhananjay Yadav
Department of Mathematical and Physical Sciences, University of Nizwa, Nizwa, Oman

Acknowledgment

We are grateful to the hardworking editorial team at CRC Press (Taylor & Francis) for their invaluable assistance in making this book, "Nanofluid Dynamics and Transport Phenomenon," a reality. Their priceless advice and steadfast assistance have been very helpful. Their insightful guidance and unwavering commitment to excellence significantly enhanced the overall quality of our work.

We also extend our heartfelt gratitude to the many reviewers and contributors whose insightful viewpoints enhanced each chapter of this extensive resource. There is no denying that their combined experience added to the content's depth and scope. The book starts off by giving readers a firm understanding of the basic ideas behind nanofluids by delving into their composition, characteristics, and synthesis processes.

The story then deftly switches to an examination of the various uses of nanofluids in thermal solutions, from food safety to renewable energy systems. The collaborative efforts of contributors and reviewers have resulted in comprehensive coverage that guarantees readers will acquire a comprehensive understanding of the subject matter.

This book is devoted to the enthusiastic and committed people whose work propels the field of nanofluid dynamics forward. As we thank them for their invaluable contributions, we hope that this work serves as a useful resource for scholars and an inspiration to others. We acknowledge and value the cooperative spirit that binds us all together in the dynamic intersection of engineering and technology as we strive for excellence and advancement.

Chapter 1

Nanofluid dynamics properties and applications

Mukesh Kumar Awasthi, Ashwani Kumar, Nitesh Dutt, and Dhananjay Yadav

1.1 INTRODUCTION AND DEFINITION OF NANOFLUIDS

Nanofluids are a class of engineered fluids consisting of a base fluid, typically a liquid such as water, oil, or ethylene glycol, and dispersed nanoparticles at the nanoscale. These nanoparticles, which can vary in material composition, shape, and size, are finely dispersed throughout the base fluid, imparting unique properties to the resulting nanofluid. The term "nanofluid" encompasses a broad range of colloidal suspensions, where the size of the dispersed nanoparticles typically falls within the nanometer scale, often ranging from 1 to 100 nanometers.

The dispersion of nanoparticles within the base fluid alters the thermophysical properties of the nanofluid, such as thermal conductivity, viscosity, and density, often enhancing these properties beyond those of the base fluid alone. This enhancement is attributed to various factors, including the high surface-area-to-volume ratio of nanoparticles, their unique surface chemistry, and interactions with the base fluid molecules. Nanofluids have garnered significant attention due to their potential applications in various fields, including thermal management, heat transfer enhancement, energy systems, and biomedical engineering, among others. Their tunable properties and versatility make nanofluids promising candidates for addressing challenges in diverse technological and scientific domains.

1.1.1 Brief history and development of nanofluids

The exploration of nanofluids traces back to the late 20th century, with seminal research efforts laying the groundwork for understanding their unique properties and potential applications. One of the earliest documented studies on nanofluids can be attributed to Choi [1] in 1995, who investigated the thermal conductivity enhancement of colloidal suspensions containing nanoparticles. Choi's pioneering work demonstrated that the addition of nanoparticles to a base fluid could significantly augment its thermal conductivity, thereby sparking interest in the field of nanofluids. This landmark study paved the

DOI: 10.1201/9781003527251-1

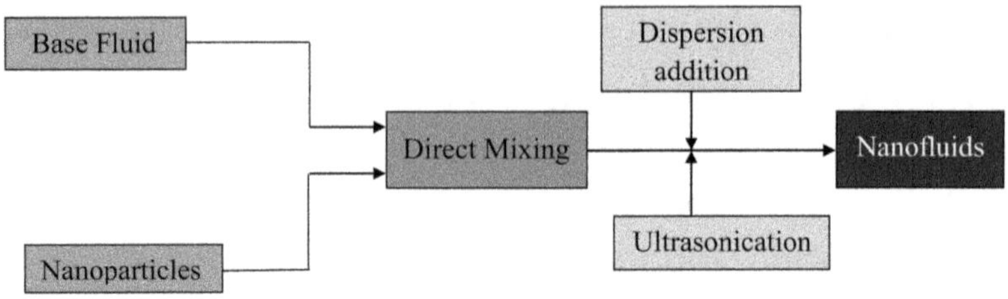

Figure 1.1 Development of nanofluids.

way for subsequent research endeavors aimed at elucidating the fundamental mechanisms governing nanofluid behavior and exploring their diverse applications. These nanoparticles encompass materials such as oxide ceramics (e.g., Al2O3, CuO), metal carbides (e.g., SiC), nitrides (e.g., AlN, SiN), or metals (e.g., Al, Cu), among others. Base fluids commonly used include water, ethylene or triethylene glycols, coolants, oils, lubricants, bio-fluids, polymer solutions, and other fluids typically encountered in various applications. The dimensions of these nanoparticles typically range from a few to about 100 nanometers. The development of nanofluids is shown in Figure 1.1.

Following Choi's seminal work, a surge of research activity ensued, encompassing both theoretical investigations and experimental studies into nanofluid dynamics and applications. Throughout the late 1990s and early 2000s, numerous researchers contributed to advancing our understanding of nanofluids, elucidating their thermal, mechanical, and rheological properties. Notable developments during this period include the elucidation of nanoparticle dispersion mechanisms within base fluids, the development of characterization techniques for nanofluids, and the exploration of novel synthesis methods for producing stable nanofluid suspensions. These collective efforts laid the foundation for the burgeoning field of nanofluidics, spurring further research into the multifaceted aspects of nanofluid behavior and its potential impact across various disciplines. Buongiorno [2] observed that the absolute velocity of nanoparticles can be conceptualized as the sum of the base fluid velocity and a relative (slip) velocity. He identified seven slip mechanisms affecting nanoparticles, including inertia, Brownian diffusion, thermophoresis, diffusiophoresis, the Magnus effect, fluid drainage, and gravity settling. Brownian diffusion and thermophoresis were noted to dominate in the absence of turbulent eddies.

Xuan and Li [3] investigated the convective heat transfer and flow characteristics of copper-water nanofluids. They found that the presence of suspended nanoparticles significantly enhances the heat transfer process, resulting in nanofluids exhibiting larger heat transfer coefficients compared to the original base liquid under similar Reynolds numbers. Moreover, the heat transfer characteristics of nanofluids increase with the volume fraction of nanoparticles. For instance, dispersing 0.3 volume percent of copper nanoparticles in ethylene glycol leads to approximately a 40% increase in thermal conductivity. Metal oxides like aluminum oxide or titanium oxide also contribute to enhanced heat transfer, albeit to a lesser extent compared to metal particles.

Keblinski et al. [4] argued that the enhanced heat transfer in nanofluids arises from a combination of factors, including Brownian motion of nanoparticles, the formation of liquid-solid interfaces, the high conductivity of the particles themselves, and the clustering of nanoparticles. Additionally, Alloui et al. [5] investigated the natural convection of nanofluids in a shallow cavity heated from below. They observed that the presence of nanoparticles reduces the strength of the flow field, particularly at low Rayleigh numbers. Moreover, temperatures on the solid boundaries are reduced (enhanced) by the presence of nanoparticles when the strength of convection is high (low).

In recent years, nanofluid research has continued to evolve, with a growing emphasis on practical applications and technological innovations. The field has witnessed significant advancements in areas such as heat transfer enhancement, thermal management systems, energy conversion, and biomedical engineering. Moreover, ongoing efforts are directed toward addressing key challenges associated with nanofluid synthesis, stability, and scalability, with the aim of translating fundamental discoveries into real-world applications. The interdisciplinary nature of nanofluid research, spanning physics, chemistry, engineering, and materials science, underscores its significance in addressing pressing societal needs and driving innovation in diverse sectors.

1.2 IMPORTANCE AND APPLICATIONS OF NANOFLUIDS IN VARIOUS FIELDS

Nanofluids have emerged as a promising class of engineered fluids with a wide range of applications across various fields due to their unique thermal, mechanical, and rheological properties. One of the primary areas where nanofluids have garnered significant attention is in thermal management and heat transfer applications. By dispersing nanoparticles in conventional heat transfer fluids such as water or oil, nanofluids exhibit enhanced thermal conductivity, enabling more efficient heat transfer in heat exchangers, cooling systems, and electronic devices. This property makes nanofluids particularly valuable in industries such as automotive, aerospace, and electronics, where efficient heat dissipation is crucial for optimal performance and reliability.

Furthermore, nanofluids find extensive applications in energy systems, where they play a pivotal role in improving the efficiency and performance of thermal energy conversion processes. For instance, in solar thermal collectors and concentrated solar power systems, nanofluids can enhance solar absorption and heat transfer rates, thereby increasing overall system efficiency. Similarly, in nuclear reactors and thermal power plants, nanofluids offer potential benefits in enhancing heat transfer and coolant performance, leading to improved safety and operational efficiency.

In addition to thermal management and energy applications, nanofluids are increasingly being explored in biomedical engineering for various purposes, including drug delivery, medical imaging, and hyperthermia treatment. The tunable properties of nanofluids allow for precise control over their behavior in biological environments, enabling targeted drug delivery to specific tissues or organs. Moreover, the high thermal conductivity of nanofluids makes them suitable for applications such as thermal ablation therapy and hyperthermia treatment, where localized heating of cancerous cells is utilized for therapeutic purposes.

Nanofluids also hold promise in environmental remediation and wastewater treatment applications. The high surface-area-to-volume ratio of nanoparticles facilitates efficient adsorption and removal of contaminants from water and air, making nanofluids effective agents for pollutant removal and purification. Additionally, the enhanced thermal conductivity of nanofluids can be harnessed for the efficient treatment of industrial wastewater through processes such as evaporation and distillation.

Furthermore, nanofluids have found applications in advanced materials science, where they are utilized in the fabrication of functional coatings, composite materials, and nano-composites with tailored properties. By incorporating nanoparticles into polymers, ceramics, or metals, nanofluids enable the development of materials with enhanced mechanical strength, electrical conductivity, and thermal stability, opening up new avenues for applications in structural engineering, electronics, and aerospace. Overall, the diverse range of applications of nanofluids underscores their importance as versatile materials with significant potential to address a wide array of technological challenges across various fields.

1.3 FUNDAMENTAL CONCEPTS OF NANOFLUID DYNAMICS

1.3.1 Explanation of fluid dynamics and its relevance to nanofluids

Fluid dynamics is a branch of physics and engineering that studies the motion of fluids (liquids and gases) and the forces acting upon them. It encompasses various phenomena, including flow behavior, turbulence, viscosity, and fluid interactions with solid surfaces. Understanding fluid dynamics is crucial for a wide range of practical applications, as it provides insights into how fluids behave in different environments and under various conditions. In the context of nanofluids, fluid dynamics plays a central role in elucidating the transport properties, flow behavior, and heat transfer characteristics of these engineered fluids.

One key aspect of fluid dynamics relevant to nanofluids is the study of fluid flow behavior. Fluid flow can occur in various regimes, ranging from laminar to turbulent, depending on factors such as fluid velocity, viscosity, and Reynolds number. In nanofluids, the addition of nanoparticles can significantly alter flow behavior, affecting factors such as flow velocity profiles, boundary layer thickness, and flow stability. Understanding how nanoparticles interact with the fluid medium and influence flow dynamics is essential for predicting and optimizing the performance of nanofluid-based systems, such as heat exchangers, microfluidic devices, and cooling systems.

Viscosity, or the resistance of a fluid to flow, is another critical aspect of fluid dynamics that is pertinent to nanofluids. The presence of nanoparticles can influence the viscosity of the base fluid, leading to changes in flow resistance and shear stress. Nanoparticles can affect viscosity through mechanisms such as particle-particle interactions, particle-fluid interactions, and particle aggregation. By characterizing the viscosity of nanofluids under different conditions, researchers can gain insights into their flow behavior and optimize their performance for specific applications. Additionally, understanding viscosity variations in nanofluids is essential for predicting pressure drop, flow stability, and energy consumption in fluid systems.

Heat transfer is a fundamental aspect of fluid dynamics that is particularly relevant to nanofluids due to their enhanced thermal conductivity. Heat transfer mechanisms, such as conduction, convection, and radiation, govern the transfer of thermal energy within nanofluidic systems. The addition of nanoparticles to the base fluid can significantly enhance heat transfer rates by increasing the effective thermal conductivity of the nanofluid. This property is exploited in various heat transfer applications, including heat exchangers, thermal management systems, and solar collectors. Understanding the mechanisms underlying heat transfer enhancement in nanofluids is crucial for optimizing their performance and maximizing energy efficiency in thermal systems.

Turbulence is another phenomenon in fluid dynamics that is pertinent to nanofluids, particularly in high-speed flow regimes. Turbulent flow is characterized by chaotic, irregular motion, leading to enhanced mixing and heat transfer rates compared to laminar flow. In nanofluids, turbulence can promote the dispersion of nanoparticles and enhance their interaction with the base fluid, thereby influencing heat transfer and flow behavior. Understanding how turbulence affects nanofluid dynamics is essential for predicting their performance in turbulent flow environments, such as jet flows, boundary layers, and turbulent shear flows.

Overall, fluid dynamics provides a comprehensive framework for understanding the behavior of nanofluids and their interactions with the surrounding environment. By leveraging principles from fluid dynamics, researchers can develop models, simulations, and experimental techniques to elucidate the transport properties, flow behavior, and heat transfer characteristics of nanofluids, enabling the design and optimization of nanofluid-based systems for various applications across engineering, science, and technology.

1.3.2 Overview of nanoparticle behavior in fluid systems

An overview of nanoparticle behavior in fluid systems provides crucial insights into the fundamental mechanisms governing the dispersion, interaction, and transport of nanoparticles within the fluid medium. Nanoparticles exhibit unique behaviors compared to their bulk counterparts due to their small size, high surface-area-to-volume ratio, and distinct surface chemistry. Understanding these behaviors is essential for elucidating the properties and performance of nanofluids in various applications.

First, the dispersion of nanoparticles within the fluid matrix is a critical aspect that significantly influences the behavior of nanofluids. Nanoparticles tend to agglomerate or form clusters due to van der Waals forces, electrostatic interactions, and Brownian motion. Effective dispersion strategies are essential to prevent or mitigate nanoparticle agglomeration and ensure uniform distribution throughout the fluid. Techniques such as ultrasonication, surfactant stabilization, and surface functionalization are commonly employed to enhance nanoparticle dispersion and stability in nanofluids.

Second, nanoparticle-fluid interactions play a pivotal role in governing the thermal, mechanical, and rheological properties of nanofluids (Figure 1.2).

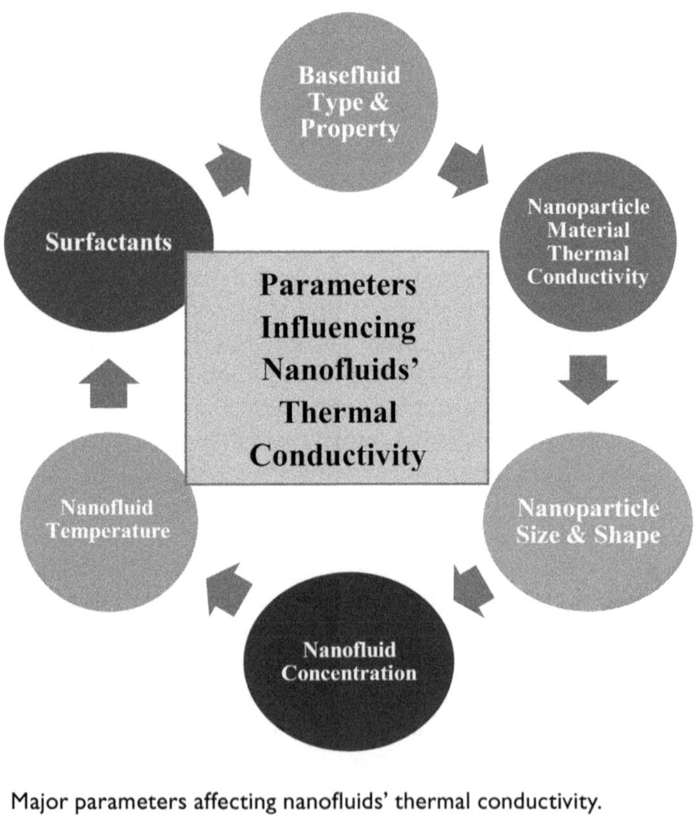

Figure 1.2 Major parameters affecting nanofluids' thermal conductivity.

At the nanoscale, surface effects become predominant, leading to altered fluid behavior and enhanced particle-fluid coupling. Nanoparticles can interact with the fluid molecules through electrostatic, steric, and hydrodynamic forces, influencing fluid viscosity, density, and thermal conductivity. Understanding these interactions is crucial for predicting and controlling the behavior of nanofluids in various flow regimes and environmental conditions.

Moreover, nanoparticle transport phenomena, including diffusion, convection, and sedimentation, dictate the spatial distribution and temporal evolution of nanoparticles within the fluid system. Brownian motion, resulting from random collisions between nanoparticles and fluid molecules, plays a significant role in nanoparticle dispersion and diffusion. Additionally, fluid flow-induced convective transport can lead to the migration and accumulation of nanoparticles in specific regions of the flow field, affecting heat and mass transfer processes. Sedimentation, driven by gravitational or centrifugal forces, can cause nanoparticle settling and stratification, influencing the stability and homogeneity of nanofluids over time.

Furthermore, nanoparticle aggregation and deposition phenomena can occur due to various factors, including particle-particle interactions, wall effects, and flow conditions (Figure 1.2). Aggregation leads to the formation of larger particle clusters, altering the rheological behavior and optical properties of nanofluids. Deposition of nanoparticles onto solid surfaces can occur through processes such as adhesion, diffusion, and electrostatic attraction, resulting in the formation of nanoparticle coatings or fouling layers. Understanding the kinetics and mechanisms of nanoparticle aggregation and deposition is crucial for optimizing the performance and durability of nanofluid-based systems in practical applications.

Overall, an overview of nanoparticle behavior in fluid systems provides valuable insights into the complex interplay between nanoparticles and the surrounding fluid medium. By elucidating the mechanisms governing nanoparticle dispersion, interaction, transport, aggregation, and deposition, researchers can develop strategies to tailor the properties and performance of nanofluids for a wide range of applications, including thermal management, energy systems, biomedical engineering, environmental remediation, and materials science.

1.3.3 Discussion on the unique properties of nanofluids, such as thermal conductivity and viscosity

Nanofluids, colloidal suspensions containing nanoparticles dispersed in a base fluid, exhibit several unique properties that distinguish them from traditional fluids and render them advantageous for various applications. Two key properties of nanofluids that have garnered significant attention are thermal conductivity and viscosity.

The thermal conductivity of nanofluids stands out as one of their most remarkable properties. Nanoparticles possess a significantly higher surface-area-to-volume ratio compared to their bulk counterparts, leading to enhanced interactions with neighboring molecules and, consequently, higher thermal conductivities. This property enables nanofluids to conduct heat more efficiently than conventional fluids, making them attractive candidates for heat transfer applications. Studies have shown that nanofluids can exhibit thermal conductivities several times higher than those of base fluids, with enhancements dependent on factors such as nanoparticle size, shape, concentration, and dispersion quality. The tunability of nanofluid thermal conductivity offers opportunities for enhancing the performance of heat exchangers, cooling systems, and thermal energy storage devices across various industries.

The viscosity of nanofluids represents another intriguing aspect of their behavior. While the addition of nanoparticles to a fluid typically increases its viscosity due to the added resistance to flow, nanofluids often exhibit unexpected rheological behavior. Research has shown that nanofluids can exhibit either Newtonian or non-Newtonian behavior depending on factors such as nanoparticle concentration, size, shape, and surface chemistry. Interestingly, some nanofluids have been observed to exhibit shear-thinning behavior, where viscosity decreases with increasing shear rate, contrary to the behavior of traditional fluids. Understanding the rheological properties of nanofluids is crucial for predicting their flow behavior in practical applications such as microfluidic devices, lubrication systems, and hydraulic systems.

Furthermore, the unique properties of nanofluids extend beyond thermal conductivity and viscosity, encompassing aspects such as optical properties, stability, and colloidal behavior. Nanoparticles dispersed in a fluid can alter its optical properties, leading to phenomena such as enhanced light absorption or scattering, which find applications in optical filters, sensors, and photonic devices. Additionally, the stability of nanofluids, characterized by their propensity to resist particle agglomeration and sedimentation, is critical for ensuring long-term performance and reliability in practical applications. Strategies for enhancing nanofluid stability include surface modification of nanoparticles, selection of appropriate surfactants or dispersants, and optimization of dispersion techniques.

In summary, nanofluids exhibit a myriad of unique properties that stem from the nanoscale characteristics of the dispersed nanoparticles and their interactions with the base fluid. Understanding and harnessing these properties are essential for unlocking the full potential of nanofluids across a diverse range of applications spanning heat transfer, fluid dynamics, optics, and materials science. Continued research efforts aimed at elucidating the underlying mechanisms governing nanofluid behavior will pave the way for the development of innovative technologies with enhanced performance, efficiency, and functionality.

1.4 GOVERNING EQUATIONS OF NANOFLUID DYNAMICS

The governing equations of nanofluid dynamics typically include the continuity equation, the momentum equation (Navier-Stokes equation), and the energy equation. These equations are augmented to account for the presence of nanoparticles in the fluid. The continuity equation describes the conservation of mass, the momentum equation accounts for the conservation of momentum, and the energy equation governs the transfer of thermal energy. Additionally, equations for nanoparticle transport and interactions may be included, such as the Brownian motion and thermophoresis equations, to capture the behavior of nanoparticles within the nanofluid.

1.4.1 Equation of continuity

$$\frac{\partial \rho}{\partial t} + \nabla \cdot (\rho \boldsymbol{q}) = 0. \tag{1.1}$$

Here, ρ denotes the density of the nanofluid and q is the nanofluid velocity. If ϕ denotes volumetric fraction of nanoparticles, the nanofluid density (ρ) is given as

$$\rho = \phi \rho_p + (1 - \phi)\rho_f. \tag{1.2}$$

Here ρ_p and ρ_f are the density of the nanoparticles and base fluid, respectively.

1.4.2 The conservation equation for the nanoparticles

$$\frac{\partial \phi}{\partial t} + (\boldsymbol{q} \cdot \nabla)\phi = -\frac{1}{\rho_p} \nabla \cdot \boldsymbol{j}_p. \tag{1.3}$$

Here, $\boldsymbol{j}_p$ represents the mass flux of nanoparticle diffusion, comprising two diffusion components (Brownian motion and thermophoresis), expressed as the sum of these terms as

$$\boldsymbol{j}_p = -\rho_p D_B \nabla \phi - \rho_p \left(\frac{D_T}{T_1} \right) \nabla T. \tag{1.4}$$

Here, D_B represents the Brownian diffusion coefficient, while D_T denotes the thermophoretic diffusion coefficient of the nanoparticles.

$$D_B = \frac{k_B T}{3\pi \mu_f d_p}; \qquad D_T = \left(\frac{\mu_f}{\rho_f} \right) \left(\frac{0.26 k_f}{2k_f + k_p} \right) \phi. \tag{1.5}$$

In this context, k_B stands for Boltzmann's constant, μ_f represents the viscosity of the base fluid, d_p signifies the nanoparticle diameter, ρ_f denotes the density of the base fluid, and k_f and k_p correspond to the thermal conductivity of the base fluid and nanoparticles, respectively.

Putting the value of $\boldsymbol{j}_p$ from equation (1.4) into equation (1.3), we get

$$\frac{\partial \phi}{\partial t}+(\boldsymbol{q}\cdot\nabla)\phi = D_B\nabla^2\phi+\left(\frac{D_T}{T_1}\right)\nabla^2 T. \tag{1.6}$$

1.4.3 The energy equation

$$(\rho c)\left(\frac{\partial T}{\partial t}+(\boldsymbol{q}\cdot\nabla)T\right)=-\nabla\cdot\boldsymbol{Q}+h_p\nabla\cdot\boldsymbol{j}_p. \tag{1.7}$$

Here, ρ refers to the density of the nanofluid, c denotes the fluid-specific heat at constant pressure, h_p denotes the specific enthalpy of nanoparticles, and $h_p\nabla\cdot\boldsymbol{j}_p$ represents the additional flow work in the energy equation resulting from Brownian motion and thermophoresis of nanoparticles relative to the flow velocities.

$$\boldsymbol{Q}=-k\nabla T+h_p\boldsymbol{j}_p. \tag{1.8}$$

In this context, k represents the nanofluid thermal conductivity. There are various models available for the thermal conductivity of the nanofluid; some of them are given in Table 1.1 as follows.

Taking the divergence of equation (1.8), we get

$$\begin{aligned}\nabla\cdot\boldsymbol{Q}&=-k\nabla^2T+\nabla\cdot\left(h_p\boldsymbol{j}_p\right)\\&=-k\nabla^2T+h_p\nabla\cdot\boldsymbol{j}_p+\boldsymbol{j}_p\cdot\nabla h_p\\&=-k\nabla^2T+h_p\nabla\cdot\boldsymbol{j}_p+\boldsymbol{j}_p\cdot\nabla\left(c_p\nabla T\right).\end{aligned} \tag{1.9}$$

Here, c_p is the specific heat of the material constituting the nanoparticle. Therefore, equation (1.7) becomes,

$$(\rho c)\left(\frac{\partial T}{\partial t}+(\boldsymbol{q}\cdot\nabla)T\right)=k\nabla^2T+\left(\rho_pc_p\right)\left[D_B\nabla\phi\cdot\nabla T+\left(\frac{D_T}{T_1}\right)\nabla T\cdot\nabla T\right]. \tag{1.10}$$

1.4.4 Equation of motion

The momentum equation for a nanofluid can be written as

$$\rho\left(\frac{\partial \boldsymbol{q}}{\partial t}+(\boldsymbol{q}\cdot\nabla)\boldsymbol{q}\right)=-\nabla p+\mu\nabla^2\boldsymbol{q}+\boldsymbol{F}, \tag{1.11}$$

Table 1.1 Various models for thermal conductivity of a nanofluid

Authors with year	*Model*	*Explanation*
Maxwell [6] (1873)	$\frac{k_{nf}}{k}=\frac{k_s+2k+2\varphi(k_s-k)}{ks_s+2k-\varphi(k_s-k)}$	The initial model, often referenced, defines the effective thermal conductivity of combined liquid and solid suspensions, with φ representing the particle sphericity.
Bruggeman [7] (1935)	$\frac{k_{nf}}{k_{bf}}=\frac{1}{4}(3\phi-1)k_{np}+(2-3\phi)k_{bf}+\frac{k_{bf}}{4}\sqrt{\Delta},$ where $\Delta=(3\phi-1)^2\left(\frac{k_{nf}}{k_{bf}}\right)^2+\left[(2-3\phi)^2+2(2+9\phi-9\phi^2)\right]\frac{k_{nf}}{k_{bf}}$	Bruggeman proposed a superior model compared to the previous one, enabling the prediction of thermal conductivity at any volume percentage and specifically applicable to spherical-shaped nanoparticles. Here, ϕ represents the particle volume concentration, k_{np} denotes the thermal conductivity of nanoparticles, and k_{bf} signifies the thermal conductivity of the base fluid.
Hamilton and Crosser [8] (1962)	$\frac{k_{nf}}{k}=\frac{k_s+(p-1)k+(p-1)\varphi(k_s-k)}{ks_s+(p-1)k-\varphi(k_s-k)}$	This model incorporates non-spherical nanoparticles and introduces a shape factor influencing the liquid-solid thermal conductivity. Here, p represents the empirical shape factor, calculated as $3/\varphi$, with φ indicating the particle sphericity.
Yu and Choi [9] (2003)	$\frac{k_{pe}}{k_p}=\frac{\left[2(1-\delta)+(1+\alpha)^3(1+2\delta)\right]\delta}{-(1-\delta)+(1+\alpha)^3+(1+2\delta)}$ $\frac{k_{nf}}{k}=\frac{k_{pe}+2k_b+2(k_{pe}-2k_p)(1+\alpha)^3\phi}{k_{pe}+2k_b-(k_{pe}-2k_p)(1+\alpha)^3\phi}$	The nano-layers at the solid/liquid interface are considered in defining the equivalent thermal conductivity, denoted as k_{pe}. Here, δ represents the ratio of nano-layer thermal conductivity to particle thermal conductivity, given by $\delta=k_{pe}/k$.
Timofeeva et al. [10] (2007)	$\frac{k_{eff}}{k}=(1+3\phi)$	The model takes into account the geometry, agglomeration state, and surface resistance of nanoparticles to determine the efficiency of thermal conductivity. k_{eff} denotes the effective thermal conductivity, where ϕ represents the particle volume concentration.

Table 1.1 (*Continued*) Various models for thermal conductivity of a nanofluid

Authors with year	*Model*	*Explanation*
Feng et al. [11] (2007)	$\frac{k_s}{k_f} = \frac{k_p + 2k_f - 2\varphi(1+\alpha)^3(k_f - k_p)}{k_p + 2k_f + \varphi(1+\alpha)^3(k_f - k_p)}$	A model that incorporates the nano-layer, denoted by α, considers the ratio of nanoparticle thickness to radius ($\alpha = t_p/r_p$) as a factor in determining the nanofluid thermal conductivity.
Vafaei et al. [12] (2017)	$\frac{k_{nf}}{k_{bf}} = 0.9787 + \exp(0.3081\phi^{0.3097} - 0.002T)$	The model relies on optimal conditions achieved through an artificial neural network and is applicable for predicting hybrid nanofluids. Here, ϕ represents the particle volume concentration, and T denotes the measured temperature.

where $\boldsymbol{F}$ represents external forces per unit volume. In this context, μ represents the nanofluid viscosity. There are various models available for the viscosity of the nanofluid; some of them are given in Table 1.2 as follows.

Table 1.2 Various models for viscosity of a nanofluid

Authors with year	*Model*	*Explanation*
Einstein [13] (1906)	$\mu_{nf} = (1 + 2.5\phi)\mu_f$	Particles with an extremely low volume fraction that are spherical in shape.
Mooney [14] (1951)	$\mu_{nf} = \mu_f e^{\frac{\xi\phi}{(1-k\phi)}}$	Suitable for self-crowding factors ranging from 1.35 to 1.91 and a fitting parameter of 2.5.
Brinkman [15] (1952)	$\mu_{nf} = (1-\phi)^{2.5}\mu_f$	Derived from the Einstein model, applicable for continuous media with particle concentrations below 4%.
Krieger and Dougherty [16] (1957)	$\frac{\mu_{nf}}{\mu_f} = \left[1 - \frac{\phi}{\phi_m}\right]^{-\eta\phi_m}$	Hard spheres, randomly distributed and monodisperse, with varying packing fractions
Neilson [17] (1970)	$\mu_{nf} = (1 + 1.5\phi)\mu_f e^{\frac{\phi}{(1-\phi_m)}}$	The power law model is applicable for particle volume fractions exceeding 0.02
Lundgren [18] (1972)	$\mu_{nf} = (1 + 2.5\phi + 6.25\phi^2 + f(\phi^3))\mu_f$	The Einstein [13] model is simplified by formulating it through a Taylor series expansion of ϕ.
Batchelor [19] (1977)	$\mu_{nf} = (1 + 2.5\phi + 6.5\phi^2)\mu_f$	Expansion of the Einstein [13] model to incorporate the effects of Brownian motion.
Chen et al. [20] (2007)	$\frac{\mu_{nf}}{\mu_f} = \left[1 - \frac{\phi_a}{\phi_m}\right]^{-2.5\phi_m}$	The Krieger-Dougherty [16] equation has been adapted to account for particle aggregates.

1.5 NUMERICAL AND ANALYTICAL METHODS IN NANOFLUID DYNAMICS

Numerical and analytical methods play a pivotal role in advancing our understanding of nanofluid dynamics, where the behavior of nanoscale particles suspended in a fluid medium is investigated. These methods provide powerful tools for studying complex phenomena such as nanoparticle transport, heat transfer enhancement, and fluid flow characteristics in various engineering and biomedical applications. Both numerical and analytical approaches offer distinct advantages and are often used complementarily to gain comprehensive insights into nanofluid behavior [21].

Numerical methods, such as finite difference, finite element, and lattice Boltzmann methods, enable the simulation of nanofluid dynamics by solving the governing equations numerically. In the context of nanofluids, these equations typically include the Navier-Stokes equations for fluid flow, the energy equation for heat transfer, and additional equations governing nanoparticle transport and interactions. For example, the conservation equations for mass, momentum, and energy can be coupled with equations describing Brownian motion and thermophoresis to model nanoparticle behavior accurately. These numerical simulations provide detailed information on flow patterns, temperature distribution, and nanoparticle concentration profiles within the nanofluid [22].

Analytical methods, on the other hand, offer closed-form solutions or approximate solutions to simplified models of nanofluid dynamics. These methods are particularly useful for gaining insights into fundamental aspects of nanofluid behavior and for validating numerical results. One common analytical approach is the homogenization method, which treats the nanofluid as a homogeneous mixture and derives effective properties such as effective thermal conductivity and effective viscosity. The Einstein model, for instance, provides an analytical expression for the effective thermal conductivity of dilute nanofluids based on the assumption of non-interacting spherical nanoparticles.

In recent years, there has been growing interest in developing hybrid numerical-analytical methods for studying nanofluid dynamics. These methods combine the strengths of both numerical and analytical approaches to achieve efficient and accurate simulations of nanofluid behavior. For example, the lattice Boltzmann method can be coupled with analytical expressions for nanoparticle transport to simulate nanofluid flow with improved computational efficiency. Similarly, numerical simulations can be used to validate analytical models and provide insights into the limitations of simplifying assumptions.

The numerical and analytical methods are indispensable tools for investigating nanofluid dynamics. While numerical methods offer the flexibility to model complex phenomena and provide detailed spatial and temporal information, analytical methods offer simplicity and insight into fundamental aspects of nanofluid behavior. By combining these methods judiciously, researchers can gain a deeper understanding of nanofluid dynamics and pave the way for the development of novel nanofluid-based technologies [23–25].

1.6 APPLICATIONS OF NANOFLUID DYNAMICS

Applications of nanofluid dynamics span a wide range of fields, from engineering and energy to medicine and environmental science. Nanofluids, consisting of nanoparticles suspended in a base fluid, exhibit unique thermal, mechanical, and rheological properties that make them valuable in various applications. Understanding the dynamics of nanofluids is crucial for optimizing their performance in these diverse applications (Figure 1.3).

In thermal management systems, nanofluids are employed to enhance heat transfer rates in heat exchangers, cooling systems, and electronic devices. The addition of nanoparticles, such as metallic or oxide nanoparticles, to the base fluid significantly improves the thermal conductivity of the nanofluid [26–29]. This enhancement in thermal conductivity enables more efficient heat dissipation, leading to smaller and more compact heat exchangers with increased heat transfer rates. The governing equations for heat transfer in nanofluids can be described by the energy equation, which incorporates

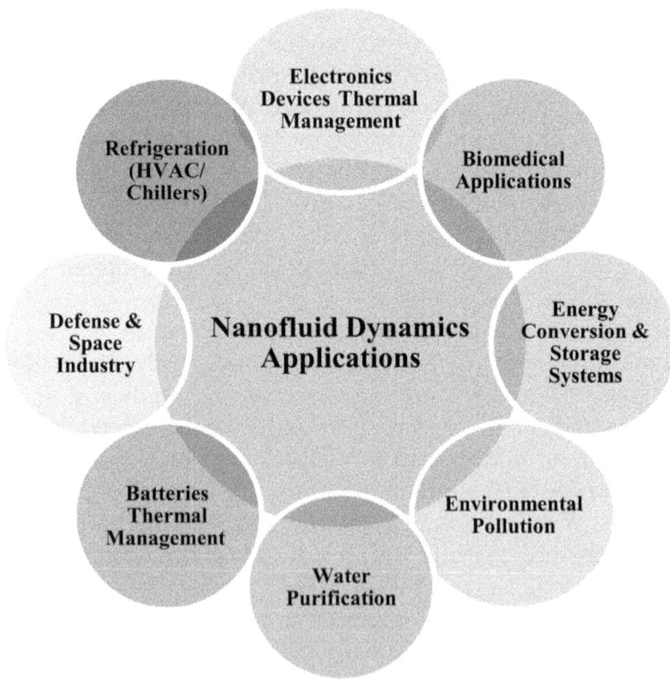

Figure 1.3 Major applications of nanofluid dynamics.

the effects of nanoparticle dispersion, Brownian motion, and thermophoresis. Mathematically, the effective thermal conductivity of nanofluids can be modeled using the Maxwell-Garnett or Bruggeman models, which account for nanoparticle concentration and size distribution [30–31].

In biomedical applications, nanofluids play a crucial role in drug delivery, imaging, and therapeutics. Magnetic nanoparticles suspended in a carrier fluid, known as magnetic nanofluids or ferrofluids, are used for targeted drug delivery and hyperthermia treatment of cancerous tumors. The dynamics of these nanofluids are governed by the Navier-Stokes equations for fluid flow coupled with the Langevin equation for describing the motion of magnetic nanoparticles under the influence of an external magnetic field. Additionally, the behavior of biological fluids, such as blood, can be modeled using nanofluid dynamics to study phenomena like hemodynamics and the effect of nanoparticles on blood flow properties [32–34].

In energy conversion and storage systems, nanofluids are utilized to improve the efficiency and performance of thermal energy storage devices, solar collectors, and thermoelectric generators. By incorporating phase change materials (PCMs) into nanofluids, the latent heat storage capacity of the fluid can be significantly enhanced. The transient behavior of nanofluids during phase change processes is described by the energy equation coupled with the phase change equation, which accounts for the energy absorbed or released during phase transition. Moreover, nanofluids containing nanoparticles with high Seebeck coefficients can be employed in thermoelectric generators to convert waste heat into electricity.

In environmental engineering, nanofluids are employed for pollution remediation, water purification, and oil spill cleanup. Nanoparticles dispersed in a carrier fluid can be used to adsorb or catalyze pollutants in contaminated water or soil, leading to their removal or degradation. The transport and dispersion of nanoparticles in porous media, such as soil or groundwater, are governed by advection, dispersion, and diffusion processes, which can be modeled using the advection-dispersion equation coupled with equations describing nanoparticle deposition and attachment (Figure 1.3) [35–37].

Overall, the applications of nanofluid dynamics are diverse and far-reaching, with implications for numerous fields including thermal management, biomedicine, energy conversion, and environmental engineering. By leveraging the unique properties of nanofluids and employing advanced mathematical models and computational techniques, researchers can continue to innovate and develop novel solutions to address complex challenges in these domains.

1.7 CHALLENGES AND FUTURE DIRECTIONS

As nanofluid dynamics continues to evolve, researchers are confronted with various challenges and opportunities that shape the future directions of this field. While nanofluids offer promising potential for enhancing heat transfer, energy conversion, and biomedical applications, several key challenges must be addressed to fully harness their benefits. One major challenge in nanofluid dynamics is the accurate characterization and modeling of nanoparticle behavior within the fluid medium. Nanoparticles exhibit complex interactions with the surrounding fluid, including Brownian motion, thermophoresis, and aggregation, which significantly influence nanofluid properties such as viscosity, thermal conductivity, and stability. The dynamics of nanoparticles can be described using mathematical models such as the Langevin equation for Brownian motion and the Smoluchowski equation for aggregation. However, accurately capturing these phenomena in numerical simulations remains a formidable task due to the computational complexity involved [38–39].

Another challenge is the development of experimental techniques for measuring nanofluid properties under realistic conditions. Traditional experimental methods often struggle to accurately characterize nanofluid behavior, particularly at high temperatures, pressures, and nanoparticle concentrations. Advanced experimental techniques such as dynamic light scattering, atomic force microscopy, and scanning electron microscopy are being developed to provide insights into nanofluid dynamics at the nanoscale. Additionally, the integration of experimental data with numerical simulations through techniques like data assimilation and uncertainty quantification can improve the accuracy and reliability of nanofluid models [25].

Furthermore, the scalability of nanofluid-based technologies poses a significant challenge for large-scale implementation. While nanofluids show promising performance enhancements at the laboratory scale, scaling up these technologies for practical applications presents numerous engineering and logistical challenges. Issues such as nanoparticle synthesis, dispersion, and stability must be addressed to ensure the uniformity and consistency of nanofluid properties across large volumes. Moreover, the economic viability and environmental sustainability of nanofluid-based technologies must be carefully evaluated to justify their widespread adoption. Looking ahead, future research directions in nanofluid dynamics will focus on addressing these challenges and exploring new avenues for innovation. Multiscale modeling approaches, which integrate molecular dynamics simulations with continuum-based models, hold promise for improving our understanding of nanofluid behavior at the nanoscale. Additionally, the development of advanced nanomaterials with tailored properties will enable the design of nanofluids with enhanced performance and stability. Collaborative efforts between researchers from diverse disciplines, including fluid mechanics, materials science, and biomedical engineering, will be essential for advancing the frontiers of nanofluid dynamics and unlocking its full potential in various applications [40–42]. Nanofluid dynamics presents exciting opportunities for enhancing heat transfer, energy conversion, and biomedical applications, and significant challenges remain to be addressed. By overcoming these challenges through interdisciplinary collaboration, innovative research, and advanced experimental and computational techniques, nanofluid dynamics will continue to evolve as a transformative field with far-reaching implications for science, engineering, and technology.

1.8 CONCLUSIONS

Traditional characterization techniques often fall short when it comes to accurately capturing the behavior of nanofluids under real-world operating conditions, particularly at high temperatures, pressures, and nanoparticle concentrations. These harsh environments can significantly influence factors like nanoparticle dispersion, interaction with the base fluid, and stability. However, a deeper understanding of nanoparticle behavior in fluid systems is crucial. By elucidating the fundamental mechanisms governing dispersion, interparticle interactions, transport phenomena, aggregation tendencies, and deposition patterns, researchers can unlock the full potential of nanofluids. This knowledge empowers the development of targeted strategies to tailor nanofluid properties for specific applications.

Nanofluids are designed to excel in thermal management for high-performance electronics or engineered for efficient energy conversion in solar cells. The possibilities extend to targeted drug delivery in biomedical engineering, contaminant removal in environmental remediation, and the creation of novel materials with superior properties in materials science. To achieve this level of control, advanced characterization techniques are essential. These techniques should provide insights into the behavior of nanoparticles at the nanoscale, offering a deeper understanding of how they interact with the surrounding fluid under various operating conditions. Overcoming the challenges of uniform, consistent, and cost-effective production of nanofluids with desired properties in large volumes remains a significant hurdle for their widespread practical implementation. Addressing these challenges will be instrumental in translating the exciting potential of nanofluids into real-world applications across diverse scientific and engineering fields.

REFERENCES

1. Choi, S. Enhancing Thermal Conductivity of Fluids with Nanoparticles. In Developments and Applications of Non-Newtonian Flows; D. A. Siginer and H. P. Wang, Editors; New York: ASME FED; Vol. 231/MD-Vol. 66, pp. 99–105, 1995.
2. Buongiorno, J. "Convective transport in nanofluids." ASME Journal of Heat Transfer 128 (2006): 240–250.
3. Xuan, Y., and Q. Li. "Investigation of convective heat transfer and flow features of nanofluids." ASME Journal of Heat Transfer 125 (2003): 151–155.
4. Keblinski, P., S. R. Phillpot, S. Choi, and J. A. Eastman. "Mechanism of heat flow in suspensions of nano-sized particles (nanofluids)." Journal of Heat and Mass Transfer 45 (2002): 855–863.
5. Alloui, Z., P. Vasseur, and M. Reggio. "Natural convection of nanofluids in a shallow cavity heated from below." International Journal of Thermal Science 50 (2011): 385–393.
6. Maxwell, J. C. A Treatise on Electricity and Magnetism: Vol. 1; Oxford: Clarendon Press, 360–366, 1873.

7. Bruggeman, D. A. G. "Calculation of various physics constants in heterogenous substances I Dielectricity constants and conductivity of mixed bodies from isotropic substances." Annalen der Physik 416, no. 7 (1935): 636–664.
8. Hamilton, R. L., and O. K. Crosser. "Thermal conductivity of heterogeneous two-component systems." Industrial & Engineering Chemistry Fundamentals 1, no. 3 (1962): 187–191.
9. Yu, W., and S. U. S. Choi. "The role of interfacial layers in the enhanced thermal conductivity of nanofluids: A renovated Maxwell model." Journal of Nanoparticle Research 5, no. 1–2 (2003): 167–171.
10. Timofeeva, E. V., A. N. Gavrilov, J. M. McCloskey, Y. V. Tolmachev, S. Sprunt, L. M. Lopatina, and J. V. Selinger. "Thermal conductivity and particle agglomeration in alumina nanofluids: Experiment and theory." Physical Review E 76, no. 6 (2007): 061203.
11. Feng, Y., B. Yu, P. Xu, and M. Zou. "The effective thermal conductivity of nanofluids based on the nanolayer and the aggregation of nanoparticles." Journal of Physics D: Applied Physics 40, no. 10 (2007): 3164–3171.
12. Vafaei, M., M. Afrand, N. Sina, R. Kalbasi, F. Sourani, and H. Teimouri. "Evaluation of thermal conductivity of MgO-MWCNTs/EG hybrid nanofluids based on experimental data by selecting optimal artificial neural networks." Physica E: Low-Dimensional Systems and Nanostructures 85 (2017): 90–96.
13. Einstein, A. "Eineneuebestimmung der moleküldimensionen." Annalen der Physik 324, no. 2 (1906): 289–306.
14. Mooney, M. "The viscosity of a concentrated suspension of spherical particles." Journal of Colloid Science 6, no. 2 (1951): 162–170.
15. Brinkman, H. C. "The viscosity of concentrated suspensions and solutions." The Journal of Chemical Physics 20, no. 4 (1952): 571.
16. Krieger, I. M., and T. J. Dougherty. "A mechanism for non-Newtonian flow in suspensions of rigid spheres." Transactions of the Society of Rheology 3, no. 1 (1957): 137–152.
17. Nielsen, L. E. "Generalized equation for the elastic moduli of composite materials." Journal of Applied Physics 41, no. 11 (1970): 4626–4627.
18. Lundgren, T. S. "Slow flow through stationary random beds and suspensions of spheres." Journal of Fluid Mechanics 51, no. 02 (1972): 273–299.
19. Batchelor, G. K. "The effect of Brownian motion on the bulk stress in a suspension of spherical particles." Journal of Fluid Mechanics 83, no. 01 (1977): 97–117.
20. Chen, H., Y. Ding, and C. Tan. "Rheological behaviour of nanofluids." New Journal of Physics 9, no. 10 (2007): 367.
21. Awasthi, M. K., A. Kumar, and N. Dutt. Modeling Rayleigh-Taylor Instability in Nanofluid Layers. In Computational Fluid Flow and Heat Transfer: Advances, Design, Control and Applications; M. K. Awasthi, A. Kumar, N. Dutt, and S. Singh, Editors; CRC Press: Boca Raton, FL, USA, 2024; Chapter 13; https://doi.org/10.1201/9781003465171-13
22. Awasthi, M. K., N. Dutt, A. Kumar, and S. Kumar. "Electrohydrodynamic capillary instability of Rivlin–Ericksen viscoelastic fluid film with mass and heat transfer." Heat Transfer53, no.1 (2024): 115–133. https://doi.org/10.1002/htj.22944.
23. Awasthi, M. K., A. K. Shukla, A. Kumar, D. Yadav, and N. Dutt. "Viscous correction to the potential flow analysis of Rayleigh-Taylor instability in a Rivlin-Ericksen viscoelastic fluid layer with heat and mass transfer." Heat Transfer (2024), https://onlinelibrary.wiley.com/doi/10.1002/htj.23076.
24. Yadav, D., M. K. Awasthi, A. Kumar, and N. Dutt. "Swirling capillary instability of Rivlin-Ericksen liquid with heat transfer and axial electric field." Physics 6, no. 2 (2024): 828–844. https://doi.org/10.3390/physics6020051.

25. Dutt, N., A. Hedau, A. Kumar, M. K. Awasthi, and V. P. Singh. "Effect of duct length variation on solar air heater performance for smooth and d-shaped roughened absorber plate." Heat Transfer (2024): 1–29, https://onlinelibrary.wiley.com/doi/10.1002/htj.23115.
26. Awasthi, M. K., N. Dutt, and A. Kumar, Editors, "Nanofluids Technology for Thermal Sciences and Engineering: Research, Development, and Applications; CRC Press, Boca Raton, FL, USA, 2024, ISBN 9781032799117.
27. Awasthi, M. K., A. Kumar, and N. Dutt. Recent Innovation, Technological Development and Futuristic Applications of Nanofluids. In Nanofluids Technology for Thermal Sciences and Engineering: Research, Development, and Applications; M. K. Awasthi, N. Dutt, and A. Kumar, Editors; CRC Press: Boca Raton, FL, USA, 2024; Chapter 01.
28. Kumar, A., S. Thangavel, M. K. Awasthi, V. P. Singh, and N. Dutt. Safety and Environmental Considerations in Nano Fluid-Based Systems. In Nanofluids Technology for Thermal Sciences and Engineering: Research, Development, and Applications; M. K. Awasthi, N. Dutt, and A. Kumar, Editors; CRC Press: Boca Raton, FL, USA, 2024; Chapter 06.
29. Awasthi, M. K., A. Kumar, N. Dutt, and S. Singh, Editors, Computational Fluid Flow and Heat Transfer: Advances, Design, Control and Applications; CRC Press: Boca Raton, FL, USA, 2024; https://doi.org/10.1201/9781003465171.
30. Awasthi, M. K., N. Dutt, A. Kumar, and A. Hedau. Introduction to Mathematical and Computational Methods. In Computational Fluid Flow and Heat Transfer: Advances, Design, Control and Applications; M. K. Awasthi, A. Kumar, N. Dutt, and S. Singh, Editors; CRC Press: Boca Raton, FL, USA, 2024; Chapter 01; https://doi.org/10.1201/9781003465171-1
31. Kumar, A., N. Dutt, and M. K. Awasthi, Editors, Heat Transfer Enhancement Techniques: Thermal Performance, Optimization and Applications; John Wiley & Sons, 2024.
32. Kumar, A., M. K. Awasthi, N. Dutt, and V. P. Singh. Recent Innovation in Heat Transfer Enhancement Techniques. In Heat Transfer Enhancement Techniques: Thermal Performance, Optimization and Applications; A. Kumar, N. Dutt, and M. K. Awasthi, Editors; John Wiley & Sons, 2024; Chapter 01.
33. Dutt, N., A. Hedau, A. Kumar, M. K. Awasthi, S. Hedau, and C. S. Meena, Thermo-hydraulic performance investigation of solar air heater duct having staggered D-shaped ribs: Numerical approach. Heat Transfer (2024), https://onlinelibrary.wiley.com/doi/10.1002/htj.22998.
34. Dutt, N., A. J. Hedau, A. Kumar, M. K. Awasthi, V. P. Singh, and G. Dwivedi. Thermo-hydraulic performance of solar air heater having discrete D-shaped ribs as artificial roughness. Environmental Science and Pollution Research (2023), https://doi.org/10.1007/s11356-023-28247-9.
35. Reddy, P. N., V. Verma, A. Kumar, and M. Awasthi. CFD simulation and thermal performance optimization of flow in a channel with multiple baffles. Journal of Heat and Mass Transfer Research 10, no. 2 (2023): 257–268, https://doi.org/10.22075/JHMTR.2023.31108.1458.
36. Singh, B., P. Kaur, A. K. Yadav, M. K. Awasthi, and A. Kumar. Design and Analysis of Solar Tracking System for PV Thermal Performance Enhancement. In Heat Transfer Enhancement Techniques: Thermal Performance, Optimization and Applications; A. Kumar, N. Dutt, and M. K. Awasthi, Editors; John Wiley & Sons, 2024; Chapter 11.
37. Li, J., and Y. Xuan. "Thermal conductivity of nanofluids: A review." International Journal of Heat and Mass Transfer 138 (2019): 1184–1208.
38. Prasher, R., and P. E. Phelan. "Predicting the thermal conductivity of nanoscale colloidal suspensions (nanofluids)." Journal of Heat Transfer 128, no. 6 (2006): 586–592.

39. Singh, V. P., A. Kumar, and M. K. Awasthi. Augmentation of Solar, Geothermal and Earth-Air Heat Exchanger in Sustainable Buildings. In Heat Transfer Enhancement Techniques: Thermal Performance, Optimization and Applications; A. Kumar, N. Dutt, and M. K. Awasthi, Editors; John Wiley & Sons, 2024; Chapter 06, 119–158.
40. Buongiorno, J., and D. Venerus. "Challenges in implementing nanofluids for thermal management of electronics." Applied Physics Letters 113, no. 17 (2018): 173101.
41. Chen, H., Y. Ding, and Y. He. "Scalable synthesis of nanofluids for heat transfer." Nano Research 13, no. 7 (2020): 1821–1830.
42. Wang, X., and Y. Zheng. "A review on stability and heat transfer performance of nanofluids." Nano Microfluidics 11, no. 1 (2017): 13.

Chapter 2

Experimental techniques for nanofluid testing and validation

Kandasamy Suganeswaran, Velu Kaliyannan Gobinath, Gunasekaran Raja, Nagarajan Nithyavathy, and Rathinasamy Parameshwaran

2.1 INTRODUCTION

Nanofluids are the homogeneous colloidal dispersion of nanoparticles (1–100 nm) into the base fluid (Figure 2.1). Nanofluids have novel properties different from base fluids that included thermophysical properties such as specific heat, density, viscosity, and thermal conductivity. This makes them potentially useful in many applications in heat transfer, including microelectronics, fuel cells, pharmaceutical processes, hybrid-powered engines, engine cooling/vehicle thermal management, domestic refrigerator, chiller, and heat exchanger; in grinding and machining; and in boiler flue gas temperature reduction. They are two-phase systems with one phase (solid phase) in another (liquid phase) [1]. Some typical nanofluids are ethylene-glycol-based copper nanofluids and water-based copper oxide nanofluids.

Nanofluids are dilute suspensions of functionalized nanoparticle composite materials developed about a decade ago with the specific aim of increasing the thermal conductivity of heat transfer fluids, which have now evolved into a promising nanotechnological area. Such thermal nanofluids

Figure 2.1 Schematic representation of nanofluids.

DOI: 10.1201/9781003527251-2

for heat transfer applications represent a class of its own difference from conventional colloids for other applications. The advent of high heat flow processes has created significant demand for new technologies to enhance the heat transfer. For example, microprocessors have continually become smaller and more powerful, and as a result, heat flow demands have steadily increased over time leading to new challenges in thermal management. Furthermore, there is increasing interest in improving the efficiency of existing heat transfer processes. An example is in automotive systems where improved heat transfer could lead to smaller heat exchangers for cooling resulting in reduced weight of the vehicle.

2.2 CHARACTERISTICS OF NANOFLUIDS

2.2.1 Enhanced thermal conductivity

One of the most significant characteristics of nanofluids is their enhanced thermal conductivity compared to the base fluid. The addition of nanoparticles significantly improves the ability of the fluid to conduct heat, making nanofluids effective in heat transfer applications.

2.2.2 Increased heat transfer rates

Due to the improved thermal conductivity, nanofluids can enhance heat transfer rates. This property is beneficial in applications such as heat exchangers, cooling systems, and electronic devices where efficient heat dissipation is crucial.

2.2.3 Enhanced convective heat transfer

Nanofluids can improve convective heat transfer, which is important in applications like automotive cooling systems and HVAC (heating, ventilation, and air conditioning) systems.

2.2.4 Stability

Achieving and maintaining stability in nanofluids is essential for practical applications. Stability refers to the ability of the nanoparticles to remain uniformly dispersed in the fluid without settling over time.

2.2.5 Viscosity

The addition of nanoparticles can influence the viscosity of nanofluids. While some nanofluids may exhibit increased viscosity, others may show a reduction, depending on factors such as nanoparticle concentration and size.

2.2.6 Brownian motion

The Brownian motion of nanoparticles in nanofluids contributes to their stability by preventing particles from agglomerating or settling. This motion is a result of the random movement of particles due to collisions with surrounding fluid molecules.

2.2.7 pH and chemical stability

The pH of nanofluids can impact their stability, and maintaining chemical stability is essential for long-term use. Researchers focus on developing nanofluids with stable chemical properties under various conditions.

2.2.8 Optical properties

Some nanofluids exhibit unique optical properties, such as increased light absorption or scattering. This characteristic can be relevant in applications like solar thermal systems.

2.2.9 Specific heat capacity

The specific heat capacity of nanofluids may differ from that of the base fluid due to the presence of nanoparticles. This property can influence the overall heat transfer performance of the nanofluid.

2.2.10 Cost and environmental considerations

The cost of producing nanofluids, including the cost of nanoparticles, and potential environmental impacts are important factors to consider in practical applications.

2.2.11 Thermal boundary layer improvement

Nanofluids can enhance the thermal boundary layer at the solid–fluid interface, improving overall heat transfer efficiency.

2.2.12 Dispersion

Achieving a uniform dispersion of nanoparticles in the fluid is critical. Agglomeration or clustering of nanoparticles can lead to reduced effectiveness and stability issues.

2.2.13 Size and shape dependence

The size, shape, and type of nanoparticles used in nanofluids can impact their properties. Small nanoparticles with a high aspect ratio tend to provide better thermal conductivity enhancement.

2.3 CHARACTERIZATION TECHNIQUES AVAILABLE FOR NANOFLUIDS

2.3.1 Transmission electron microscopy (TEM)

It has represented as a very powerful instrument for studying and researching about the structure of nanomaterials in the material science world. The quantitative measures of particle size, grain size, size distribution, size homogeneity, lattice type, morphological information, crystallographic details, chemical composition of phases distribution, and parameters can obtain by transmission electron micrographs. So, TEM is the best technique for characterization of the nanomaterials such as nanoparticles and nanocomposites. Moreover, this study shows that electron diffraction pattern via the TEM is a perfect procedure for determining the structure of materials, including perfect crystals, defect structure, and phases.

2.3.2 Scanning electron microscopy (SEM)

SEM is another imaging technique that provides detailed surface morphology information. It is useful for examining the surface characteristics and distribution of nanoparticles.

2.3.3 Dynamic light scattering (DLS)

It is a fast and non-invasive tool used to measure particle size, size distribution, and stability in solutions or suspensions. DLS is a non-destructive spectroscopic analysis technique that allows measuring corresponding to (i) light intensity scattered by the nanoparticles and (ii) size of particles suspended in a liquid from 1 to 1000 nm. The detection of scattered light from light-matter interactions provides information related to the physical properties of the sample. It measures the Brownian motion of nanoparticles in a fluid, providing information about their size distribution. This technique is valuable for assessing the stability of nanofluids.

2.3.4 Zeta potential

It is defined as electric potential in the interfacial double layer, which is between the stationary layer of fluid attached to the dispersed particles and the dispersion medium. Subsequently, electric surface charges are generated around nanoparticles that form an electrostatic field affecting the ions in the solution. Both thermal motion of ions and electrostatic field can screen the electric surface charge when the net surface charge is equal to the resultant electric charge in magnitude but opposite in polarity. Zeta potential is an effective parameter to evaluate nanofluid stability in the synthesis process of nanofluids.

2.3.5 X-ray diffraction (XRD)

XRD is a rapid analytical technique primarily used for phase identification of a crystalline material and can provide information on unit cell dimensions. The analyzed material is finely ground and homogenized, and average bulk composition is determined.

2.3.6 Fourier transform infrared (FTIR) spectroscopy

FTIR method is a type of spectroscopy that can detect changes in the total composition of biomolecules by determining changes in functional groups. The vibration and rotation of molecules influenced by infrared radiation at a particular wavelength are measured using FTIR. This method identifies structural differences in molecular binding between entities, which can reveal details about the existence of their interactions. Transmittance FTIR, attenuated total reflectance (ATR–FTIR), and micro-spectroscopy FTIR are the most popular FTIR-based methods for characterization.

2.3.7 UV-Visible (UV-Vis) spectroscopy

It measures the extinction (scatter + absorption) of light passing through a sample. Nanofluids have unique optical properties that are sensitive to the size, shape, concentration, agglomeration state, and refractive index near the nanoparticle surface, which makes UV-Vis a valuable tool for identifying, characterizing, and studying nanomaterials.

2.3.8 Thermal conductivity measurement

Techniques such as the transient hot-wire method or the laser flash method can be used to measure the thermal conductivity of nanofluids. This is crucial for understanding their heat transfer capabilities.

2.3.9 Differential scanning calorimetry (DSC)

It is a thermal analytical technique that measures the energy absorbed or emitted by a sample as a function of temperature. DSC equipment can be used to determine phase diagram, enthalpy, entropy, and specific heat.

2.3.10 Raman spectroscopy

It becomes a prominent analysis technique among other optical metrology techniques, such as those involving absorption of infrared light, particularly when water and other useful polar solvents were present, because these media typically strongly absorb light in the infrared region.

2.3.11 Atomic force microscope (AFM)

It allows for 3D characterization of nanoparticles with sub-nanometer resolution. Nanoparticle characterization using atomic force microscopy has a number of advantages over DLS, electron microscopy, and optical characterization methods. AFM is a high-resolution technique used to study the surface topography of nanoparticles in nanofluids. It complements other imaging techniques like TEM and SEM.

2.3.12 Nuclear magnetic resonance (NMR) spectroscopy

It is an another important analytical technique in the quantitative and structural determination of nanoscale materials. It is based on the NMR phenomenon exhibited by nuclei that possess non-zero spin when placed in a strong magnetic field, which causes a small energy difference between the 'spin-up' and 'spin-down' states. Transitions between these states can be probed by electromagnetic radiation in the radio wave range. NMR is typically used to study the interactions or coordination between the ligand and the surface of diamagnetic or antiferromagnetic nanoparticles.

2.3.13 Viscosity measurement

Various viscometers can be employed to measure the viscosity of nanofluids, providing insights into the fluid's flow behavior.

2.4 FLUID FLOW ANALYSIS

It is the process of examining the effects of adding nanoparticles on the entropy generation of water–Al_2O_3 and ethylene glycol–Al_2O_3 nanofluid flows through a circular pipe under uniform wall heat flux thermal boundary condition in both laminar and turbulent regimes. The nanofluid flow is studied in both laminar and turbulent regimes, and nanoparticles' volume concentration up to 5% is considered. Figure 2.2 displays the effects of adding Al_2O_3 nanoparticles to water on both fluid flow irreversibility number (NS,Δ_P) and heat transfer irreversibility number (NS,Δ_T). As it is expected, increase of Reynolds (Re) number causes ascent of NS,Δ_P and descent of NS,Δ_T. As shown in Figure 2.2 fluid flow irreversibility of water–Al_2O_3 laminar flow is too smaller than the heat transfer irreversibility ($NS,\Delta_P/NS,\Delta_T \approx 0.001$). Adding nanoparticles raises the fluid flow irreversibility (by increasing the viscosity of the nanofluid) and lowers the heat transfer irreversibility (by increasing the thermal conductivity and skin friction coefficient). Since the fluid flow irreversibility is negligible compared to the heat transfer irreversibility,

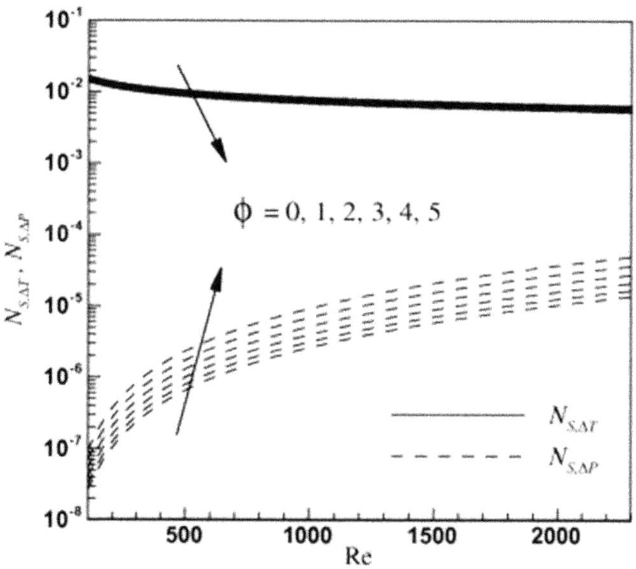

Figure 2.2 Heat transfer and pressure drop irreversibility numbers of water–Al_2O_3 nanofluid in laminar flow [2].

it has no effect on the total entropy generation. Therefore, in laminar flow of the water–Al_2O_3 nanofluid, adding nanoparticles causes irreversibility and entropy generation to decline.

Figure 2.3 shows the effect of adding nanoparticles and Re number on the entropy generation of the ethylene glycol–Al_2O_3 nanofluid for laminar flow. It is shown that only in very low Re numbers (Re < 11) heat transfer irreversibility is dominant, so adding nanoparticles results in reduction of the entropy generation. The minimum entropy generation points for each specified volume concentration are represented by triangular symbols in Figure 2.3. For higher Re numbers (Re > 14), the fluid flow irreversibility is dominant and increasing nanoparticles volume concentration leads to entropy generation raise. In laminar flow of the ethylene glycol–Al_2O_3 nanofluid, it is found that the maximum reduction of the entropy generation is about 33% which occurs at Re = 5.4 for ϕ = 5% and the highest increase of the entropy generation occurs at Re = 2300 with ϕ = 5% and is around 325%. These results show that using nanoparticles in a viscose fluid is only efficient for low Re numbers.

Figure 2.4 illustrates the effect of Re number variations and adding nanoparticles on the augmentation entropy generation number in turbulent flow of ethylene glycol–Al_2O_3 nanofluid. Since the fluid flow irreversibility is dominant even in laminar flow, it is expected that in turbulent flow, fluid flow irreversibility will be dominant again and adding nanoparticles has adverse

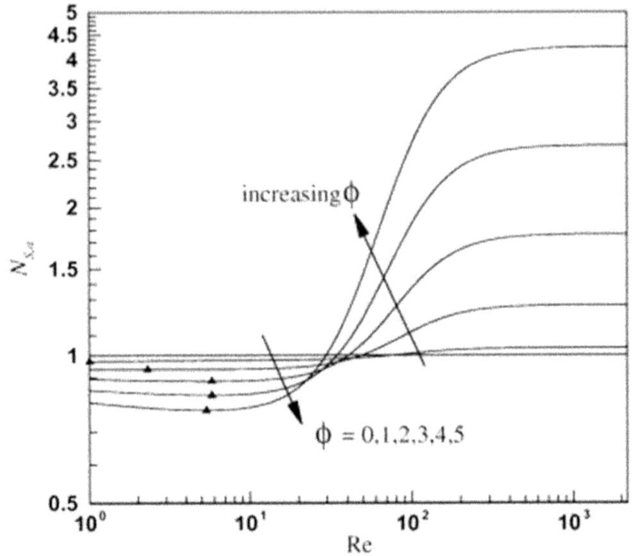

Figure 2.3 Augmentation entropy generation number of ethylene glycol–Al_2O_3 nanofluid in laminar flow [2].

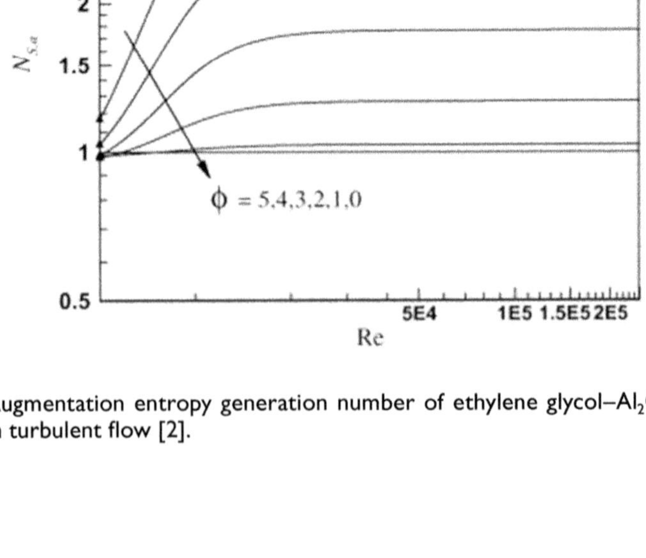

Figure 2.4 Augmentation entropy generation number of ethylene glycol–Al_2O_3 nanofluid in turbulent flow [2].

effects on the thermodynamic performance. As it can be seen in Figure 2.4, adding Al_2O_3 nanoparticles in turbulent flow of the ethylene glycol causes increase in the entropy generation. It is found that the largest augmentation entropy generation number is about 4.25 and occurs for $\phi = 5\%$ at Re numbers greater than about 40,000.

Another research deals with the investigation of magnetohydrodynamics (MHD) convective nanofluid flow over a vertical stretching sheet considering the thermal radiation and buoyancy effects. Three different base fluids and four different nanoparticles were examined as nanofluid. The governing nonlinear partial differential equations along with the boundary conditions were transformed into a dimensionless form; afterwards, they were solved numerically by the fourth-order Runge–Kutta method with the shooting technique. Parametric investigation was conducted, and the influence of various physical parameters, such as magnetic parameter (Mn), buoyancy parameter, nanoparticle size, nanoparticle concentrations, and radiation parameter on velocities and temperature profiles, skin fraction coefficient, and reduced Nusselt number, is reported.

The effect of particle volume fraction on velocities and temperature profiles of TiO_2–ethylene glycol 50% is shown in Figure 2.5. It is shown that the

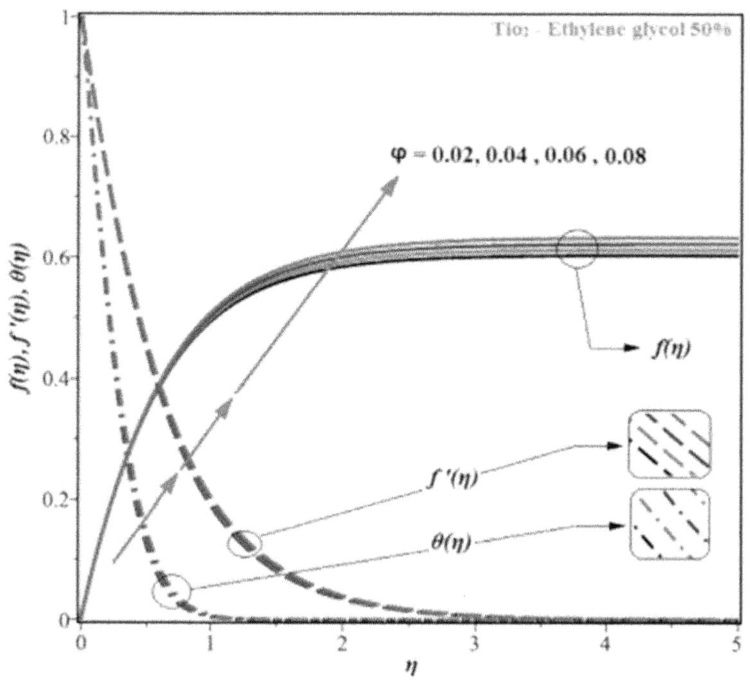

Figure 2.5 Effect of particle volume fraction on velocities and temperature profiles of TiO_2–ethylene glycol 50% when dp = 30 nm, N = 2, Mn = 2, Pr = 24.4, and $\lambda = 1$ [3].

velocity profiles and the temperature profile increase with the increasing the particle volume fraction.

Figure 2.6a–c represents the effect of buoyancy parameter on reduced Nusselt number, skin friction coefficient, velocity profiles and temperature profiles for TiO_2–ethylene glycol 50%. Since the momentum boundary layer thickness increases and temperature profile decreases with increasing values of buoyancy parameter enabling more flow, the rising values of buoyancy parameter result in thinning of the thermal boundary layer and hence produce an increase in the heat transfer rate and decrease the skin friction coefficient.

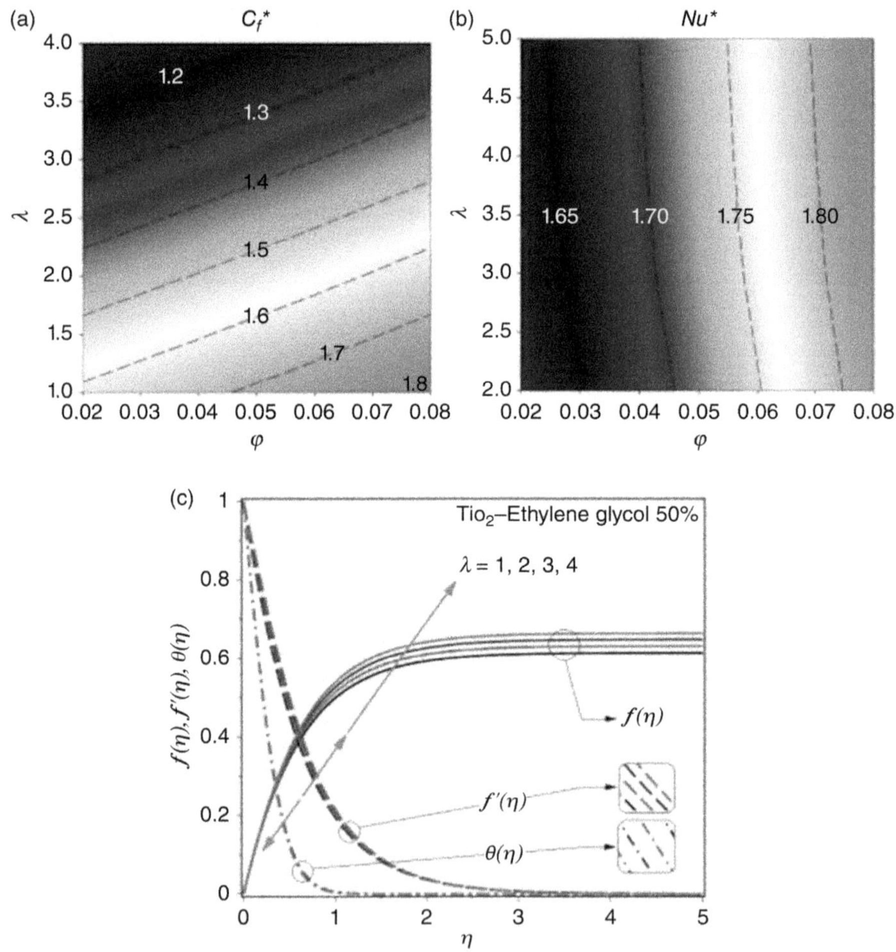

Figure 2.6 Effect of λ on (a) skin friction coefficient; (b) reduced Nusselt number when dp = 30 nm, N = 2, and Mn = 2; and (c) velocities and temperature profiles of TiO_2–ethylene glycol 50% when dp = 30 nm, N = 2, ϕ = 0.04, Pr = 24.4, and Mn = 2 [3].

The effect of Mn on the velocity profiles and the temperature profile of TiO_2–ethylene glycol 50% is shown in Figure 2.7a–c. It is depicted that the velocity profile decreases with the Mn and the temperature profile increases with the increasing of Mn. This is due to the fact that with the increasing Mn, the Lorentz force associated with the magnetic field increases and it produces more resistance to the flow and the nanoparticles by increasing Cf*. But the magnetic field enhances the temperature at all points leading to increase in thermal boundary layer thickness and decrease in the Nu* as it is shown graphically.

Cu–water nanofluid flow analysis between two parallel palates is also investigated using a differential transformation method (DTM) and

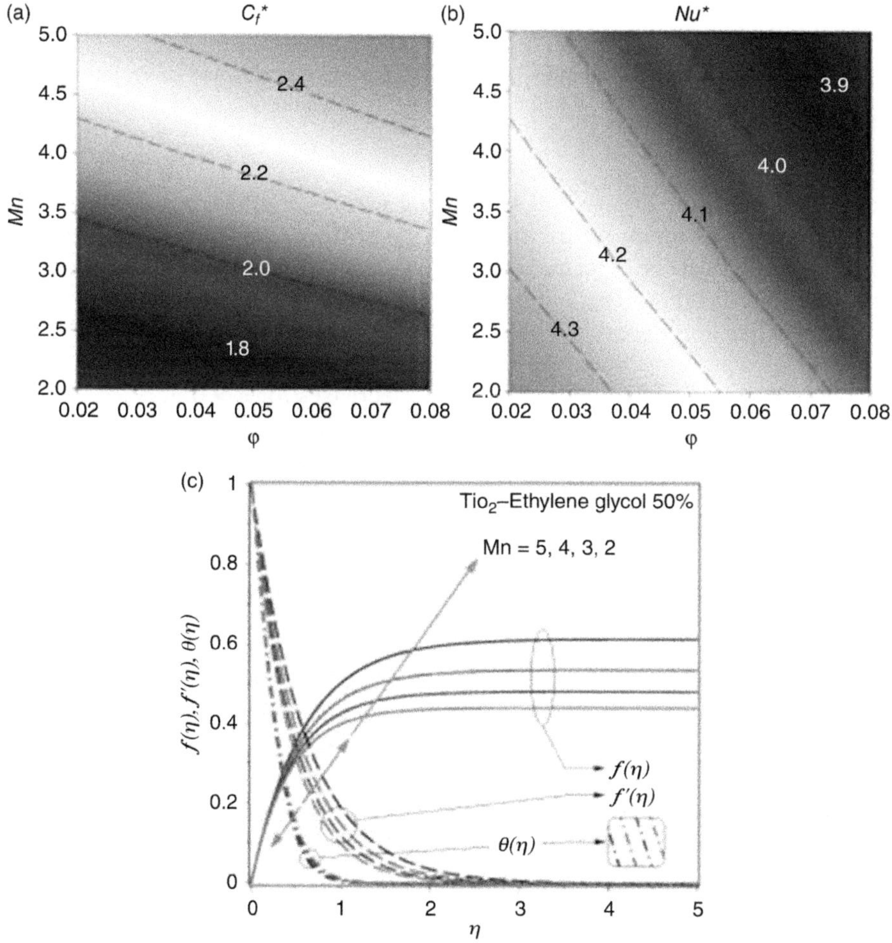

Figure 2.7 Effect of Mn on (a) skin friction coefficient; (b) reduced nusselt number when dp = 30 nm, N = 2, and λ = 1; and (c) velocities and temperature profiles of TiO_2–ethylene glycol 50% when dp = 30 nm, N = 2, ϕ = 0.04, Pr = 24.4, and λ = 1 [3].

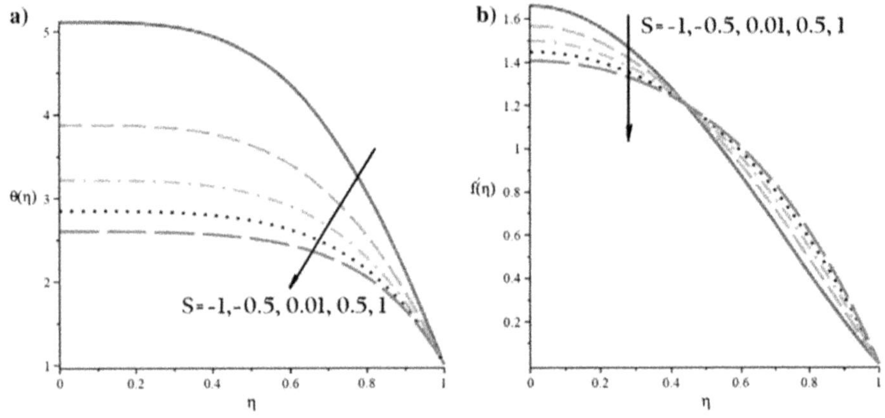

Figure 2.8 Effect of squeeze number (s) on a) temperature profile and b) velocity profile when Pr = 6.2, Ec = 0.5, δ = 0.1, and ϕ = 0.06 [4].

numerical method. Figure 2.8 shows the effect of squeeze number (S) on non-dimensional temperature and velocity profiles, respectively. When two plates move together, thermal boundary layer thickness increases as the absolute magnitude of the S enhances.

The positive and negative S has different effects on the velocity profile. For the case of squeezing flow, the velocity increases due to an increase in the absolute value of S when $\eta < 0.5$, while it decreases for $\eta > 0.5$.

In other words, when S is negative, α is negative too and by decreasing its magnitude, its behavior is like squeezing flow so the temperature profile is decreased. The same treatment is observed in the velocity profile. By increasing in positive values for the S number, the space between the two plates decreased and consequently the velocity profile near the lower plate is decreased. The effect of nanoparticles volume fraction on velocity and temperature profiles is shown in Figure 2.9. Adding nanoparticles into the base fluid leads to increase in thermal boundary layer thickness while it has no significant effect on velocity boundary layer thickness. It's obvious that when nanoparticles were added to the base fluid, heat transfer will increase due to their high thermal conductivity, so temperature profiles will decrease.

Figure 2.10 depicts the effect of the Eckert number (Ec) on temperature profile. As seen in this figure, by increasing the Ec, the thermal boundary layer is decreased and the maximum nanofluid temperature between two plates is increased significantly which is due to the increase in the viscous dissipation effect.

Another work is to elaborate the characteristics of heat transport and MHD finite film flow of human blood with carbon nanotubes (CNTs) nanofluids over a stretchable upright cylinder. Two kinds of CNTs nanoparticles, namely

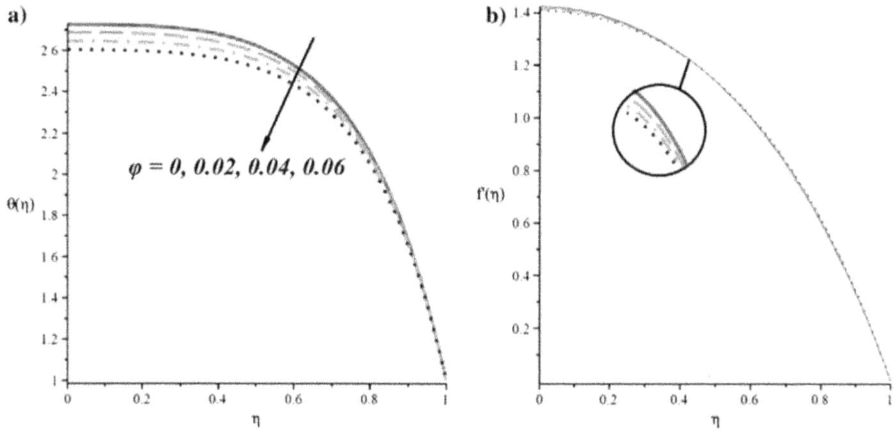

Figure 2.9 Effect of nanoparticles volume fraction (φ) on a) temperature profile and b) velocity profile when Pr = 6.2, Ec = 0.5, δ = 0.1, and S = 1.0 [4].

(i) SWCNTs (single-walled CNTs) and (ii) MWCNTs (multi-walled CNTs), are used with human blood as a base liquid. In addition, a uniform magnetic field (B) has been conducted perpendicularly to the motion of nanoliquid.

Figure 2.11 depicts the behavior of df (η)/dη (variation in the velocity field) by varying the Mn and thin film thickness parameter α of both CNTs (SWCNTs and MWCNTs) nanofluids. The impacts of these quantities on df (η)/dη are very clear during the flow of both CNTs nanofluids. It can be noted that the larger magnitude of M reduced the fluid motion in both cases

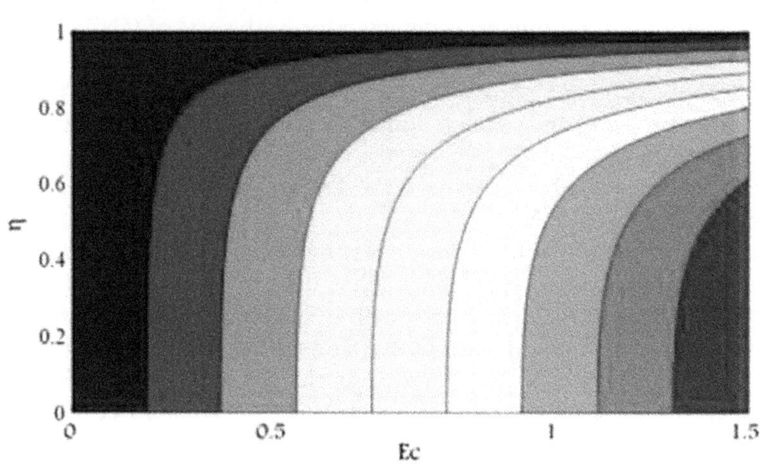

Figure 2.10 Effect of Eckert number (Ec) on temperature profile when Pr = 6.2, ϕ = 0.06, δ = 0.1, and S = 1.0 [4].

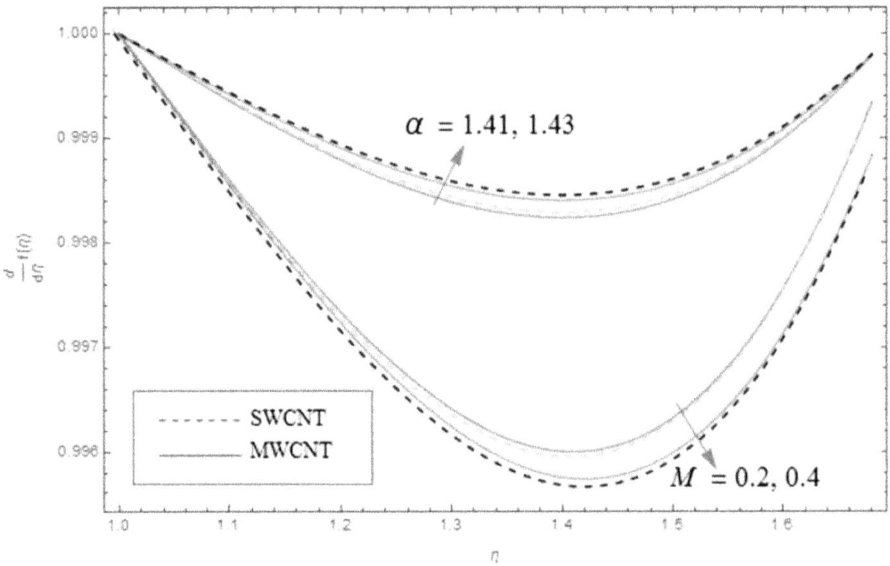

Figure 2.11 df (η)/dη distribution for varying M and α [5].

(SWCNTs and MWCNTs). Physically, such a situation arises as a result of a constantly applied magnetic field β_o that can be induced current in inductive nanoliquid. It creates resistant forces called Lorentz forces, which reduce the liquid velocity. Finally, it is clear that β_o is used to govern the boundary layer separation. Comparatively, a rapid fall in the velocity field is perceived in the case of SWCNTs as related to the MCWNTs. In Figure 2.11, the effect of a (thin film nanofluid parameter) is depicted for both sorts of CNTs nanofluids. It can be observed that by increasing the value of α, the fluid motion decelerates because in this case, the mass of the fluid is enhanced. Actually, the tiny size of film accelerates the velocity and less energy is required for fluid motion, for example, the flow in the pipe is much easier and faster than the flow in sea water. Moreover, in the case of MWCNTs, the velocity field is dominant when compared to SWCNTs.

Figure 2.12 presents the velocity distribution df (η)/dη for several values of Grashof number (Gr_ using SWCNTs- and MWCNTs-based nanofluid. The velocity df (η)/dη elevates for both CNTs by maximizing the value of Gr. Similarly, the velocity field df (η)/dη shows the slowing behavior for both CNTs (SWCNTs and MWCNTs), reducing the value of Gr. Actually, the ratio of the thermal buoyancy force in the direction of viscous force is termed the Gr. Therefore, the basic reason for this is that in the absenteeism of buoyancy force, there is no motion of fluid. The present outline indicates that motion of liquid is occurring due to the buoyancy force and the liquid is stationary in

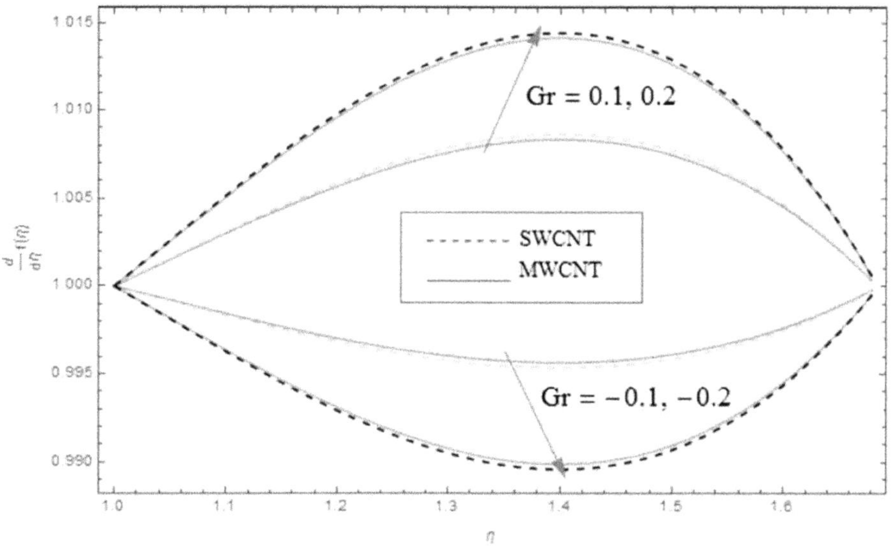

Figure 2.12 df (η)/dη distribution for varying Gr [5].

the absence of this force. In addition, it is clear from the figure that SWCNTs are more dominant than MWCNTs.

Investigation of heat transfer behavior of hybrid nanofluid (HyNF) flow through the tubular heat exchanger is experimentally studied. In this analysis, the effects of thermal characteristics of forced convection, thermal conductivity, and heat transfer coefficient were explored. The nanofluid was prepared by dispersing the copper–titanium hybrid nanocomposite (HyNC) in the water. The experiments were performed for various nanoparticle volume concentrations added in the base fluid ranging from 0.1% to 1.0%.

The thermal conductivity of HyNF as a function of the particle volume concentration and temperature is depicted in Figure 2.13. The result showed that the thermal conductivity of HyNF increased with the nanocomposite volume concentration and the nanofluid temperature. With the increase of fluid temperature, the movement between the particles increased, raising the thermal conductivity [6]. The probable cause of increase in thermal conductivity of the nanofluids not only depended on particle concentration and temperature but also due to the influence of other parameters like size and shape. It is usual that the tiny dissimilarity of the trends may perhaps due to the particle configuration and clustering [7].

The copper nanoparticles on the titania surface act as the thermal boundary between the base fluid and the wall surfaces which induced the enhancement

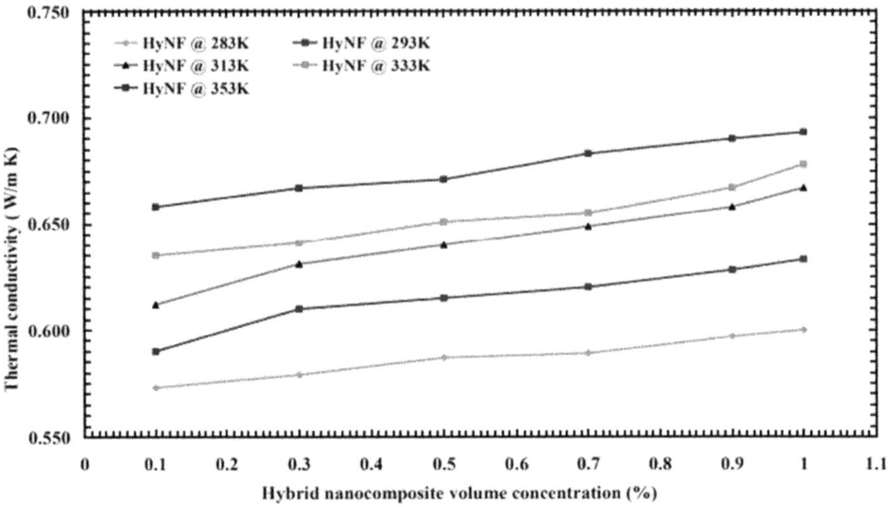

Figure 2.13 Variation of thermal conductivity of hybrid nanofluid (HyNF) with various volume concentration of hybrid nanocomposite [8].

of thermal conductivity of nanofluid. Titania material is one of the better heat transfer material and economical for the heat transfer applications typically suitable for nanofluid. From this perception, the creation of highly crystalline and heat-conductive copper nanoparticles on the surface of titania particles was distinct and significant, which collectively enhanced the capabilities of nanofluids in capturing and dissipating thermal energy from the heat source and dissipating it to the surrounding fluid.

The addition of nanoparticle concentration of HyNC led to the formation of compactly packed thermal interfaces, which in turn contributed to accomplish a better thermal conductivity of the nanofluids. This can be considered to be an indispensable factor for utilizing HyNC in the base fluid, as a mean to achieve improved thermophysical characteristics, and the heat transfer potential of nanofluids.

Variations of convective heat transfer coefficient versus Re numbers for various concentrations of nanoparticles are illustrated in Figure 2.14. Based on the results, it was clear that at the lower volume concentrations of less than 0.7% HyNC nanoparticle, it showed better heat transfer behavior than at the higher volume concentrations of greater than 0.7%. For the lesser particle sizes, the phonon heat transfer was adequate to raise the temperature of the particle to that of heat source [9]. It was seen that the nanoparticles suspended in base fluid increases the heat transfer coefficient even for a very low particle weight fraction of 0.1% and the maximum heat transfer enhancement is obtained for nanofluids with 0.7 vol. % of nanoparticle concentration. At 1.0 vol. % concentration of nanoparticles, compared to the base fluid the

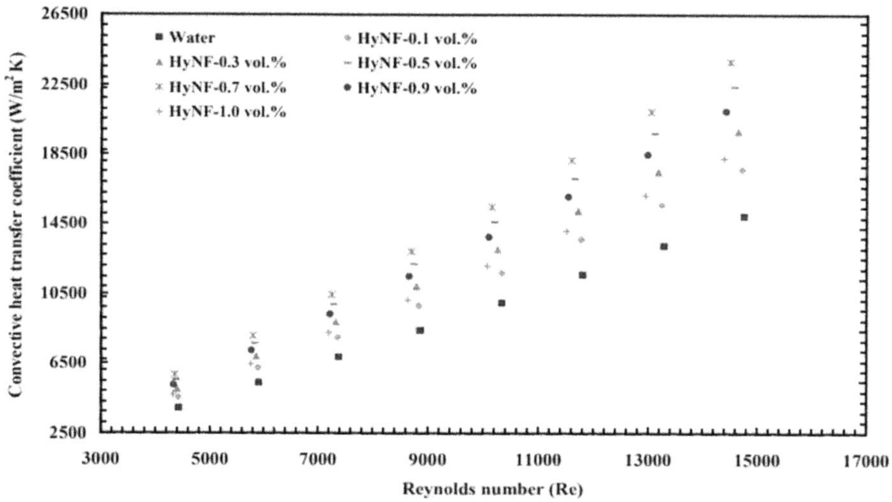

Figure 2.14 Variation of convective heat transfer coefficient with Re number [8].

heat transfer coefficient was increased from 20.8% to 24.4% for the various ranges of Re number starting from 4315 to 14,384.

Further, the heat transfer coefficient increased by 48.4% compared to the base fluid at 0.7% volume concentration for the Re number of 4348. The reason for the increase in heat transfer was due to the motion of nanoparticles which caused microconvection in fluid [10]. From this study, it was known that the heat transfer enhancement of nanofluids was more at high Re numbers. The higher values of Re numbers caused turbulence due to the higher velocity of nanoparticles which in turn enhanced the heat transfer. It can be concluded from the figure that other than the thermal conductivity some of the other mechanisms also played the role in enhancement of heat transfer of nanofluids.

2.5 RHEOLOGICAL EVALUATION

It is the method of examining the effects of parameters such as nanoparticle size and shear rate on rheological behavior of nanofluids. Rheological behavior of nanofluids influences on their pressure drop. On the other hand, rheological behavior of nanofluids can provide the idea of making materials with optimal rheological structure, which is useful to predict their thermal conductivity. To determine the rheological behavior of nanofluids, some researchers used rheometer [11–13] and some other viscometers [14, 15]. The viscometer is a device for measuring the viscosity of fluids and it is not able to measure the fluids that are flowing. For this purpose, the rheometer is used to measure the flow properties of the fluids and similar pseudo-solids and measure the amount of stress in the substrate. The rheometer also has the ability to measure the rheological properties of different materials in a wide range of shear rates and temperatures [16, 17].

It is observed that when the TiO_2 nanoparticles are added to the SiO_2/water nanofluid, the rheological behavior of the nanofluid is changed from Newtonian fluid to non-Newtonian one [14]. Adding TiO_2 nanoparticles to the ethylene glycol base fluid even at high shear rates leads to Newtonian behavior. The nanofluid MWCNTs have both Newtonian and non-Newtonian behavior based on the type of fluid. The combination of MWCNTs in the base water, oil, and resin fluid exhibits a shear-thinning behavior [18]. Generally, nanofluids containing MWCNTs with high volume fractions exhibit non-Newtonian behavior. They show Newtonian behavior at low volume fractions [19, 20]. SiO_2 nanoparticles with different base fluids exhibit Newtonian behavior [21]. Al_2O_3 nanoparticles with a water-base fluid has non-Newtonian behavior [22]. The relative viscosity of nanofluids containing CNTs significantly increases by adding CNTs. In fact, cylindrical shape and large length of MWCNTs reduce the ability of movement between the fluid layers [23].

Another study presents the viscosity of pure coconut oil/Al_2O_3 nanofluids with different solid volume fraction of 0.2–1.2%. The experiments were conducted at different temperature varying from 303 to 413 K with shear rate ranging from 10 to 50 per second. It is found that the viscosity of the nanofluids as well as base fluid (coconut oil) declines with an increase in temperature. They found that, in the direction of increasing shear, the molecular structure of fluid varies temporarily and slowly align themselves, as the spindle rotates at the low speed in the fluid. Therefore, at low shear rate, a less resistance in fluid or a reduction in viscosity of fluid takes place. Moreover, it is also observed that, as shear rate for nanofluids with different solid volume fraction increased, the viscosity of high-solid-volume-fraction-based nanofluid showed higher shear thinning compared to that of low solid volume fraction same nanofluids. This showed that, at higher solid volume fraction, both nanoparticle and base fluid have a significant effect on shear thinning. The solid volume fraction has a remarkable effect on the viscosity for the shear rate ranging from 10 to 50 per second. Meanwhile, as the shear rate declines, the viscosities of nanofluids are increased.

2.6 CONCLUSION

In this chapter, various experimental techniques for analyzing the nanofluids is discussed. Moreover, different characteristics and its corresponding testing methods are identified. The effects of adding nanoparticles on the flow behavior with respect to different profiles and the effects of nanoparticle size (SiO_2,

TiO_2, Al_2O_3, and CNT) and shear rate on rheological behavior of nanofluids are examined using the appropriate testing method. According to these characteristics, nanofluids with excellent thermal qualities is effectively used for heat transfer applications.

REFERENCES

[1]. W. Yu, H. Xie, A review on nanofluids: Preparation, stability mechanisms, and applications, Journal of Nanomaterials, 2012 (2012) 1–17.
[2]. M. Moghaddami, A. Mohammadzade, S.A.V. Esfehani, Second law analysis of nanofluid flow, Energy Conversion and Management, 52 (2011) 1397–1405.
[3]. O. Pourmehran, M. Rahimi-Gorji, D. Ganji, Heat transfer and flow analysis of nanofluid flow induced by a stretching sheet in the presence of an external magnetic field, Journal of the Taiwan Institute of Chemical Engineers, 65 (2016) 162–171.
[4]. G. Domairry, M. Hatami, Squeezing Cu–water nanofluid flow analysis between parallel plates by DTM-Padé method, Journal of Molecular Liquids, 193 (2014) 37–44.
[5]. A.S. Alsagri, S. Nasir, T. Gul, S. Islam, K. Nisar, Z. Shah, I. Khan, MHD thin film flow and thermal analysis of blood with CNTs nanofluid, Coatings, 9 (2019) 175.
[6]. M. Elias, I.M. Mahbubul, R. Saidur, M. Sohel, I. Shahrul, S. Khaleduzzaman, S. Sadeghipour, Experimental investigation on the thermo-physical properties of Al2O3 nanoparticles suspended in car radiator coolant, International Communications in Heat and Mass Transfer, 54 (2014) 48–53.
[7]. H. Zhu, C. Zhang, S. Liu, Y. Tang, Y. Yin, Effects of nanoparticle clustering and alignment on thermal conductivities of Fe3O4 aqueous nanofluids, Applied Physics Letters, 89 (2006).
[8]. D. Madhesh, S. Kalaiselvam, Experimental analysis of hybrid nanofluid as a coolant, Procedia Engineering, 97 (2014) 1667–1675.
[9]. V. Karthik, S. Sahoo, S. Pabi, S. Ghosh, On the phononic and electronic contribution to the enhanced thermal conductivity of water-based silver nanofluids, International Journal of Thermal Sciences, 64 (2013) 53–61.
[10]. Y. Ren, H. Xie, A. Cai, Effective thermal conductivity of nanofluids containing spherical nanoparticles, Journal of Physics D: Applied Physics, 38 (2005) 3958.
[11]. Y. He, Y. Jin, H. Chen, Y. Ding, D. Cang, H. Lu, Heat transfer and flow behaviour of aqueous suspensions of TiO2 nanoparticles (nanofluids) flowing upward through a vertical pipe, International Journal of Heat and Mass Transfer, 50 (2007) 2272–2281.
[12]. H. Chen, Y. Ding, C. Tan, Rheological behaviour of nanofluids, New Journal of Physics, 9 (2007) 367.
[13]. H. Chen, Y. Ding, Y. He, C. Tan, Rheological behaviour of ethylene glycol based titania nanofluids, Chemical Physics Letters, 444 (2007) 333–337.
[14]. W. Richmond, R. Jones, P. Fawell, The relationship between particle aggregation and rheology in mixed silica–titania suspensions, Chemical Engineering Journal, 71 (1998) 67–75.
[15]. W.J. Tseng, K.-C. Lin, Rheology and colloidal structure of aqueous TiO2 nanoparticle suspensions, Materials Science and Engineering: A, 355 (2003) 186–192.
[16]. B. Ambasta, Chemistry for engineers, Laxmi Publications, 2008.

[17]. T. Mezger, The rheology handbook: For users of rotational and oscillatory rheometers, European Coatings, 2020.
[18]. G.H. Ko, K. Heo, K. Lee, D.S. Kim, C. Kim, Y. Sohn, M. Choi, An experimental study on the pressure drop of nanofluids containing carbon nanotubes in a horizontal tube, International Journal of Heat and Mass Transfer, 50 (2007) 4749–4753.
[19]. P. Pötschke, T. Fornes, D. Paul, Rheological behavior of multiwalled carbon nanotube/polycarbonate composites, Polymer, 43 (2002) 3247–3255.
[20]. Y. Yang, E.A. Grulke, Z.G. Zhang, G. Wu, Thermal and rheological properties of carbon nanotube-in-oil dispersions, Journal of Applied Physics, 99 (2006) 114307-1–114307-8.
[21]. J. Chevalier, O. Tillement, F. Ayela, Rheological properties of nanofluids flowing through microchannels, Applied Physics Letters, 91 (2007) 233103-1–233103-3.
[22]. W.J. Tseng, C.H. Wu, Aggregation, rheology and electrophoretic packing structure of aqueous A12O3 nanoparticle suspensions, Acta Materialia, 50 (2002) 3757–3766.
[23]. A.A. Nadooshan, H. Eshgarf, M. Afrand, Evaluating the effects of different parameters on rheological behavior of nanofluids: A comprehensive review, Powder Technology, 338 (2018) 342–353.

Chapter 3

The application of nanofluids in thermal management systems

Rashi Bhargava, Ankit Agrawal, and Kanchan Bhardwaj

3.1 INTRODUCTION

In recent years, the quest for efficient thermal management solutions has led to the exploration of nanotechnology-based approaches among which nanofluids have garnered significant attention. Nanofluids are engineered colloidal suspensions containing nanoparticles dispersed in a base fluid, offering improved thermal conductivity compared to conventional heat transfer fluids (Choi & Eastman, 1995; Eastman et al., 2001). This chapter provides an overview of the application of nanofluids in diverse thermal management systems, elucidating their potential benefits and challenges.

3.2 ENHANCED THERMAL CONDUCTIVITY OF NANOFLUIDS

The remarkable enhancement in thermal conductivity exhibited by nanofluids can be attributed to various mechanisms at the nanoscale level. Brownian motion and particle-particle interactions play crucial roles in augmenting thermal transport properties (Keblinski et al., 2002; Murshed et al., 2008). Additionally, the high surface-area-to-volume ratio of nanoparticles facilitates efficient heat transfer between the solid phase and the base fluid. Furthermore, the selection of appropriate nanoparticles and base fluids, along with optimized dispersion techniques, influences the thermal conductivity enhancement of nanofluids (Wang et al., 2007; Wen & Ding, 2004).

3.3 METHODS OF NANOFLUID PREPARATION

The one-step method in nanofluid filtration involves the direct incorporation of nanoparticles into the filtration medium solution in a single stage, thereby eliminating the need for additional processing steps

DOI: 10.1201/9781003527251-3

(Choi & Eastman, 1995). This approach is effective because it can achieve the intended filtration outcomes with good efficiency. Nanoparticles are dispersed within the filtration matrix enhancing its performance by altering surface properties. One-step methods have been explored extensively in recent research to develop advanced filtration systems with improved capabilities for various applications. For instance, Chen and Li (2020) demonstrated the efficacy of a one-step method utilizing graphene oxide nanoparticles for water purification highlighting its potential for sustainable filtration solutions.

The two-step method in nanofluid filtration involves a sequential process where nanoparticles are synthesized separately before being incorporated into the filtration medium (Sadasivuni et al., 2018). This method consists of two distinct stages: the preparation of nanoparticles with desired properties followed by their integration into the filtration system. The first step may include the synthesis of nanoparticles through various techniques such as chemical precipitation, sol-gel or hydrothermal methods along with surface modification for improved compatibility with the filtration matrix. In the second step, the synthesized nanoparticles are introduced into the filtration medium where they can enhance filtration efficiency by modifying pore size distribution. Two-step methods offer flexibility in tailoring nanoparticle properties and optimizing filtration performance but may involve more complex procedures compared to one-step approaches (Chopkar et al., 2006). For instance Sadasivuni et al. (2018) employed a two-step method involving the synthesis of silver nanoparticles followed by their incorporation into polymeric membranes for antibacterial water filtration applications. Figure 3.1 shows the method of nanofluid preparation, and Table 3.1 shows the difference between both methods.

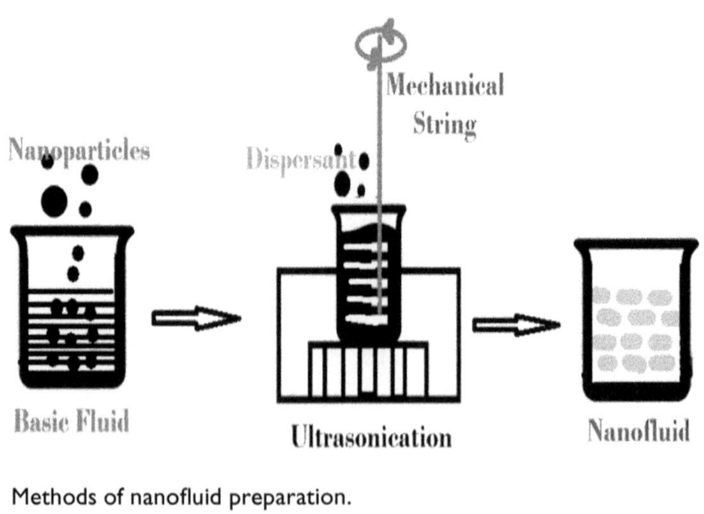

Figure 3.1 Methods of nanofluid preparation.

Table 3.1 Difference between one- and two-step methods of nanofluids preparation

Aspect	*One-step method*	*Two-step method*
Process complexity	Simple	Moderate to high
Number of stages	Single	Two
Synthesis of nanoparticles	Nanoparticles are directly incorporated during filtration medium preparation	Nanoparticles are synthesized separately before integration into filtration medium
Surface modification	Typically limited	Can be extensively tailored during synthesis
Filtration medium	Nanoparticles dispersed within the medium	Nanoparticles integrated into the medium afterward
Control over properties	Limited	Enhanced control over nanoparticle properties
Advantages	Simplicity, cost-effectiveness, efficiency	Flexibility in nanoparticle properties, optimization of filtration performance
Disadvantages	Limited control over nanoparticle properties	Increased complexity, potentially higher production costs

3.4 APPLICATION OF NANOFLUIDS IN ELECTRONICS COOLING

The cooling of electronic devices is essential to maintain optimal operating temperatures and prevent overheating-induced malfunctions. Figure 3.2 depicts the application of nanofluids in cooling. Nanofluids offer a promising

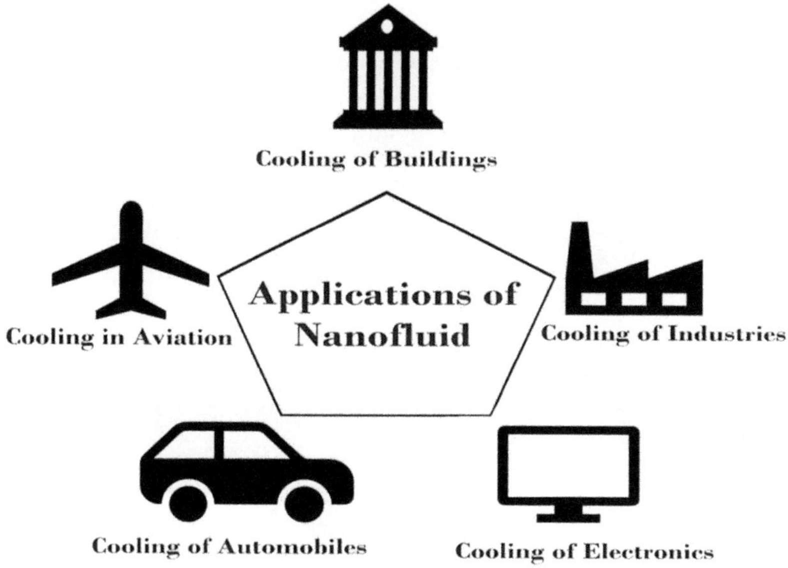

Figure 3.2 Application of nanofluids in cooling.

solution for enhancing heat transfer in electronics cooling systems due to their superior thermal properties (Yu et al., 2008). By circulating nanofluid coolant through microchannels or heat sinks within electronic devices, efficient heat dissipation can be achieved, thereby improving device reliability and longevity.

3.5 NANOFLUID-BASED HEAT EXCHANGERS AND MECHANISM

Heat exchangers play a vital role in various industrial processes facilitating the exchange of heat between fluid streams. Nanofluids have demonstrated potential for enhancing the performance of heat exchangers by increasing heat transfer rates and reducing thermal resistance (Khanafer et al., 2003). Incorporating nanofluid coolant in heat exchanger systems enhances convective heat transfer coefficients leading to improved overall heat transfer efficiency and reduced energy consumption (Esfe et al., 2017).

The enhanced thermal conductivity of nanofluids can be attributed to the increased surface-area-to-volume ratio of nanoparticles compared to their bulk counterparts facilitating more efficient heat transfer at the fluid-solid interface (Mahian et al., 2013). The Brownian motion of nanoparticles within the fluid leads to enhanced convection and reduced thermal boundary layers, further promoting heat transfer. Moreover, the presence of nanoparticles alters the phonon dynamics within the fluid resulting in enhanced phonon scattering and thus higher thermal conductivity (Xie et al., 2002). Figure 3.3 shows nanofluids-based heat exchangers. These mechanisms collectively contribute to the overall improvement in thermal conductivity observed in nanofluids (Wen & Ding, 2004).

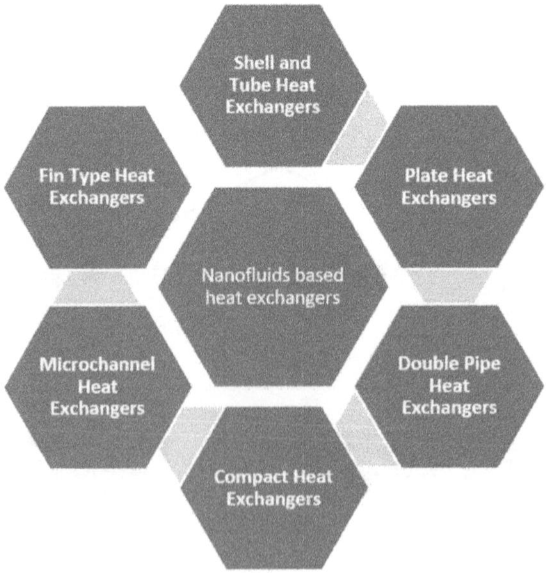

Figure 3.3 Nanofluids-based heat exchangers.

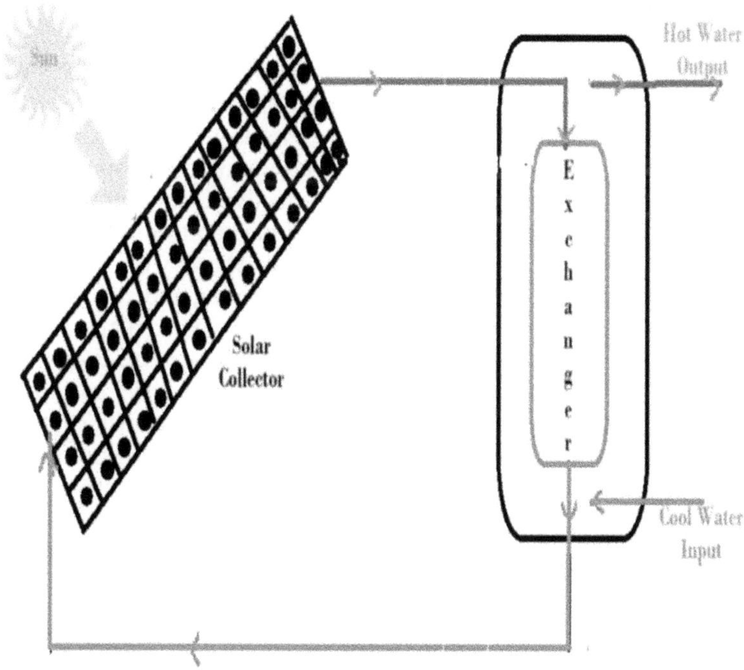

Figure 3.4 Utilization of nanofluids in solar collectors.

3.6 UTILIZATION OF NANOFLUIDS IN SOLAR COLLECTORS

Solar collectors harness solar energy for various applications, including electricity generation and water heating (Wang et al., 2007) (Figure 3.4). Nanofluids offer a compelling option for enhancing the thermal performance of solar collectors by increasing the absorption of solar radiation and improving heat transfer characteristics. By circulating nanofluid through the absorber tubes of solar collectors, greater heat absorption and transfer efficiencies can be achieved, thereby enhancing overall system efficiency and energy output.

3.7 RECENT ADVANCEMENTS AND CHALLENGES

Recent advancements in nanofluid research have focused on tailoring nanoparticle properties and optimizing nanofluid formulations for specific applications. However, several challenges persist in the practical implementation of nanofluids, including nanoparticle stability, long-term performance and cost-effectiveness (Kole & Dey, 2013; Zhang et al., 2016). Addressing these challenges requires continued research efforts to elucidate fundamental mechanisms and develop robust nanofluid formulations tailored to different thermal management applications.

3.8 CONCLUSION

The application of nanofluids in thermal management systems presents a promising avenue for enhancing heat transfer efficiency and addressing challenges associated with conventional heat transfer fluids. Through the utilization of colloidal suspensions containing nanoparticles dispersed in a base fluid, nanofluids offer a unique opportunity to overcome limitations in thermal conductivity and heat dissipation. The remarkable enhancement in thermal properties exhibited by nanofluids has paved the way for their integration into diverse applications including electronics cooling and solar collectors.

By leveraging the mechanisms underlying the enhanced thermal conductivity of nanofluids and employing advanced synthesis and dispersion techniques, researchers can tailor nanofluid formulations to meet specific thermal management requirements. The ability to enhance heat transfer rates, reduce thermal resistance and improve overall system efficiency positions nanofluids as a valuable asset in the quest for sustainable and energy-efficient thermal management solutions.

Despite the considerable progress made in nanofluid research, several challenges remain, including long-term performance and cost-effectiveness. Addressing these challenges necessitates ongoing interdisciplinary collaboration and innovative approaches to overcome barriers hindering the practical implementation of nanofluids in real-world applications.

As research in nanofluid technology continues to evolve, future advancements hold the potential to revolutionize thermal management practices across various industries leading to more efficient and environmentally sustainable systems. By harnessing the unique properties of nanofluids and translating scientific discoveries into practical solutions, we can unlock new opportunities for enhancing heat transfer efficiency and meeting the growing demands of modern thermal management challenges.

REFERENCES

Chen, L., Wang, W., Fang, Q., Zuo, K., Hou, G., Ai, Q., ... & Lou, J. (2020). High performance hierarchically nanostructured graphene oxide/covalent organic framework hybrid membranes for stable organic solvent nanofiltration. Applied Materials Today, 20, 100791.

Choi, S. U., & Eastman, J. A. (1995). *Enhancing thermal conductivity of fluids with nanoparticles* (No. ANL/MSD/CP-84938; CONF-951135-29). Argonne National Lab. (ANL), Argonne, IL.

Chopkar, M., Sudarshan, S., Das, P. K., & Manna, I. (2006). *Synthesis and stability of nanofluids—A review. Bulletin of Materials Science*, 29(6), 569–583. https://doi.org/10.1007/BF02705293

Eastman, J. A., Choi, S. U. S., Li, S., Yu, W., & Thompson, L. J. (2001). Anomalously increased effective thermal conductivities of ethylene glycol-based nanofluids containing copper nanoparticles. *Applied Physics Letters*, *78*(6), 718–720.

Esfe, M. H., Saedodin, S., Rostamian, H., Toghraie, D., & Afrand, M. (2017). A review on applications and challenges of nanofluids. *Renewable and Sustainable Energy Reviews*, 70, 1088–1101. https://doi.org/10.1016/j.rser.2016.12.085

Keblinski, P., Phillpot, S. R., Choi, S. U. S., & Eastman, J. A. (2002). Mechanisms of heat flow in suspensions of nano-sized particles (nanofluids). *International Journal of Heat and Mass Transfer*, *45*(4), 855–863.

Khanafer, K., Vafai, K., & Lightstone, M. (2003). Buoyancy-driven heat transfer enhancement in a two-dimensional enclosure utilizing nanofluids. *International Journal of Heat and Mass Transfer*, *46*(19), 3639–3653.

Kole, M., & Dey, T. K. (2013). Enhanced thermophysical properties of copper nanoparticles dispersed in gear oil. *Applied thermal engineering*, 56(1–2), 45–53. https://doi.org/10.1016/j.applthermaleng.2013.03.022

Mahian, O., Kianifar, A., Sahin, A. Z., Wongwises, S., & Pop, I. (2013). A review of entropy generation in nanofluid flow. *International Journal of Heat and Mass Transfer*, 65, 514–532. https://doi.org/10.1016/j.ijheatmasstransfer.2013.06.010

Murshed, S. M. S., Leong, K. C., & Yang, C. (2008). Investigations of thermal conductivity and viscosity of nanofluids. *International Journal of Thermal Sciences*, 47(5), 560–568. https://doi.org/10.1016/j.ijthermalsci.2007.05.004

Sadasivuni, K. K., Rattan, S., Waseem, S., Brahme, S. K., Kondawar, S. B., Ghosh, S., ... & Mazumdar, P. (2019). Silver nanoparticles and its polymer nanocomposites—Synthesis, optimization, biomedical usage, and its various applications. Polymer nanocomposites in biomedical engineering, 331–373.

Wang, L.& Wei, X. (2009). Nanofluids: Synthesis, heat conduction, and extension. *Journal of Heat Transfer*, 131(3), 033102 (7 pages). https://doi.org/10.1115/1.3056597

Wen, D., & Ding, Y. (2004). Experimental investigation into convective heat transfer of nanofluids at the entrance region under laminar flow conditions. *International Journal of Heat and Mass Transfer*, 47(24), 5181–5188.

Xie, H., Wang, J., Xi, T., Liu, Y., & Ai, F. (2002). Thermal conductivity enhancement of suspensions containing nanosized alumina particles. *Journal of Applied Physics*, 91(7), 4568–4572. https://doi.org/10.1063/1.1454184 (not accessible as of August 8, 2024)

Yu, W., France, D. M., Routbort, J. L., & Choi, S. U. S. (2008). Review and comparison of nanofluid thermal conductivity and heat transfer enhancements. *Heat Transfer Engineering*, 29(5), 432–460. https://doi.org/10.1080/01457630701850851

Zhang, J., Kong, Q., Yang, L., & Wang, D. Y. (2016). Few layered Co (OH) 2 ultrathin nanosheet-based polyurethane nanocomposites with reduced fire hazard: from eco-friendly flame retardance to sustainable recycling. Green Chemistry, 18(10), 3066–3074. https://doi.org/10.1039/C5GC03048J

Chapter 4

Optimizing SAH performance

A computational approach with nanofluid integration

Arun Kumar Singh Gangwar, Mohd Anas, and Nitesh Dutt

4.1 INTRODUCTION

The world needs sustainable energy sources to meet its growing demands. Renewable energy offers a promising alternative to conventional systems. Solar energy, in particular, is a frontrunner due to its versatility. Solar systems can convert sunlight directly into electricity using photovoltaic panels. They can also generate heat indirectly through concentrated solar power systems. Solar air heaters (SAHs) take yet another approach, utilizing the greenhouse effect to capture solar energy as heat. Researchers are actively exploring the potential of different solar technologies. A recent study analyzed highly cited publications containing keywords related to various solar systems as shown in Figure 4.1. Photovoltaic systems received the most attention, followed by solar thermal systems [1–2]. Hybrid and concentrated

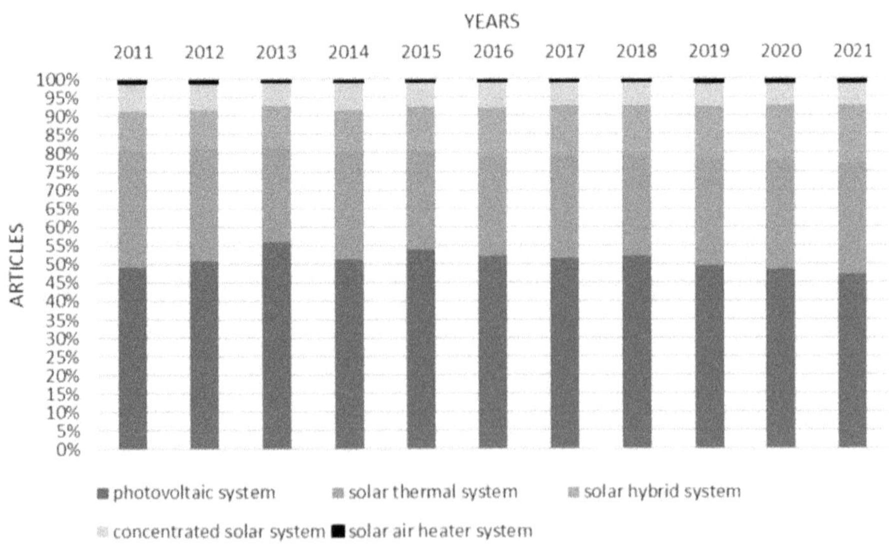

Figure 4.1 Distribution of field-wise articles published related to solar air heaters.

DOI: 10.1201/9781003527251-4

solar systems garnered moderate interest, while SAH systems saw the least research focus based on the number of publications. A significant portion of research on solar air heating systems focuses on energy fuels (30.58%) and thermodynamics (15.33%). Studies show that various components significantly impact the system's thermal performance. These include the absorber material, the cover material, ducts, insulation, enclosure material, airflow rate, and tilt angle. To improve performance, researchers explore different methods like optimizing equipment dimensions, testing various collector shapes and sizes, incorporating heat storage, concentrating solar radiation, and even using photovoltaic panels in the collection area [30–32]. Green sustainable science and technology (10.54%) is another crucial area of research. This is because solar air heating systems must be environmentally friendly, safe, and durable to meet environmental requirements [33, 34]. It's important to note that the percentages in Figure 4.2 represent the proportion of keywords within a total of 2485 keywords assigned to each category. The "other" category (12.47%) encompasses categories where keywords accounted for less than 2% of the total [3–4].

SAHs play a crucial role in sustainable heating applications. Optimizing their design for efficiency and performance is vital. Computational methods offer a powerful tool for this purpose, providing virtual testing grounds before physical prototypes are built. Here's a breakdown of prominent techniques employed for SAH analysis, as follows.

4.1.1 Computational fluid dynamics (CFD)

CFD solves the governing equations of fluid mechanics (Navier-Stokes equations) within a discretized geometry of the SAH. This allows for detailed analysis of airflow patterns, heat transfer processes, and pressure drops. CFD

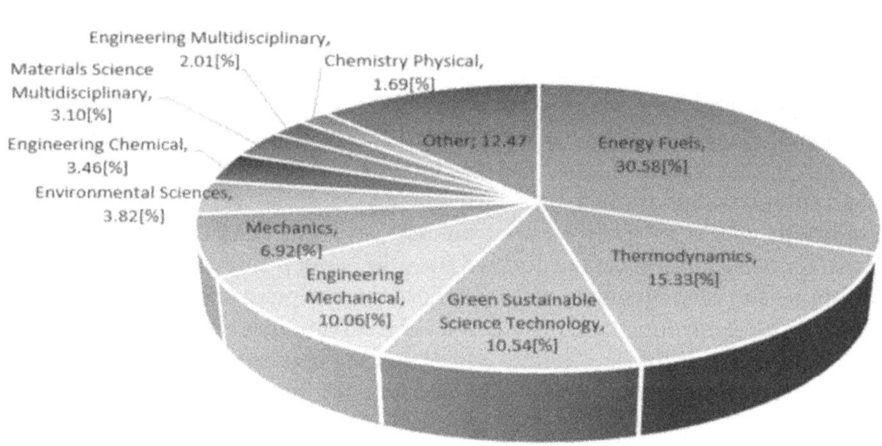

Figure 4.2 Topic-wise active multidisciplinary subjects used in study of solar air heaters.

excels at capturing complex flow phenomena like turbulence within the SAH channels. It can predict the impact of design modifications on thermal performance, such as incorporating fins or ribs on the absorber plate. Accurate modeling of turbulence requires advanced turbulence models, which can be computationally expensive. CFD simulations rely on user-defined boundary conditions, and their accuracy can affect the final results. SAHs represent a simple yet effective technology for harnessing solar energy. They function by transferring solar radiation absorbed by the collector plate to the flowing air, achieving thermal energy gain [5–7]. Despite its strengths, CFD simulations for SAHs present certain challenges. Accurate representation of complex geometries, including fins or artificial roughness on the absorber plate, requires careful mesh generation. Additionally, experimental validation of CFD results is essential to ensure the accuracy of the simulations. This can be achieved by comparing predicted temperature profiles or outlet air temperatures with measured data. In conclusion, CFD simulations serve as a valuable tool for unveiling the complex flow and heat transfer phenomena within SAHs. By enabling detailed analysis and design optimization, CFD paves the way for developing more efficient and effective solar air heating systems [8, 9].

4.1.2 Finite element method (FEM)

FEM breaks down the SAH geometry into smaller elements (finite elements). It then applies heat transfer equations to each element to determine temperature distribution throughout the system. FEM is adept at handling complex geometries of the SAH, including irregular shapes and discontinuities. It is a versatile tool that can be coupled with CFD simulations for a comprehensive analysis of heat transfer through conduction, convection, and radiation. Similar to CFD, FEM simulations can be computationally intensive, especially for intricate SAH designs. The accuracy of FEM results depends on the mesh size (element size) used in the discretization process. SAHs are a simple and cost-effective technology for harnessing solar energy. Evaluating their thermal performance is crucial for optimizing design and efficiency. FEMs offer a powerful tool for this purpose. FEM discretizes the SAH geometry into small elements, allowing for detailed temperature distribution analysis. The governing equations for heat transfer, including conduction, convection, and radiation, are applied to each element. Material properties, boundary conditions (such as solar irradiation and ambient temperature), and heat transfer coefficients are incorporated into the model. By solving the system of equations for each element, FEM predicts the temperature profile across the entire SAH. This approach provides several advantages over traditional analytical methods. FEM can handle complex SAH geometries with fins, baffles, and varying flow paths. It can also account for non-uniform solar irradiation and airflow patterns, leading to more realistic performance predictions. Additionally, FEM allows for sensitivity analysis, where the impact of design changes (fin shape, material selection) on thermal performance can be readily evaluated. This facilitates the optimization of SAH design for specific applications [10, 11].

4.1.3 Artificial neural networks (ANNs)

ANNs are a type of machine learning (ML) algorithm inspired by the structure of the human brain. When paired with CFD simulations, ANNs can be trained on a large dataset of SAH performance data. This training enables them to predict the thermal behavior of new SAH designs without running complex CFD simulations every time. This approach offers a fast and efficient way to evaluate SAH performance for various operating conditions. It can be particularly valuable during the initial design exploration stage. The accuracy of ANN predictions hinges on the quality and quantity of training data. ANNs might not be able to generalize well for SAH designs that deviate significantly from the data used for training. In conclusion, computational methods like CFD, FEM, ANNs, and their combinations offer valuable tools for optimizing SAH design. CFD provides detailed insights into fluid flow and heat transfer, while FEM excels at handling complex geometries. ANNs, coupled with CFD, can accelerate the design process by offering rapid performance predictions. By leveraging these methods, researchers and engineers can create more efficient and effective SAHs for a sustainable future and ANNs for performance prediction and optimization. SAHs are a simple yet effective technology for harnessing solar energy to heat air. Optimizing their performance is crucial for maximizing energy gains [12, 13]. Additionally, selecting the appropriate ANN architecture (number of layers and neurons) and training algorithms are crucial for optimal performance. Validation with experimental data is essential to ensure the ANN's generalizability to real-world conditions. In conclusion, ANNs offer a powerful approach for performance prediction and optimization in SAHs. Their ability to learn complex relationships, handle non-linearity, and adapt to diverse scenarios makes them a valuable tool for researchers and engineers in the field of solar thermal energy. By addressing data quality, network architecture, and validation techniques, ANNs can play a significant role in designing and operating high-performance SAHs [14, 15].

4.1.4 Machine learning with CFD

The quest for ever-more efficient SAHs has led to the exciting integration of ML with CFD simulations. This powerful coupling offers an intelligent design approach, optimizing SAH configurations for targeted performance metrics. CFD simulations play a crucial role in analyzing airflow patterns, heat transfer characteristics, and pressure drops within SAHs. However, these simulations require significant computational resources and expertise to define the geometry and set up the boundary conditions. Here's where ML steps in. ML algorithms can be trained on a vast database of CFD simulations coupled with real-world SAH performance data. This training empowers the ML model to identify key design features – like fin geometry, channel spacing, and absorber plate material – that significantly influence SAH performance. The integration process typically involves two key stages. First, a large number of CFD simulations are conducted for various SAH designs. These simulations capture the complex thermo-fluidic interactions within the heater [16, 17]. Additionally, the ML model can be continually refined with new data, leading to continuous improvement in design accuracy and performance prediction capabilities. In conclusion, the integration of ML with CFD offers a powerful approach to intelligent design for SAHs. By leveraging the strengths of both techniques, engineers can create highly optimized SAH configurations that deliver superior thermal performance while minimizing design time and resource consumption. This paves the way for a new generation of efficient and cost-effective solar air heating systems [18, 19].

4.1.5 Application of nanofluid in SAH

SAHs are a promising technology for space heating and drying applications. However, their efficiency can be limited by the thermal conductivity of the heat transfer fluid (HTF). Here, nanofluids, suspensions of nanoparticles in a base fluid (water, ethylene glycol, etc.), offer a potential solution. Several types of nanoparticles have been explored for SAH applications. Metal oxides, like aluminum oxide (Al_2O_3) and silicon dioxide (SiO_2), are popular choices due to their good thermal conductivity and stability. Carbon-based nanoparticles, such as carbon nanotubes (CNTs) and graphene, are also gaining interest for their exceptional thermal properties and high solar absorptivity, which can improve heat capture within the collector as shown in Figure 4.3. Research is

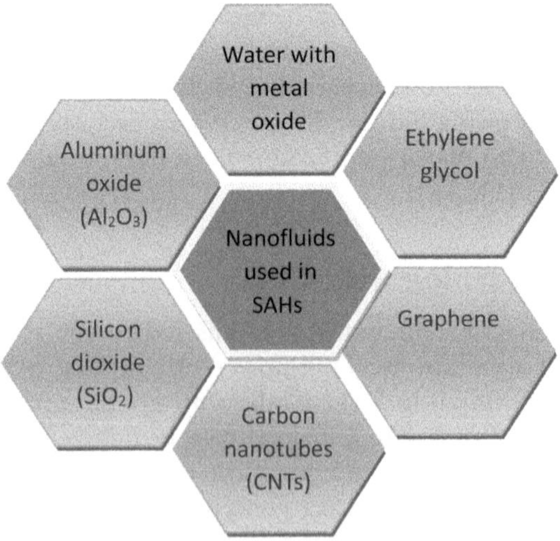

Figure 4.3 Different types of nanofluids used in solar air heaters application.

ongoing to optimize the nanoparticle type, concentration, and surface modification for achieving the desired balance between thermal performance, stability, and cost-effectiveness in SAHs [20–23].

SAHs are technology for utilizing solar energy to heat air for various applications. However, their efficiency can be limited by the heat transfer capabilities of conventional fluids like air or water. Here, nanofluids, suspensions of nanoparticles in a base fluid, offer a potential solution. Nanoparticles possess high thermal conductivity, which can significantly improve the heat transfer properties of the base fluid. Integrating nanofluids into SAHs can lead to increased thermal efficiency, meaning more heat gets transferred to the air being heated. Evaluating the performance improvement of SAHs using nanofluids requires a multi-pronged approach. Temperature sensors are placed at key locations within the SAH, such as the absorber plate inlet and outlet, and the air inlet and outlet. These sensors monitor the temperature rise of both the air and the nanofluid. Additionally, airflow rate through the SAH is measured. By analyzing these parameters, researchers can calculate the thermal efficiency of the SAH with nanofluids and compare it to the efficiency achieved with conventional fluids. Furthermore, pressure drop across the SAH can be measured to assess the pumping power required to circulate the nanofluid. This helps determine the optimal balance between enhanced heat transfer and increased pumping costs [24–28].

4.1.6 Improvement in effectiveness of SAH with nanofluids

SAHs are a simple and efficient technology for harnessing solar energy to heat air for various applications. However, a key challenge lies in improving their thermal performance, particularly heat transfer between the absorber plate and the air stream. Here's where nanofluids come into play. Studies have demonstrated increased collector efficiency and outlet air temperature when using nanofluids compared to conventional air- or water-based HTFs. The improved thermal conductivity of nanofluids facilitates faster heat absorption by the absorber plate and better transfer to the air stream. This can lead to significant reductions in heat loss and improved overall system performance [29–31]. However, challenges like long-term stability, pumping power requirements, and cost-effectiveness of nanofluids need further investigation for their widespread adoption in SAH technology as described in Table 4.1.

The visualization appears to depict a network analysis of co-authorship among researchers in a specific field over a four-year period, likely from 2020 to 2023. The nodes, labeled with country names, are connected by lines, while circles around the nodes connect researchers within the same country. The size of the circles and nodes may indicate publication output or centrality within the network. Some of the most prominent central nodes include the United States, China, India, and Germany. It is important to note that the

Table 4.1 Different type of nanofluids and their impact on performance of solar air heaters

Type of nanofluid	*Impacts on performance of SAHs*
Aluminum oxide (Al_2O_3) Nanofluid	Enhanced thermal conductivity and collector efficiency
Silicon carbide (SiC) nanofluid	Improved thermal efficiency and reduced heat loss
Titanium dioxide (TiO_2) nanofluid	Increased heat transfer rate and collector performance
Copper oxide (CuO) nanofluid	Elevated thermal conductivity and collector heat gain
Zinc oxide (ZnO) nanofluid	Improved thermal performance and antibacterial properties for air purification

specific field this network represents cannot be determined from the image alone [32–36].

4.2 METHODOLOGY

4.2.1 CFD simulations methodology

CFD plays a crucial role in analyzing and optimizing the performance of SAHs. At the heart of these simulations lie the governing equations that represent the fundamental physical principles of fluid flow, heat transfer, and mass conservation within the system. These equations are typically expressed in a time-dependent form, but for steady-state simulations of SAHs, the time derivatives become zero [37–39]. Here's a closer look at the three key governing equations:

Conservation of mass equation ensures that within a control volume, the mass entering must equal the mass leaving, accounting for any mass source or sink within the volume. Mathematically, it's represented as follows:

$$\partial \rho / \partial t + \nabla(\rho u) = 0,$$

where ρ is the air density, t is time, and u is the velocity vector. The divergence (∇) operator represents the net flow of mass out of a differential control volume. Incompressible flow, a common assumption for SAHs due to low operating pressures, simplifies this equation to $\nabla(u) = 0$, signifying constant air density throughout the domain.

4.2.2 Momentum equation (Conservation of momentum)

This equation expresses Newton's second law of motion, stating that the rate of change of momentum of a fluid element is equal to the sum of forces acting on it. It accounts for inertia, pressure forces, viscous forces, and body forces (like gravity). In CFD simulations of SAHs, the momentum equation is typically expressed in two or three dimensions depending on the chosen model geometry [40, 41]. Here's a simplified form for incompressible flow in x-direction:

$$\partial(\rho u) / \partial t + \nabla(\rho uu) = -\partial p / \partial x + \partial(\mu\ \partial u / \partial x) + \partial(\tau xx) + F,$$

where p is the pressure, μ is the dynamic viscosity, τxx is the viscous stress tensor component, and F represents any external body forces. Solving this equation along with the continuity equation provides the velocity field within the air domain.

4.2.3 Energy equation (conservation of energy)

This equation ensures that the total energy within a control volume remains constant. It considers energy transfer due to convection, conduction, and heat sources within the domain. In SAHs, the heat source is the absorbed solar radiation by the absorber plate. The equation can be expressed as follows:

$$\partial(\rho E) / \partial t + \nabla(\rho uE) = \nabla(k\ \nabla T) + Q,$$

where E is the total energy per unit mass, k is the thermal conductivity, T is the temperature, and Q represents the volumetric heat source due to solar radiation absorption. Solving this equation along with the continuity and momentum equations provides the temperature distribution of the air flowing through the SAH. These governing equations, along with appropriate turbulence models (if applicable) and boundary conditions, are discredited and solved numerically within the chosen CFD software. The resulting solutions provide valuable insights into the airflow patterns, temperature distribution, and overall thermal performance of the SAH, aiding in design optimization for better efficiency [42, 43].

4.2.4 Turbulence modeling in CFD

Turbulent flow is a dominant characteristic within SAHs. Accurately capturing this turbulence in CFD simulations is crucial for predicting the heater's thermal performance. However, selecting the appropriate turbulence model presents a significant challenge. Here, we explore into the factors influencing this choice and explore suitable models for SAH simulations. A critical consideration is the balance between model complexity and computational cost. Popular Reynolds-Averaged Navier-Stokes (RANS) models like the standard k-ε model offer a reasonable compromise. These models solve transport equations for turbulent kinetic energy (k) and dissipation rate (ε) to account for the influence of turbulence on the mean flow. However, their reliance on empirical constants can limit accuracy, especially in complex flow regimes with strong separations or secondary flows. SAHs often involve thin boundary layers adjacent to the absorber plate. Accurately resolving these layers is essential for capturing heat transfer mechanisms. Enhanced wall functions can be employed with RANS models to address this. These functions bridge the gap between the viscous sublayer (where the model cannot resolve) and the fully turbulent region [44–46]. Selecting an optimal turbulence model for SAH CFD simulations requires careful consideration of flow complexity, computational resources, and desired accuracy. RANS models with enhanced wall functions often provide a good starting point. For complex geometries or the need for high accuracy, exploring advanced models like shear stress transport (SST) k-ω or detached Eddy simulation (DES) might be necessary. Ultimately, validation and comparison with experimental data are crucial steps in ensuring the reliability of the CFD results [47, 48].

4.2.5 Geometry creation and meshing strategy for CFD simulations of solar air heaters

Accurately representing the geometry and implementing an appropriate meshing strategy are crucial aspects of performing successful CFD simulations for SAHs. Here's a detailed breakdown of these considerations. The geometry of the SAH needs to capture all relevant features that influence airflow and heat transfer. This includes the collector casing, absorber plate, fins (if present), inlet and outlet channels, and any internal flow structures. Software like SolidWorks or Ansys SpaceClaim can be used to create a precise 3D model of the SAH. Depending on the complexity of the fin geometry, simplifications might be necessary. For instance, thin fins can be modeled as solid walls with adjusted thermal conductivity to account for heat transfer through the fins. Another simplification involves assuming periodic boundary conditions for situations with multiple flow passages. This reduces the computational domain while maintaining the representative nature of the simulation. Meshing refers to discretizing the computational domain into smaller elements (cells) over which the governing equations of fluid flow and heat transfer are solved. Structured meshes, with uniformly sized and aligned elements, are often preferred for SAH simulations due to their simplicity and efficiency. However, unstructured meshes offer more flexibility for complex geometries with curved surfaces. A key aspect of meshing is ensuring mesh refinement in critical regions. This includes areas near the absorber plate where thermal gradients are high and the boundary layer develops. Refining the mesh in these areas improves the accuracy of capturing heat transfer phenomena. Conversely, overly refined meshes throughout the entire domain can lead to unnecessary computational cost [49–51]. Popular options include Ansys Fluent, OpenFOAM, and STAR-CCM+. User expertise in mesh generation and CFD simulation setup is critical for achieving reliable results. Understanding the interplay between geometry, mesh, turbulence models, and boundary conditions is essential for an accurate representation of the physical phenomena occurring within the SAH [52–54].

4.2.6 Boundary conditions and material properties in CFD simulations of solar air heaters

The inlet boundary defines the incoming air's mass flow rate, temperature, and turbulence characteristics. Mass flow rate can be specified directly or derived from the desired Reynolds number. Inlet temperature should reflect ambient conditions or pre-heating stages. Selecting a suitable turbulence model (e.g., k-ε or RNG k-ε) and specifying the turbulence intensity and eddy viscosity ratio at the inlet are vital for capturing turbulent flow behavior. The outlet boundary defines the pressure conditions at the SAH exit. Typically, a fixed static pressure corresponding to ambient conditions is applied. Alternatively, a pressure outlet with a backflow condition can be used if the flow rate is unknown. Simulating solar radiation absorption by the absorber plate is essential. A uniform or non-uniform heat flux can be applied on the top surface, representing the absorbed solar energy. Alternatively, a radiative heat transfer model can be coupled with a solar radiation source to account for angular dependence and spectral effects [55–57]. By carefully defining these boundary conditions and material properties, CFD simulations can provide valuable insights into the performance of SAHs, aiding in design optimization and efficiency improvement [58–61].

4.2.7 Solution algorithms and convergence criteria

CFD simulations play a crucial role in optimizing the design and performance of SAHs. These simulations rely on robust solution algorithms and well-defined convergence criteria to ensure accurate and reliable results. Segregated solvers tackle pressure-velocity coupling iteratively. Common approaches include Semi-Implicit Method for Pressure Linked Equations (SIMPLE) and its variants. These are efficient for low-speed, incompressible flows prevalent in SAHs. Coupled solvers handle pressure and velocity equations simultaneously, offering better stability for complex flows. However, they can be computationally expensive for SAH applications. Discretization schemes approximate the governing equations (continuity, momentum, and energy) into algebraic equations for each computational cell. Popular choices include finite volume method (FVM), which is most widely used in SAH simulations. It conserves variables across each cell, ensuring mass and energy balance. Finite difference method (FDM) is simpler to implement but may struggle with complex geometries present in SAH designs. Turbulent flow is common in SAHs due to wind and buoyancy effects. Turbulence models account for these complexities. Common approaches include RANS equations with turbulence models like k-ε or k-ω SST [62, 63]. This ensures the results are independent of mesh discretization and not merely artifacts of the chosen grid size. Selecting appropriate solution algorithms and convergence criteria requires careful consideration. While stricter criteria ensure higher accuracy, they can significantly increase computational time. CFD software often offers automatic convergence control, but user expertise is crucial for fine-tuning settings and interpreting results within the context of SAH design optimization [64–65].

4.3 LIMITATIONS: LIMITATIONS OF TRADITIONAL EXPERIMENTAL METHODS FOR SOLAR AIR HEATER ANALYSIS

SAHs represent a promising technology for harnessing solar energy for space and water heating applications. However, optimizing their performance relies heavily on thorough analysis. Traditional experimental methods, while valuable, suffer from several limitations that hinder comprehensive SAH evaluation [66–68].

- **Time-consuming nature:** Traditional SAH experiments involve constructing physical prototypes, instrumenting them with sensors, and conducting tests under varying operating conditions. This process can be incredibly time-consuming. Each test configuration requires meticulous setup, data acquisition, and analysis. Iterating through different design parameters to identify optimal configurations further extends the research timeline.
- **Costly endeavors:** Building SAH prototypes requires materials, sensors, and data acquisition equipment. These costs can be significant, especially for complex designs or those requiring specialized materials. Additionally, maintaining a controlled testing environment (e.g., simulating different solar radiation levels) often necessitates expensive facilities.
- **Limited parameter exploration:** Traditional experiments typically focus on a limited set of parameters due to time and cost constraints. Investigating the combined effects of multiple design variables (e.g., absorber plate material, flow channel geometry, and glazing type) becomes challenging. This limited scope may lead to overlooking potentially beneficial design combinations.
- **Addressing the limitations:** Computational modeling and simulation techniques offer a promising alternative to address the limitations of traditional experimental methods. These tools allow researchers to perform several analysis: Virtual SAH models can be subjected to various weather patterns, flow rates, and material properties, enabling a more comprehensive exploration of the design space, etc. Simulations are significantly faster and less expensive than physical experiments. This allows for rapid iteration and optimization of SAH designs.

Computational models can account for the interplay between multiple design variables, providing valuable insights into their combined effects on performance. While traditional experimental methods remain crucial for validating SAH models and performance, their limitations hinder comprehensive analysis. Computational modeling offers a valuable complement, enabling researchers to explore a wider design space, reduce costs, and gain deeper understanding of SAH behavior. Integrating these approaches will lead to more efficient and optimal SAH designs, accelerating the development of this promising solar technology.

4.4 SAH APPLICATIONS

SAHs have seen significant progress in recent years, driven by the desire for energy-efficient and sustainable heating solutions. Here's a closer look at some key advancements and areas requiring further exploration. Traditional metal absorbers with black paint are being challenged by selective surfaces with high solar absorptance and low thermal emittance. Research on nanocomposite coatings and photonic crystals promises even more efficient heat capture. Additionally, exploring absorber geometries like fins and corrugated channels is improving heat transfer to the air stream. One area of focus is reducing heat loss from the collector box. Glazing materials with high transmittance and low thermal conductivity are being developed. Evacuated collectors further minimize heat loss by reducing convection within the collector [69, 70].

4.4.1 A sustainable solution for space and crop drying

SAHs are ingenious devices that harness the power of sunlight for thermal applications. Unlike solar photovoltaic cells that convert light directly into electricity, SAHs focus on capturing solar energy and transforming it into heated air. This heated air can then be utilized for various purposes, making SAHs a valuable tool in the realm of renewable energy. The core functionality of a SAH lies in its deceptively simple design. The fundamental components include an absorber plate, a transparent cover, insulation, and air channels. The absorber plate, typically made of metal, is coated with a selective surface that efficiently absorbs solar radiation and converts it into heat. The transparent cover, often constructed of glass, allows sunlight to pass through while minimizing heat loss back into the atmosphere. Insulation on the backside of the absorber plate further restricts heat from escaping, ensuring optimal thermal performance. Finally, air channels situated behind the absorber plate allow air to flow and become heated as it comes in contact with the hot surface. The significance of SAHs stems from their ability to tap into a clean and abundant source of energy – the sun. By harnessing solar power, SAHs contribute to reducing dependence on conventional fossil fuels for space heating and crop drying. This translates into a multitude of advantages. From an environmental standpoint, SAHs promote sustainability by curbing greenhouse gas emissions associated with fuel combustion. Moreover, SAHs offer economic benefits by lowering reliance on conventional energy sources, potentially leading to cost savings. In the domain of space heating, SAHs can be integrated into buildings to provide a sustainable source of warmth. The heated air generated by the SAH can be directly distributed within the building or channeled through a thermal storage system to ensure warmth even during periods of low sunlight. This application is particularly advantageous in regions with ample sunshine and mild winters. Another significant application of SAHs lies in the agricultural sector. Crop drying is a crucial postharvest process that ensures the preservation and quality of agricultural products. SAHs can provide a clean and cost-effective solution for drying crops like grains, fruits, and spices. The heated air from the SAH can efficiently remove moisture from the crops, extending their shelf life and minimizing losses. In conclusion, SAHs offer a promising avenue for harnessing renewable energy for space heating and crop drying applications. Their simple design, coupled with their ability to utilize a clean and abundant energy source, makes them a valuable tool in promoting sustainability and energy independence. As research and development in SAH technology progress, we can expect even greater efficiency and wider adoption of these innovative solar thermal systems [71–73].

4.4.2 Socioeconomic benefits of solar air heaters

SAHs offer a compelling solution for harnessing renewable energy to deliver significant socioeconomic benefits, particularly in regions like Bareilly, India, where there's ample sunlight. Here's a detailed exploration of these advantages:

- **Reduced energy costs and improved energy security:** SAHs significantly reduce dependence on conventional energy sources like electricity or natural gas for space heating and water heating. This translates to lower energy bills for households and businesses. Moreover, in areas with unreliable grid access, SAHs provide a dependable source of heat, enhancing energy security and mitigating the disruptions associated with conventional power outages.
- **Economic opportunities and job creation:** The adoption of SAHs stimulates economic activity by creating new job opportunities in installation, maintenance, and manufacturing sectors. As the technology matures and demand increases, local economies benefit from the establishment of SAH production facilities and associated service industries.
- **Improved public health and well-being:** SAHs contribute to improved public health by reducing indoor air pollution. Traditional heating methods, particularly those reliant on burning solid fuels, release harmful pollutants that can exacerbate respiratory illnesses. SAHs, by contrast,

generate clean heat, creating a healthier indoor environment, especially in poorly ventilated spaces.

- **Enhanced quality of life:** SAHs can play a vital role in improving the quality of life, particularly in rural areas. They can provide access to hot water for bathing and washing, contributing to better hygiene standards. Additionally, SAHs can be used for space heating, creating a more comfortable living environment during cold weather conditions.
- **Environmental sustainability:** One of the most significant socioeconomic benefits of SAHs is their contribution to environmental sustainability. By harnessing solar energy, a clean and renewable resource, SAHs reduce reliance on fossil fuels and consequently help mitigate greenhouse gas emissions. This not only combats climate change but also improves air quality on a local level. In conclusion, SAHs offer a multitude of socioeconomic benefits, making them a valuable technology for regions like Bareilly, India. From reduced energy costs and improved energy security to enhanced public health, economic opportunities, and environmental sustainability, SAHs have the potential to significantly improve the lives of people and contribute to a more sustainable future.

REFERENCES

1. Ghritlahre, H. K., Chandrakar, P., & Ahmad, A. (2021). A comprehensive review on performance prediction of solar air heaters using artificial neural network. Annals of Data Science, 8, 405–449.
2. Korpale, V. S., Deshmukh, S. P., Mathpati, C. S., & Dalvi, V. H. (2020). Numerical simulations and optimization of solar air heaters. Applied Thermal Engineering, 180, 115744.
3. Kabeel, A. E., Hamed, M. H., Omara, Z. M., & Kandeal, A. W. (2017). Solar air heaters: Design configurations, improvement methods and applications–A detailed review. Renewable and Sustainable Energy Reviews, 70, 1189–1206.
4. Gopi, R., Ponnusamy, P., Arokiaraj, A. F., & Raji, A. (2021). Performance comparison of flat plate collectors in solar air heater by theoretical and computational method. Materials Today: Proceedings, 39, 823–826.
5. Gawande, V. B., Dhoble, A. S., Zodpe, D. B., & Chamoli, S. (2016). A review of CFD methodology used in literature for predicting thermo-hydraulic performance of a roughened solar air heater. Renewable and Sustainable Energy Reviews, 54, 550–605.
6. Sharma, S., Das, R. K., & Kulkarni, K. (2021). Computational and experimental assessment of solar air heater roughened with six different baffles. Case Studies in Thermal Engineering, 27, 101350.
7. Ghritlahre, H. K., & Verma, M. (2021). Accurate prediction of exergetic efficiency of solar air heaters using various predicting methods. Journal of Cleaner Production, 288, 125115.
8. Ghritlahre, H. K., Chandrakar, P., & Ahmad, A. (2020). Application of ANN model to predict the performance of solar air heater using relevant input parameters. Sustainable Energy Technologies and Assessments, 40, 100764.
9. Esen, H., Ozgen, F., Esen, M., & Sengur, A. (2009). Modelling of a new solar air heater through least-squares support vector machines. Expert Systems with Applications, 36(7), 10673–10682.

10. Abulkhair, H., Alsaiari, A. O., Ahmed, I., Almatrafi, E., Madhukeshwara, N., & Sreenivasa, B. R. (2023). Heat transfer and air flow friction in solar air heaters: A comprehensive computational and experimental investigation with wire-roughened absorber plate. Case Studies in Thermal Engineering, 48, 103148.
11. Saxena, A., & El-Sebaii, A. A. (2015). A thermodynamic review of solar air heaters. Renewable and Sustainable Energy Reviews, 43, 863–890.
12. Potgieter, M. S. W., Bester, C. R., & Bhamjee, M. (2020). Experimental and CFD investigation of a hybrid solar air heater. Solar Energy, 195, 413–428.
13. Sharma, S. K., & Kalamkar, V. R. (2017). Experimental and numerical investigation of forced convective heat transfer in solar air heater with thin ribs. Solar Energy, 147, 277–291.
14. Sharma, S., & Maithani, R. (2024). A computational study on thermal and sustainability analysis of solar air heater with s and airfoil tabulators. Evergreen, 11, 295–305.
15. Alta, D., Bilgili, E., Ertekin, C., & Yaldiz, O. (2010). Experimental investigation of three different solar air heaters: Energy and exergy analyses. Applied Energy, 87(10), 2953–2973.
16. Yadav, A. S., & Bhagoria, J. L. (2013). A CFD (computational fluid dynamics) based heat transfer and fluid flow analysis of a solar air heater provided with circular transverse wire rib roughness on the absorber plate. Energy, 55, 1127–1142.
17. Raj, A. K., Srinivas, M., & Jayaraj, S. (2019). A cost-effective method to improve the performance of solar air heaters using discrete macro-encapsulated PCM capsules for drying applications. Applied Thermal Engineering, 146, 910–920.
18. Lanjewar, A., Bhagoria, J. L. L., & Sarviya, R. M. (2011). Heat transfer and friction in solar air heater duct with W-shaped rib roughness on absorber plate. Energy, 36, 4531–4541.
19. Awasthi, M. K., Kumar, A., & Dutt, N. Modeling Rayleigh-Taylor Instability in Nanofluid Layers. In Computational Fluid Flow and Heat Transfer: Advances, Design, Control and Applications; M. K. Awasthi, A. Kumar, N. Dutt, & S. Singh, Editors; CRC Press: Boca Raton, FL, USA, 2024; Chapter 13, https://doi.org/10.1201/9781003465171-13
20. Awasthi, M. K., Dutt, N., Kumar, A., & Kumar, S. (2024). Electrohydrodynamic capillary instability of Rivlin–Ericksen viscoelastic fluid film with mass and heat transfer. Heat Transfer, *53*(1), 115–133, https://doi.org/10.1002/htj.22944.
21. Lanjewar, A. M., Bhagoria, J. L., & Sarviya, R. M. (2011). Performance analysis of W-shaped rib roughened solar air heater. Journal of Renewable and Sustainable Energy, 3, 043110.
22. Kumar, A., Saini, R. P., & Saini, J. S. (2012). Experimental investigation on heat transfer and fluid flow characteristics of air flow in a rectangular duct with multi v-shaped rib with gap roughness on the heated plate. Solar Energy, *86*(6), 1733–1749.
23. Sahu, M. K., & Prasad, R. K. (2016). Exergy based performance evaluation of solar air heater with arc-shaped wire roughened absorber plate. Renewable Energy, *96*, 233–243.
24. Awasthi, M. K., Shukla, A. K., Kumar, A., Yadav, D., & Dutt, N. (2024). Viscous correction to the potential flow analysis of Rayleigh-Taylor instability in a Rivlin-Ericksen viscoelastic fluid layer with heat and mass transfer, Heat Transfer, https://onlinelibrary.wiley.com/doi/10.1002/htj.23076.
25. Yadav, D., Awasthi, M. K., Kumar, A., & Dutt, N. (2024). Swirling capillary instability of Rivlin-Ericksen liquid with heat transfer and axial electric field. Physics, *6*(2), 828–844. https://doi.org/10.3390/physics6020051

26. Verma, V., Thangavel S., Dutt N., Kumar A., Weerasinghe R., Editors, Highly Efficient Thermal Renewable Energy Systems: Design, Optimization and Applications, Pages 1–341, 2024, CRC Press: Boca Raton, FL, USA, https://doi.org/10.1201/9781003472629.
27. Deo, N. S. N. S., Chander, S., & Saini, J. S. (2016). Performance analysis of solar air heater duct roughened with multigap v-down ribs combined with staggered ribs. Renewable Energy, 91, 484–500.
28. Skullong, S., Promthaisong, P., Promvonge, P., Thianpong, C., & Pimsarn, M. (2018). Thermal performance in solar air heater with perforated-winglet-type vortex generator. Solar Energy, *170*, 1101–1117.
29. Awasthi, M. K., Dutt, N., & Kumar, A., Editors. "Nanofluids Technology for Thermal Sciences and Engineering: Research, Development, and Applications" CRC Press, Boca Raton, FL, USA, ISBN 9781032799117, 2024.
30. Awasthi, M. K., Kumar, A., & Dutt, N. Recent Innovation, Technological Development and Futuristic Applications of Nanofluids. In Nanofluids Technology for Thermal Sciences and Engineering: Research, Development, and Applications; M. K. Awasthi, N. Dutt, & A. Kumar, Editors; CRC Press: Boca Raton, FL, USA, 2024; Chapter 01.
31. Kumar, A., Thangavel, S., Awasthi, M. K., Singh, V. P., & Dutt, N. Safety and Environmental Considerations in Nano Fluid-Based Systems. In Nanofluids Technology for Thermal Sciences and Engineering: Research, Development, and Applications; M. K. Awasthi, N. Dutt, & A. Kumar, Editors; CRC Press: Boca Raton, FL, USA, 2024; Chapter 06.
32. Jain, S. K., Misra, R., Kumar, A., & Agrawal, G. D. (2022). Thermal performance investigation of a solar air heater having discrete v-shaped perforated baffles. International Journal of Ambient Energy, *43*(1), 243–251.
33. Awasthi, M. K., Kumar, A., Dutt, N., & Singh, S., Editors. Computational Fluid Flow and Heat Transfer: Advances, Design, Control and Applications; CRC Press: Boca Raton, FL, USA, 2024, https://doi.org/10.1201/9781003465171.
34. Awasthi, M. K., Dutt, N., Kumar, A., & Hedau, A. Introduction to Mathematical and Computational Methods. In Computational Fluid Flow and Heat Transfer: Advances, Design, Control and Applications; M. K. Awasthi, A. Kumar, N. Dutt, & S. Singh, Editors; CRC Press: Boca Raton, FL, USA, 2024; Chapter 01, https://doi.org/10.1201/9781003465171-1.
35. Kumar, A., Dutt, N., & Awasthi, M. K., Editors. Heat Transfer Enhancement Techniques: Thermal Performance, Optimization and Applications, John Wiley & Sons, 2024.
36. Kumar, A., Awasthi, M. K., Dutt, N., & Singh, V. P.. Recent Innovation in Heat Transfer Enhancement Techniques. In Heat Transfer Enhancement Techniques: Thermal Performance, Optimization and Applications; A. Kumar, N. Dutt, M. K. Awasthi, Editors; John Wiley & Sons, 2024; Chapter 01, pages 1–38.
37. Manjunath, M. S., Karanth, K. V., & Sharma, N. Y. (2017). Numerical analysis of the influence of spherical turbulence generators on heat transfer enhancement of flat plate solar air heater. Energy, *121*, 616–630.
38. Ravi, R. K., & Saini, R. P. (2016). Experimental investigation on performance of a double pass artificial roughened solar air heater duct having roughness elements of the combination of discrete multi V shaped and staggered ribs. Energy, *116*, 507–516.
39. Kumar, A., Saini, R. P., & Saini, J. S. (2012). Experimental investigation on heat transfer and fluid flow characteristics of air flow in a rectangular duct with multi v-shaped rib with gap roughness on the heated plate. Solar Energy, *86*(6), 1733–1749.

40. Verma, V., Thangavel, S., Dutt, N., Kumar, A., & Weerasinghe, R., Editors. Highly Efficient Thermal Renewable Energy Systems: Design, Design, Optimization and Applications; CRC Press: Boca Raton, FL, USA, 2024; pages 1–341, https://doi.org/10.1201/9781003472629.
41. Dutt, N., Hedau, A., Kumar, A., Awasthi, M. K., Hedau, S., & Meena, C. S. (2024). Thermo-hydraulic performance investigation of solar air heater duct having staggered D-shaped ribs: Numerical approach. Heat Transfer, https://onlinelibrary.wiley.com/doi/10.1002/htj.22998.
42. Dewangan, A. K., Moinuddin, S. Q., Cheepu, M., Sajjan, S. K., & Kumar, A. Thermal Energy Storage: Opportunities, Challenges and Future Scope. In Thermal Energy Systems: Design, Computational Techniques and Applications; A. Kumar, V. P. Singh, C.S. Meena, & N. Dutt, Editors; CRC Press: Boca Raton, FL, USA, 2023; Chapter 2; pages 17–28, http://dx.doi.org/10.1201/9781003395768-2.
43. Pandey, N. K., & Bajpai, V. K. (2016). Experimental investigation of heat transfer augmentation using multiple arcs with gap on absorber plate of solar air heater. Solar Energy, *134*, 314–326.
44. Ahmadizadeh, M., Heidari, M., Thangavel, S., Naamani, E. A., Khashehchi, M., Verma, V., & Kumar, A. Technological Advancements in Sustainable and Renewable Solar Energy Systems. In Highly Efficient Thermal Renewable Energy Systems: Design, Design, Optimization and Applications; V. Verma, S. Thangavel, N. Dutt, A. Kumar, & R. Weerasinghe Editors; CRC Press: Boca Raton, FL, USA, 2024; Chapter 02; pages 23–39, https://doi.org/10.1201/9781003472629-2.
45. Verma, V., Thangavel, S., Dutt, N., Kumar, A., & Weerasinghe, R. Recent Development of Introduction to Thermal Energy Storage: Solar, Geothermal and Hydrogen Energy. In Highly Efficient Thermal Renewable Energy Systems: Design, Design, Optimization and Applications; V. Verma, S. Thangavel, N. Dutt, A. Kumar, & R. Weerasinghe, Editors; CRC Press: Boca Raton, FL, USA, 2024; Chapter 01; pages 1–22, https://doi.org/10.1201/9781003472629-1.
46. Parsa, H., Saffar-Avval, M., & Hajmohammadi, M. R. (2021). 3D simulation and parametric optimization of a solar air heater with a novel staggered cuboid baffles. International Journal of Mechanical Sciences, *205*, 106607.
47. Chitt, M., Thangavel, S., Verma, V., & Kumar, A. Green Hydrogen Productions: Methods, Designs and Smart Applications. In Highly Efficient Thermal Renewable Energy Systems: Design, Design, Optimization and Applications; V. Verma, S. Thangavel, N. Dutt, A. Kumar, & R. Weerasinghe, Editors; CRC Press: Boca Raton, FL, USA, 2024; Chapter 16; pages 261–276, https://doi.org/10.1201/9781003472629-16.
48. Khashehchi, M., Thangavel, S., Rahmanivahid, P., Heidari, M., Moazzeni, T., Verma, V., & Kumar, A. Solar Desalination Techniques: Challenges and Opportunities. In Highly Efficient Thermal Renewable Energy Systems: Design, Design, Optimization and Applications; V. Verma, S. Thangavel, N. Dutt, A. Kumar, & R. Weerasinghe, Editors; CRC Press: Boca Raton, FL, USA, 2024; Chapter 19; pages 305–329, https://doi.org/10.1201/9781003472629-19.
49. Dutt, N., Hedau, A. J., Kumar, A., Awasthi, M. K., Singh, V. P., & Dwivedi, G. (2023). Thermo-hydraulic performance of solar air heater having discrete D-shaped ribs as artificial roughness. Environmental Science and Pollution Research, https://doi.org/10.1007/s11356-023-28247-9.
50. Klein, S. A., Beckman, W. A., & Duffie, J. A. (1977). A design procedure for solar air heating systems. Solar Energy, *19*, 509–512.
51. Ho, C. D., Hsiao, C. F., Chang, H., & Tien, Y. E. (2017). Investigation of device performance for recycling double-pass V-corrugated solar air collectors. Energy Procedia, 105, 28–34.

52. Reddy, P. N., Verma, V., Kumar, A., & Awasthi, M. (2023). CFD simulation and thermal performance optimization of flow in a channel with multiple baffles. Journal of Heat and Mass Transfer Research, *10*(2), 257–268, https://doi.org/10.22075/JHMTR.2023.31108.1458.
53. Singh, B., Kaur, P., Yadav, A. K., Awasthi, M. K., & Kumar, A. Design and Analysis of Solar Tracking System for PV Thermal Performance Enhancement. In Heat Transfer Enhancement Techniques: Thermal Performance, Optimization and Applications; A. Kumar, N. Dutt, & M. K. Awasthi, Editors; John Wiley & Sons, 2024; Chapter 11, pages 251–258.
54. Singh, V. P., Kumar, A., & Awasthi, M. K. Augmentation of Solar, Geothermal and Earth-Air Heat Exchanger in Sustainable Buildings. In Heat Transfer Enhancement Techniques: Thermal Performance, Optimization and Applications; A. Kumar, N. Dutt, & M. K. Awasthi, Editors; John Wiley & Sons, 2024; Chapter 06, pages 119–158.
55. Fudholi, A., Sopian, K., & Ruslan, M. H. (2011). Thermal efficiency of double pass solar collector with longitudinal fins absorbers. American Journal of Applied Science, *8*, 254–260.
56. Kundu, A., Kumar, A., Awasthi, M. K., Dwivedi, G., & Shukla, A. K. (In Press). An explorative study of earth-air heat exchanger for sustainable building applications. International Journal of Energy for a Clean Environment. https://doi.org/10.1615/InterJEnerCleanEnv.2024052513
57. Heidari, M., Thangavel, S., Ghafri, K. A., & Kumar, A. Future Trends and Emerging Research in Nano Fluids for Aerospace Applications. In Nanofluids Technology for Thermal Sciences and Engineering: Research, Development, and Applications; M. K. Awasthi, N. Dutt, & A. Kumar, Editors; CRC Press: Boca Raton, FL, USA, 2024; Chapter 15.
58. Dutt, N., Hedau, A., Kumar, A., Awasthi, M. K., & Singh, V. P. (2024). Effect of duct length variation on solar air heater performance for smooth and D-shaped roughened absorber plate. Heat Transfer, 1–29. https://onlinelibrary.wiley.com/doi/10.1002/htj.23115
59. Jaideep, P. T., & Kumar, A. CFD Numerical Investigation of Thermal Performance of Diamond Shape Micro Rectangular Heat Exchanger. In Heat Transfer Enhancement Techniques: Thermal Performance, Optimization and Applications; A. Kumar, N. Dutt, & M. K. Awasthi, Editors; John Wiley & Sons, 2024; Chapter 07, pages 159–192.
60. Hernández, A. L., & Quiñonez, J. E. (2013). Analytical models of thermal performance of solar air heaters of double-parallel flow and double-pass counter flow. Renewable Energy, *55*, 380–391.
61. Kumar, A., Singla, Y., & Namboodri, T. Globalization and International Issues in Sustainable Manufacturing. In Sustainability in Smart Manufacturing: Trends, Scope, and Challenges; S. Shah, H. Nautiyal, G. Gugliani, A. Kumar, T. Namboodri, & Y.K. Singla, Editors; CRC Press: Boca Raton, FL, USA, 2024; Chapter 01; pages 1–18, http://dx.doi.org/10.1201/9781003467496-1.
62. Karwa, R., & Maheshwari, B. K. K. (2009). Heat transfer and friction in an asymmetrically heated rectangular duct with half and fully perforated baffles at different pitches. International Communications in Heat and Mass Transfer, *36*, 264–268.
63. Kwak, J. S., & Shin, S. (2008). Effect of hole shape on the heat transfer in a rectangular duct with perforated blockage walls. Journal of Mechanical Science and Technology, *22*, 1945–1951.
64. Alam, T., Saini, R. P., & Saini, J. S. (2014). Effect of circularity of perforation holes in v-shaped blockages on heat transfer and friction characteristics of rectangular solar air heater duct. Energy Conversion and Management, *86*, 952–963.

65. Bahuguna, R., Mer, K. K. S., Kumar, M., & Chamoli, S. (2021). Entropy generation analysis in a tube heat exchanger integrated with triple blade vortex generator inserts. Energy Sources Part A: Recovery, Utilization, and Environmental Effects, 1–19. https://doi.org/10.1080/15567036.2021.1918291
66. Chamoli, S., & Thakur, N. S. (2013). Heat transfer enhancement in solar air heater with V-shaped perforated baffles. Journal of Renewable Sustainable Energy, *5*, 023122.
67. Phu, N. M., Bao, T. T., Hung, H. N., Tu, N. T., & van Hap, N. (2020). Analytical predictions of exergoeconomic performance of a solar air heater with surface roughness of metal waste. Journal of Thermal Analysis and Calorimetry, *144*, 1727–1740.
68. Phu, N. M., Thao, P. B., & van Hap, N. (2021). Effective efficiency assessment of a solar air heater having baffles spaced with different successive ratios. Case Studies in Thermal Engineering, *28*, 101486.
69. Gupta, B., Manikpuri, G. P., Waiker, J. K., & Pandya, M. (2013). Experimental investigation of double pass solar air heater using different type of porous media. International Journal of Current Engineering and Technology, *3*, 2006–2009.
70. Klein, S. A., Duffie, J. A., & Beckman, W. A. (1974). Transient considerations of flat-plate solar collectors. Journal of Engineering for Gas Turbines and Power, *96*, 109–113.
71. Bliss, R. W. (1959). The derivations of several 'Plate-efficiency factors' useful in the design of flat-plate solar heat collectors. Solar Energy, *3*, 55–64.
72. Mullick, S. C., & Nanda, S. K. (1989). An improved technique for computing the heat loss factor of a tubular absorber. Solar Energy, *42*, 1–7.
73. Shetty, S. P., Paineni, A., Kande, M., Madhwesh, N., Sharma, N. Y., & Karanth, K. V. (2020). Experimental investigations on a cross flow solar air heater having perforated circular absorber plate for thermal performance augmentation. Solar Energy, *197*, 254–265.

Chapter 5

Carbon nanotubes for evacuated tube solar collector

Seyed-Mostafa Moussavi, Ali Ahadi, and Mohsen Sheikholeslami

5.1 INTRODUCTION

Better energy utilization and storage are needed to keep up with population expansion and ever-increasing development. Fossil fuels account for more than 80% of the energy generated globally today (British Petroleum, 2013). It increases the atmosphere's carbon concentration and causes global warming. Instead of declining, the use of fossil fuels is rising (by 2% yearly on average) (Team, 2012). So, green energies to overcome the aforementioned problem with fossil fuel sources with higher efficiency are preferred (Pachauri et al., 2014). Scientists worldwide are experimenting with their national energy policies to develop renewable energy in response to recent worries about the depletion of non-renewable energy they emit. Solar energy is a big area of renewable energy that has purposefully attracted a lot of attention. Since there have always been affordable options for energy sources, solar energy has never actually been used on a large scale. This was tied to a cost-effective technique of utilizing solar energy and not only because there were cheap resources available. More specifically, though solar energy is completely free, it was predicted that sustaining such systems would be more expensive than developing the necessary hardware and systems (Plante, 2014). With a drop in the overall cost of solar system parts, recent technical developments, and the potential use of energy storage devices, there is an increased tendency to employ solar energy applications. Therefore, both industrialized and developing nations are expected to rely heavily on solar energy in the future (Adaramola, 2014). According to the International Energy Agency (IEA), solar array installations will be able to supply around 45% of the world's energy needs by 2050 (Mekhilef et al., 2011). The energy situation across the world is significantly influenced by the UK, the USA, China, India, and Russia. China and India both import energy and rely largely on fossil fuels to meet their energy needs. China saw a 25% increase in renewable energy in 2004 compared to a 7–9% increase in global power consumption, which made over 40% of the collecting area worldwide (Asif and Muneer, 2007). After China and Japan, India uses the third-most energy in Asia (Kumar and Vimala, 2016). The majority of India's electricity is produced by traditional

DOI: 10.1201/9781003527251-5

fuels like coal and mineral oil, which significantly increase glasshouse gas emissions (Lalwani and Singh, 2010). For practically all solar applications, absorbent materials are required to both absorb solar radiation and perform temperature control techniques. It is always crucial to use absorbent materials to take in solar radiation and to implement heat management strategies. This transmitter is then fed to a liquid to raise its temperature. Due to their wide absorption spectrum in solar thermal systems, it has been demonstrated that carbon nanotubes (CNTs) are efficient sun absorbers. Sumio Iijima published the first study on CNTs in 1991 (Iijima, 1991). Researchers have been drawn to CNTs because of their distinct optical, electrical, and mechanical characteristics. Previous studies have demonstrated that CNTs effectively and opaquely absorb light at all wavelengths, from visible to infrared. The base fluid has been mixed with CNT nanoparticles to improve the mixture's thermal performance. The research team of Xie et al. (2003), who may have been the first to study CNT, included nanofluids. Multi-walled carbon nanotube (MWCNT) nanoparticles were infused into poly-alpha olefin (PAO) oil at quantities of up to 1 vol%. The maximal thermal conductivity ratio (and/k_{bf}) was shown to be more than 2.5. Since then, the literature has published a large number of papers on CNT nanofluids. More research is done in this section on the features, properties, and effects of using CNT nanofluid in evacuated tube solar collectors (ETSCs).

5.2 APPLICATION OF CNT IN SOLAR ENERGY

CNTs are a new material with strong thermal and optical properties that may be used in a variety of solar systems. We examine the usage of CNTs as solar absorbers in this section because of their innate capability for high-bandwidth light absorption. CNT-based solar absorbers have the potential to improve power generation in solar and improve solar water heating (SWH) device thermal efficiency.

They are useful in a variety of solar technologies, thanks to their changed optical and thermal properties in contrast to the currently used conventional fluids.

5.2.1 Solar collector

The solar thermal collector is one type of heat exchanger that takes in the incoming solar irradiance and produces usable heat for use in a variety of applications (Sabiha et al., 2015a). According to the fluids' intended uses (such as air, water, and oil), the absorber, a part of the solar collector, transfers heat to them. Materials used as absorbers must have lower emissivity and upper absorptivity.

Various solar collector types are shown in the article according to the location, the weather, and the intended use (Kalogirou, 2004).

Solar technology has long used solar collectors to gather solar power. In actuality, the solar collector was created for the first time in 1767 (Butti and Perlin, 1980). The first ETSC was created by Emmet in 1911 (Emmet, 1911), and the first flat-plate solar collector (FPSC) with a tank was created by Bailey in 1909 (Crawford, 1975). Modern FPSC and ETSC have similar designs, except for materials and heat transfer fluid (HTF) used to improve system performance.

5.2.1.1 Flat plate collector

FPSCs are a recently created idea that has the largest several applications, e.g. contain water and space heating, home usage, and commercial, residential, and industrial use (Struckmann, 2008). It has a solid reputation for being the cheapest and simplest to build, install, and maintain. The surface area of a FPSC for solar radiation absorption is often about as large as the actual collection. There are two types of collectors: liquid-type collectors, which use liquid as the medium for transferring heat, and air-type collectors, which use air as the media (Amrutkar et al., 2012; Klevinskis and Bučinskas, 2011). The FPC, invented in 1958, is the first kind of solar collector (Hottel and Whillier, 1955). In order to develop, examine, and eventually improve this collector's thermal efficiency, many investigations were afterward carried out on it.

Since 2012, Yousefi and his colleagues have been experimenting with using nanofluid, notably CNT nanofluid, which serves as the HTF in FPSC (Yousefi et al., 2012a) (Yousefi et al., 2012b). They discovered that FPSC efficiency is significantly influenced by the pH of their manufactured nanofluids (Yousefi et al., 2012a). This is because the stability of the nanofluid and pH are strongly connected. With the same types of nanofluid and experimental setup, the same scientists carried out additional research, but in this example, they studied the effect of including a surfactant at different concentrations (Yousefi et al., 2012b). According to their findings, the FPSC efficiency was lower when nanofluid was employed at lower concentrations than water without the addition of surfactant. Intriguingly, a greater concentration of nanofluid without surfactant was shown to boost the effectiveness of the collector.

5.2.1.2 Evacuated tube collector

ETSCs make up the bulk of solar collectors with the maximum efficiency. It takes up less room and may be utilized on chilly, windy, and gloomy days (Kalogirou, 2003). Evacuated tube collector (ETC) consists of two cylindrical tubes; the outer tube is made of glass with a solar absorber and a vacuum space between it (Bhatia, 2014). The vacuum is created above the selective coating in order to reduce heat losses through convection and conduction. The heat that the selective coating absorbs is transferred to the water running through the tube or the heat pipe, which is a sealed copper pipe containing the refrigerant (Kalogirou, 2016). The ETSC's thermal efficiency is increased because the thermal losses are reduced by enclosing the absorber tube in a vacuum.

Numerous experts have undertaken studies on the effectiveness of ETSCs and studied the thermal performance of an ETSC using water/single-walled carbon nanotube (SWCNT) nanofluid (Sabiha et al., 2015b). The impact amount of nanoparticles and density rate are taken into account in their experimental investigation. Both of these characteristics were seen to have an impact on the solar collector efficiency, which peaked at 93.43% at 0.2 vol% SWCNTs and 0.025 kg/s flow rate, 71.84% higher than water. The thermal conductivity of the SWCNT nanofluid was measured at three-volume concentrations of 0.05, 0.1, and 0.2 vol%. SWCNTs nanofluid has a thermal conductivity that is significantly greater than that of water, and that rises with temperature and volume fraction. In contrast to water, SWCNT nanofluid may, therefore, absorb more heat from the heat pipe as it passes through the collector's manifold. Water and SWCNT nanofluids were used to examine how well an ETSC performed. Prior to using nanofluids as HTF, it was determined that they were stable for more than a month.

ETCs will make up 68.8% of the market, FPCs make a share of 24.7%, and unglazed water collectors make a share of 6.3% (Newell et al., 2021).

5.3 FEATURES OF CNT

5.3.1 Thermal conductivity

Contrasted with their basic fluids, CNT nanofluids have superior thermal conductivities, which get even better as the temperature rises. Researchers used a variety of CNT nanofluid types and measurement methods to assess the thermal conductivity of their nanofluids. Since Paul et al. (2010) have previously covered several of the techniques used to evaluate the thermal conductivity of nanofluids, we won't go into great detail about them. However, it was found that the method most frequently employed for figuring out the thermal conductivity of CNT nanofluids was the transient hot-wire approach. According to stability procedures, CNT type, base HTF, and other factors, the improved thermal conductivity of the nanofluids varied.

5.3.2 Temperature effects

At different temperatures, CNT nanofluids' thermal conductivity varies in tandem with their CNT content. A nanofluid using MWCNT and gum arabic (GA) as surfactants was made (Ding et al., 2006). Researchers later discovered that at 1 wt% MWCNT, thermal conductivity significantly improved from 10% to 80%. They created a nanofluid and examined its thermal conductivity at a range of temperatures (15–35°C). They found that the CNT nanofluids' percentage enhancement in thermal conductivity considerably rises by more than 5% at temperatures greater than 24°C (Ding et al., 2006; Garg et al., 2009). According to (Chen and Xie, 2010), temperatures have a significant impact on how well MWNT nanofluids stabilized by cationic gemini surfactants transmit heat. The thermal conductivity improvement of the nanofluid containing 0.6 vol% of MWCNTs increases from 5.6% to 34.3% when the temperature is elevated from 5°C to 65°C. At temperatures ranging from 15°C to 55°C, respectively, the thermal conductivities of the functionalized multi-walled carbon nanotubes (f-MWCNT)/glycol nanofluid (Meng et al., 2012) increased from 0.296 to 0.306 at 4.0% of f-MWCNTs and MWCNT nanofluids with 0.5 volume percent. When the temperature rose from 20°C to 60°C, they saw that the rate of thermal conductivity improvement increased practically linearly, from 10.64% to 20.21%. The thermal conductivity of an ethylene glycol (EG)-based nanofluid containing up to 0.3 vol% f-MWCNTs improved from 12.8% to 55.4%, according to a study (Zhang et al., 2015). Karami et al. (2014) produced f-MWCNTs up to 100 ppm in a water-to-EG (70%:30%) mixture and assessed them at various temperatures. Upon increasing the temperature from 20°C to 60°C, at 100 ppm of f-MWCNTs, they obtained an increase in thermal conductivity ranging from 10.0% to 54.4%. In addition to the studies mentioned above, numerous more research has also shown that nanofluids' thermal conductivities increase as a result of rising temperatures (Abubakar et al., 2016; Alawi et al., 2016; Ny et al., 2016). The bulk of studies have demonstrated that the thermal conductivity of nanofluids considerably rises at high temperatures. The behavior is partially explained by the fact that the thermal resistance between the base fluid and the CNTs diminishes at high temperatures.

5.3.3 Effects of concentrations

As previously mentioned, compared to their base fluids, CNT nanofluids exhibit higher thermal conductivity increases with temperature.

Numerous investigations have also demonstrated that the enhanced thermal conductivities in nanofluids are a direct consequence of the increasing CNT concentrations. The increase in thermal conductivity with CNT concentration in base fluid is assumed to be caused by the additional CNTs' better thermal conductivity relative to the base fluid. Interestingly, a recent study (Maré et al., 2015). found that nanofluids with MWCNT and sodium dodecyl benzene sulfonate (SDBS) as surfactants had remarkably high thermal conductivity even at a small amount of CNT. In their work, various MWCNT percentages and temperatures were used to test the thermal conductivity. They were able to boost thermal conductivity at 20°C by 13%.

5.3.4 Geometries effects

There are two different types of CNT shapes: SWCNT and MWCNT. The graphene sheet is conceptually rolled into a seamless cylinder. As time passed, additional graphene sheets were folded up to create MWCNT (Zhang and Li, 2009).

Due to the inclusion of additional CNT layers over the previous layer, MWCNT has a greater diameter than SWCNT. The effect of CNT diameters on the conductivity of CNT nanofluids has been considerable, according to a small number of studies. CNTs' length might also be altered in addition to their diameter. CNTs can have their length physically reduced by ball milling or ultrasonication, as well as chemically, during the covalent functionalization process. However, using shorter CNT might lessen the CNT nanofluids' increased heat conductivity. Following mechanical ball milling, Chen et al. (2009) studied the impact of MWCNT on the conductivity of nanofluids, including f-MWCNT, and found that with 1 vol% at 63.9°C, they improved thermal conductivity by 29.5%.

5.3.5 Consequences of physical and chemical treatment techniques

For the manufacture of CNT nanofluids, a two-step procedure is frequently used. This method's approach involves spreading CNT into any form of base fluid in the required amount. However, this process has to be further stabilized using certain steps, which often involve chemical and physical treatment techniques. The ultrasonicator device's homogenization of the suspensions is the easiest method for achieving the aforementioned goal. An efficient approach that has been utilized often recently is before being uniformly distributed into the base fluids via the ultrasonication process; the CNT surfaces are chemically modified using acids. Another efficient method combines CNT with the addition of a surfactant, then homogenizing the suspensions using ultrasonic equipment. One of the chemical processes for treating CNTs is covalent functionalization utilizing a combination of acids. This method has been shown to improve CNT nanofluids. Chemically treating the CNT is also a possibility with non-covalent. Non-covalent surfactants coat the CNTs' surfaces and render them both repellent to other CNTs and hydrophilic with base fluids. High thermal conductivity of the suspensions can only be attained by having homogeneous nanoparticle states in solvents. Higher power ultrasonication provides enough energy to separate clusters of particles with high impact, assisting in the creation of homogenous suspensions. The homogeneous condition of CNT nanofluids leads to a rise in the conductivity of the suspensions. The above-discussed chemical treatment procedures have been employed in a variety of ways to increase the conductivity of CNT nanofluids while maintaining their long-term stability. There are, nevertheless, relative disparities between them in terms of enhancing conductivity, as evidenced by a number of literature reports, and for the aforementioned aim, most research leans more toward covalent functionalization than non-covalent functionalization.

5.4 INVESTIGATION OF CNT IN NANOFLUIDS

5.4.1 Preparation methods of CNT nanofluids

A CNT nanofluid may be created using either a one-step methodology or a two-step procedure, according to the literature. The synthesis of nanofluids was performed simultaneously in a one-step method. The study by a team of researchers (Teng et al., 2011) examined this approach to make nanofluid containing CNT having the shape of flakes, which is not CNT and was the closest that was discovered. The two-step procedure produced the nanotube suspension by first fabricating nanotubes and then dispersing them into a base HTF or solvent in a second manufacturing phase. Traditional base fluids have constrained thermal and physical qualities, such as water, oil, and EG, and are often utilized as coolants and heat storage. The features of thermophysics in the combination might be further improved when particles are added to it. When these nanoparticles are disseminated in various base fluids or for a particular purpose, they give rise to new terminology, such as heat-storage or nano coolants ionanofluids, nanofluids, nano lubricants, nano greases, and nano refrigerants.

CNTs are crucial for several technical applications. The exterior of virgin CNTs, however, is chemically inert to the base HTF and hydrophobic (Chen et al., 2005). In addition, the high amount of van der Waals contact caused by the CNTs' enormous surface area causes aggregation during synthesis. These elements are harmful because they may lessen the CNTs' ideal qualities. The best dispersibility and stability of CNTs in various based fluids are crucial for realization. As was already mentioned, there are two broad techniques that may be utilized to ensure CNT stability in the base fluids: chemical and mechanical treatment. Covalent and non-covalent functionalization are two subsets of the chemical process known as functionalization. In the literature, functional groups can also be added electrophilically while being exposed to microwave radiation, wet-mechanochemical reactions that use acid treatment before ball milling, and plasma treatments. After functionalization, the procedure's efficacy is frequently assessed to verify its existence. Fourier transform infrared (FTIR) can be used for this. To measure the level of functionalization, other techniques, including Raman spectroscopy and thermogravimetric analysis (TGA), are also employed. Contrarily, mechanical treatment uses ultrasonication to mix the nanoparticle with the solvent physically. In reality, whether or not the CNTs were functionalized, it is common practice to disperse the nanoparticles via a physical mechanical method.

5.4.2 Advantages of CNT in nanofluids

To satisfy the requirements of several high-tech industries and complex thermal management systems, the creation of fluids with excellent heat transfer and transport capabilities is essential. This is due to the fact that fluid thermal characteristics are essential for the creation of machinery. Due to the fact that even microparticles are insufficient for microsystems, compared to suspensions of millimeter- or micrometer-sized particles, nanofluids are preferred. Additional problems with such suspensions include the rapid settling of these particles, occlusion of the flow channel, and increased fluid pressure loss. On the other hand, nanoparticles are orders of magnitude smaller than microsystems, remain in suspension, and reduce erosion and clogging. As a result, nanofluids may pass easily through small or microchannels, thanks to evenly scattered nanoparticles. The mobility of the nanoparticles is an additional advantage, which could create dynamic nanoscale processes and improve the efficiency of heat transmission in nanofluids. Owing to nanoparticles' small size, they are less in weight and less likely to sediment, which increases the stability of nanofluids. Since the heat transfer efficiency of heat exchangers or cooling devices is vital in many sectors, the impact of nanofluid technology is projected to be considerable. When properly engineered, CNT nanofluids have many advantages over conventional fluids and suspensions with particles that are millimeter- or micron-sized due to dispersed CNT's high aspect ratio; it has strong heat transfer and stability, without clogging, flow and cooling, streamlined systems, and a loss of pumping strength (Das et al., 2006; Murshed et al., 2008). Because of their greater stability, CNT will avoid heat transfer device wall clogging and quick settling. More energy efficiency, improved performance, and lower operational expenses are all associated with the high conductivity of nanofluids, convective, and boiling heat transfers. They can make solar energy systems more effective and apparatus while reducing the energy needed to pump HTFs. Nanofluids can be employed in systems that are smaller in size and have lower fluid stocks. Thermal systems may be more lightweight and compact. Vehicles with fewer parts have higher gas mileage, use less fuel, emit fewer pollutants, and are better for the environment. The ability to regulate characteristics and applications by adjusting the CNT's size concentration, synthesizing any number of walls, and surface treatment is a major benefit of employing CNT applications.

5.5 CNT APPLICATION IN SEVERAL EVACUATED TUBE SOLAR COLLECTOR (ETSC) TYPES

5.5.1 U-tube ETSCs

The phenomena of heat transfer at ETCs were initially studied by Eberlein. Using air as the working medium, we examine the performance of ETCs; he developed a mathematical model. The circular design of the ETC, according to the inventor, considerably reduced heat loss. The U-pipe type of ETSCs, which is regarded as an evaporator draught, is progressively being used for solar heat pump systems (space heating). The HTF's proposed temperature range is often lower than the surrounding ambient air (Nie et al., 2017). One of the most popular heat exchanger designs for ETCs is the U-pipe, which excels in cost-effective designs due to its straightforward geometry and outstanding thermal efficiency.

A few years ago, there were several experiment-based research on this kind of collector. Different U-pipe setups of the ETSC were taken into account by Liang et al. (2014) in uniform boundary conditions. Additionally, three distinct varieties of U-pipe ETSC were examined in terms of their thermal performance using a theoretical and experimental method. The outcome showed that three U-pipe solar collector efficiency could reach a maximum of 0.82%, which is much higher than other types. In research, O'Neil and Sobhansarbandi (2022) studied experimentally the comparison of the thermal performance of a specially constructed U-tube ETSC vs. accessible through commerce heat pipe evacuated tube collector (HPETC). The performance evaluation with and without a thermal storage medium was done over a number of days. Results of tests without a thermal storage medium revealed that U-Pipe ETC was 13% more efficient than HPETC because it was able to generate greater peak water temperatures, a maximum of 31°C, than HPETC. During tests with storage materials, the U-Pipe ETC's water tank and fin attained their highest temperatures at 47°C and 38.7°C, respectively, whereas the water tank and heat pipe's fin both achieved their highest temperatures of 36.7°C and 60°C, respectively. PCM72 with an additional 1 wt% of MWCNT nanoparticles was integrated into the U-Pipe ETC system to see whether it may improve performance any further. Peak temperatures for the U-Pipe fin and HTF tank were 63.9°C and 43.3°C, respectively, throughout this experiment. According to the study's findings, the U-tube system was able to reach HTF temperatures that were greater than those of the HPETC arrangement. The water's ability to store latent heat was extended by the use of NEPCM, which enabled slower water heat transfer.

5.5.2 Thermosyphon ETSCs

A horizontal tank is immediately linked to the thermosyphon collector, which is composed of 15–40 submerged single-ended tubes. The tubes are made of two concentric glass tubes that are vacuum-sealed together at one end. The outer surface of the inner tube has a selective surface coating. Here, water is naturally convicting over the single-ended tube as the heat transmission mechanism. Additionally, water flow is heated by sun radiation, which increases it during its ascent via the tube to the storage. Finally, cooler water is used in place of the original water.

Numerous academics were planning to conduct numerical analyses and experimental evaluations of SWH performance throughout the previous decade. The effects of flow in the thermosyphon's ETSC were studied by Morrison et al. (2005)using both computational and experimental approaches. The findings demonstrate that the flow rate of spontaneous convection in the tube is high enough to interfere with the tank's design. Additionally, the temperature of the tank has a significant impact on the pace of circulation through the tubes. The natural circulation flow of the diffuse reflector of the thermosyphon ETSC was examined. The circulation flow rate was investigated by elements such as tank HTF, collector tilt, aspect ratio, and solar input (Budihardjo and Morrison, 2009; Tang and Yang, 2014).

5.5.3 Heat pipe ETSCs

The heat pipe is a double-phase device used for heat transfer that has a very high thermal conductivity. It is made using a capillary wick and contains just a little vaporizable fluid. HTF movement results from a circulatory process between two phases, which transports the necessary heat. The heat pipes used in solar applications were said to possess critical characteristics, including heat transfer mechanisms and upper conductivity (Hayek et al., 2011; Wang et al., 2015). The usage of heat pipes in ETSCs has lately been the subject of several investigations. U-tube ETC and HPETC were the main subjects of the study. Both types of systems under consideration, numerical simulation and experimental data, showed excellent agreement. Every sort of collector has benefits. The U-tube has a standardized and higher thermal performance in cloudy conditions (Tong and Cho, 2015).

5.6 CONCLUSIONS AND FUTURE SCOPE

In conclusion, CNTs have shown great potential in improving the efficiency of ETSCs due to their high thermal conductivity and other unique properties. The application of CNTs in solar energy, particularly in ETSCs, has been found to enhance heat transfer and overall performance. The investigation of CNTs in nanofluids has shown promising results, with various preparation methods and advantages identified. The use of CNTs in different types of ETSCs, such as U-tube, thermosyphon, and heat pipe ETSCs, has demonstrated improved heat transfer capabilities and efficiency. Future research could focus on further optimizing the use of CNTs in ETSCs, exploring new preparation methods for CNT nanofluids, and investigating the long-term durability and stability of CNT-enhanced solar collectors. Additionally, studying the effects of different concentrations, geometries, and treatment methods of CNTs on the performance of ETSCs could provide valuable insights for further advancements in this field. Overall, the integration of CNT technology in ETSCs holds great promise for enhancing solar energy utilization and contributing to sustainable energy solutions.

REFERENCES

Abubakar, S., Azwadi, C. N., & Ahmad, A. (2016). The use of Fe3O4-H2O4 nanofluid for heat transfer enhancement in rectangular microchannel heatsink. J. Adv. Res. Mater. Sci, 23, 15–24.

Adaramola, M. (Ed.). (2014). Solar Energy: Application, Economics, and Public Perception. CRC Press.

Alawi, O. A., Mohammed, H. A., & Sidik, N. C. (2016). Mixed convective nanofluids flow in a channel having forward-facing step with baffle. Journal of Advanced Research in Applied Mechanics, 24(1), 1–21.

Amrutkar, S. K., Ghodke, S., & Patil, K. N. (2012). Solar flat plate collector analysis. IOSR Journal of Engineering, 2(2), 207–213.
Asif, M., & Muneer, T. (2007). Energy supply, its demand and security issues for developed and emerging economies. Renewable and Sustainable Energy Reviews, 11(7), 1388–1413.
Bhatia, S. C. (Ed.). (2014). Advanced Renewable Energy Systems (Part 1 and 2) (1st ed.). WPI Publishing.
British Petroleum. (2013). BP Statistical Review of World Energy. British Petroleum.
Budihardjo, I., & Morrison, G. L. (2009). Performance of water-in-glass evacuated tube solar water heaters. Solar Energy, 83(1), 49–56.
Butti, K., & Perlin, J. (1980). A Golden Thread: 2500 Years of Solar Architecture and Technology. Cheshire Books, Palo Alto.
Chen, L., & Xie, H. (2010). Properties of carbon nanotube nanofluids stabilized by cationic gemini surfactant. Thermochimica Acta, 506(1–2), 62–66.
Chen, L. F., Xie, H., Yu, W., & Li, Y. (2009, January). Water Based Multiwalled Carbon Nanotube Nanofluids with Optimized Thermal Conductivity. In International Conference on Nanochannels, Microchannels, and Minichannels (Vol. 43499, pp. 917–920).
Chen, X. H., Chen, C. S., Xiao, H. N., Chen, X. H., Li, W. H., & Xu, L. S. (2005). Lipophilic functionalization of multi-walled carbon nanotubes with stearic acid. Carbon, 43(8), 1800–1803.
Crawford, F. R.. (1975). Solar heater. (No. US 3859980). Solar heater (Patent) | OSTI. GOV. https://www.osti.gov/biblio/5122668
Das, S. K., Choi, S. U., & Patel, H. E. (2006). Heat transfer in nanofluids—a review. Heat Transfer Engineering, 27(10), 3–19.
Ding, Y., Alias, H., Wen, D., & Williams, R. A. (2006). Heat transfer of aqueous suspensions of carbon nanotubes (CNT nanofluids). International Journal of Heat and Mass Transfer, 49(1–2), 240–250.
Emmet, W. L. (1911). Apparatus for utilizing solar heat (No. US 980505). US980505A - Apparatus for utilizing solar heat. - Google Patents. https://patents.google.com/patent/US980505A/en
Garg, P., Alvarado, J. L., Marsh, C., Carlson, T. A., Kessler, D. A., & Annamalai, K. (2009). An experimental study on the effect of ultrasonication on viscosity and heat transfer performance of multi-wall carbon nanotube-based aqueous nanofluids. International Journal of Heat and Mass Transfer, 52(21–22), 5090–5101.
Hayek, M., Assaf, J., & Lteif, W. (2011). Experimental investigation of the performance of evacuated-tube solar collectors under eastern mediterranean climatic conditions. Energy Procedia, 6, 618–626.
Hottel, H. C., and A. Whillier. (1955). Evaluation of Flat-Plate Solar Collector Performance. Transactions of the Conference on the use of Solar Energy - The Scientific Basis, Vol. 2, Tucson, AZ, October 31–November 1, 74–104.
Iijima, S. (1991). Helical microtubules of graphitic carbon. Nature, 354(6348), 56–58.
Kalogirou, S. (2003). The potential of solar industrial process heat applications. Applied Energy, 76(4), 337–361.
Kalogirou, S. A. (2004). Solar thermal collectors and applications. Progress in Energy and Combustion Science, 30(3), 231–295.
Kalogirou, S. A. (2016). Nontracking solar collection technologies for solar heating and cooling systems. In Advances in Solar Heating and Cooling (pp. 63–80). Woodhead Publishing.
Karami, M., Bahabadi, M. A., Delfani, S., & Ghozatloo, A. (2014). A new application of carbon nanotubes nanofluid as working fluid of low-temperature direct absorption solar collector. Solar Energy Materials and Solar Cells, 121, 114–118.
Klevinskis, A., & Bučinskas, V. (2011). Analysis of a flat-plate solar collector. Mokslas–Lietuvos ateitis/Science–Future of Lithuania, 3(6), 39–43.

Kumar, V. R., & Vimala, M. (2016). Energy consumption in India-recent trends. Asia Pacific Journal of Research, 1(XXXVI), 140–51.

Lalwani, M., & Singh, M. (2010). Conventional and renewable energy scenario of India: Present and future. Canadian Journal on Electrical and Electronics Engineering, 1(6), 122–140.

Liang, R., Zhang, J., Zhao, L., & Ma, L. (2014). Research on the universal model of filled-type evacuated tube with U-tube in uniform boundary condition. Applied Thermal Engineering, 63(1), 362–369.

Maré, T., Halelfadl, S., Van Vaerenbergh, S., & Estellé, P. (2015). Unexpected sharp peak in thermal conductivity of carbon nanotubes water-based nanofluids. International Communications in Heat and Mass Transfer, 66, 80–83.

Mekhilef, S., Saidur, R., & Safari, A. (2011). A review on solar energy use in industries. Renewable and Sustainable Energy Reviews, 15(4), 1777–1790.

Meng, Z., Wu, D., Wang, L., Zhu, H., & Li, Q. (2012). Carbon nanotube glycol nanofluids: Photo-thermal properties, thermal conductivities and rheological behavior. Particuology, 10(5), 614–618.

Morrison, G. L., Budihardjo, I., & Behnia, M. (2005). Measurement and simulation of flow rate in a water-in-glass evacuated tube solar water heater. Solar Energy, 78(2), 257–267.

Murshed, S. M. S., Leong, K. C., & Yang, C. (2008). Thermophysical and electrokinetic properties of nanofluids–a critical review. Applied Thermal Engineering, 28(17–18), 2109–2125.

Newell, R., Raimi, D., Villanueva, S., & Prest, B. (2021). Global energy outlook 2021: Pathways from Paris. Resources for the Future, 8, 39.

Nie, X., Zhao, L., Deng, S., & Lin, X. (2017). Experimental study on thermal performance of U-type evacuated glass tubular solar collector with low inlet temperature. Solar Energy, 150, 192–201.

Ny, G., Barom, N., Noraziman, S., & Yeow, S. (2016). Numerical study on turbulent-forced convective heat transfer of Ag/Heg water nanofluid in pipe. Journal of Advanced Research in Materials Science, 22(1), 11–27.

O'Neil, T. J., & Sobhansarbandi, S. (2022). Thermal performance investigation of energy storage based U-pipe evacuated tube solar collector: An experimental study. Sustainable Energy Technologies and Assessments, 52, 102146.

Pachauri, R. K., Allen, M. R., Barros, V. R., Broome, J., Cramer, W., Christ, R., ... & van Ypserle, J. P. (2014). Climate change 2014: Synthesis report. In Contribution of Working Groups I, II and III to the Fifth Assessment Report of the Intergovernmental Panel on Climate Change (p. 151). IPCC.

Paul, G., Chopkar, M., Manna, I., & Das, P. K. (2010). Techniques for measuring the thermal conductivity of nanofluids: A review. Renewable and Sustainable Energy Reviews, 14(7), 1913–1924.

Plante, R. H. (2014). Solar Energy, Photovoltaics, and Domestic Hot Water: A Technical and Economic Guide for Project Planners, Builders, and Property Owners. Academic Press.

Sabiha, M. A., Saidur, R., Hassani, S., Said, Z., & Mekhilef, S. (2015a). Energy performance of an evacuated tube solar collector using single walled carbon nanotubes nanofluids. Energy Conversion and Management, 105, 1377–1388.

Sabiha, M. A., Saidur, R., Mekhilef, S., & Mahian, O. (2015b). Progress and latest developments of evacuated tube solar collectors. Renewable and Sustainable Energy Reviews, 51, 1038–1054.

Struckmann, F. (2008). Analysis of a flat-plate solar collector. Heat and Mass Transport, Project Report, 2008MVK160.

Tang, R., & Yang, Y. (2014). Nocturnal reverse flow in water-in-glass evacuated tube solar water heaters. Energy Conversion and Management, 80, 173–177.

Team, G. E. A. W. (2012). Global Energy Assessment. Cambridge University Press.

Teng, T. P., Cheng, C. M., & Pai, F. Y. (2011). Preparation and characterization of carbon nanofluid by a plasma arc nanoparticles synthesis system. Nanoscale Research Letters, 6, 1–11.
Tong, Y., & Cho, H. (2015). Comparative study on the thermal performance of evacuated solar collectors with U-tubes and heat pipes. International Journal of Air-Conditioning and Refrigeration, 23(03), 1550019.
Wang, Y., Zhu, Y., Chen, H., Zhang, X., Yang, L., & Liao, C. (2015). Performance analysis of a novel sun-tracking CPC heat pipe evacuated tubular collector. Applied Thermal Engineering, 87, 381–388.
Xie, H., Lee, H., Youn, W., & Choi, M. (2003). Nanofluids containing multiwalled carbon nanotubes and their enhanced thermal conductivities. Journal of Applied Physics, 94(8), 4967–4971.
Yousefi, T., Shojaeizadeh, E., Veysi, F., & Zinadini, S. (2012a). An experimental investigation on the effect of pH variation of MWCNT–H2O nanofluid on the efficiency of a flat-plate solar collector. Solar Energy, 86(2), 771–779.
Yousefi, T., Veisy, F., Shojaeizadeh, E., & Zinadini, S. (2012b). An experimental investigation on the effect of MWCNT-H2O nanofluid on the efficiency of flat-plate solar collectors. Experimental Thermal and Fluid Science, 39, 207–212.
Zhang, M., & Li, J. (2009). Carbon nanotube in different shapes. Materials Today, 12(6), 12–18.
Zhang, Y., Li, C., Jia, D., Zhang, D., & Zhang, X. (2015). Experimental evaluation of the lubrication performance of MoS2/CNT nanofluid for minimal quantity lubrication in Ni-based alloy grinding. International Journal of Machine Tools and Manufacture, 99, 19–33.

Chapter 6

Optimization algorithm for CNT application in solar system

Seyed-Mostafa Moussavi, Hossein Khademi, and Mohsen Sheikholeslami

6.1 INTRODUCTION

With the fast depletion of fossil fuels, also known as non-renewable sources of energy (petroleum, natural gas, etc.), the necessity for sustainable techniques to extract energy without affecting our climate is in high demand and has included substantial study in earlier ages. Wind is a well-known example of a renewable energy source, along with solar and tidal energy. Solar energy is significantly preferred due to its high levels of radiation and abundance, making it effective in providing energy. This is the primary driver for the increased global dominance of solar installations. Solar energy, which has been obtained from the sun directly for generations, is one of several sustainable energy sources. There are two types of solar energy: active and passive solar systems. Water heaters and concentrated solar power (CSP) systems are included in active systems. Passive systems include building orientation towards the sun, material choice, and light dispersion characteristics (De Groote and Verboven, 2019). Solar energy has started to be used to meet the power needs of useful gadgets in recent years (Mohan and Senthilkumar, 2022; Senthilkumar et al., 2022a; Senthilkumar et al., 2022b). Many nations offer incentives to encourage the usage of green energy sources together with various technologies and systems for solar and wind energy. In order to minimise construction costs, solar energy systems are installed atop the structures (Araújo et al., 2019). As the planet's temperature rises daily, it is imperative that efforts be made to use renewable energy sources to limit CO_2 emissions. Evidently, a method with set sustainability for classifying solar parameter values is examined (Choudhary and Srivastava, 2019). Due to its endless supply, environmental friendliness, and the possibility for solar energy collection systems to have high efficiency of power conversion, solar energy is an essential source for energy (Gwon et al., 2014) (Dincer, 2000). The manufacture of photovoltaic (PV) devices is influenced by research on improving the efficiency of power conversion. The equipment gets more affordable with respect to conventional energy sources as power efficiency improves (Gwon et al., 2014) (Ye et al., 2013) (Raut et al., 2011). Many people are interested in employing innovative optoelectronic devices, such PV solar cells, with heterojunction designs that make use of

 DOI: 10.1201/9781003527251-6

carbon elements (Zhu et al., 2009). The usage of carbon nanotubes (CNTs) in PV solar cells will be examined, with a focus on organic, dye-sensitized, and silicon-based solar cells. Numerous studies have looked at various substances, especially translucent CNT-based conductive films, semiconducting polymers, and graphene that may easily be placed on silicon wafers used in industry (Shi et al., 2012). In order to increase the output of these ecologically friendly energy generating techniques, several tests have foreseen new developments in PV cells. However, in order to accomplish semiconductor intersection, the maximal solar emission must still be allowed in order to make affordable optical coatings. This problem may be largely resolved by covering the device with an anti-reflective coating (ARC) layer that has an improved refractive index (RI) and thickness (Mathew et al., 2014; Benzekkour et al., 2009). The majority of coatings are made of a fundamentally porous silica-based substance with a strong attachment to the surface and adjustable RI and thickness. SiO_2 porous coatings for solar cells can be useful; however, according to some studies, pure SiO_2 porous coatings are not stable in water because they are too soluble to resist wear from water. The porosity that builds up on the material's surface amplifies this impact. The pores' hydrophilic surfaces can be filled with water at simply moderate humidity due to their small size, which significantly reduces the antireflective properties while raising the RI (Faustini et al., 2010). However, the presence of holes that may be reached in the nanoscale range aids in the adsorption of environmental contaminants. This is significant because it raises the antireflective layer's RI. Making the pores hydrophobic can greatly reduce the amount of water that is taken in (Faustini et al., 2010; Garnett and Yang, 2010). In order to make surfaces hydrophobic, chemicals need to be utilised. Pollutants cannot escape when this is done, with the exception of when non-open porosities are in the presence of contaminants. The coordination of a photocatalytic species' integration would serve the purpose of cleaning the porous surface. In the optical layer, this would be arranged alternately (Zhang and Resasco, 2009).

6.2 MODELING OF SOLAR SYSTEMS

The global production of electricity from solar energy is enormous and is growing quickly every day. The cost of solar panels, batteries, and inverters is decreasing significantly on the alternative side. These reasons are causing numerous nations to modify their policies towards solar energy. Many academics have recently developed an interest in simulating solar cells. Different kinds of comparable circuits are utilised in solar cell modelling depending on the quantity of diodes in that specific circuit. Some solar cell models are (Senthilkumar et al., 2020) as follows:

- Single-diode model (SDM)
- Double-diode model (DDM)

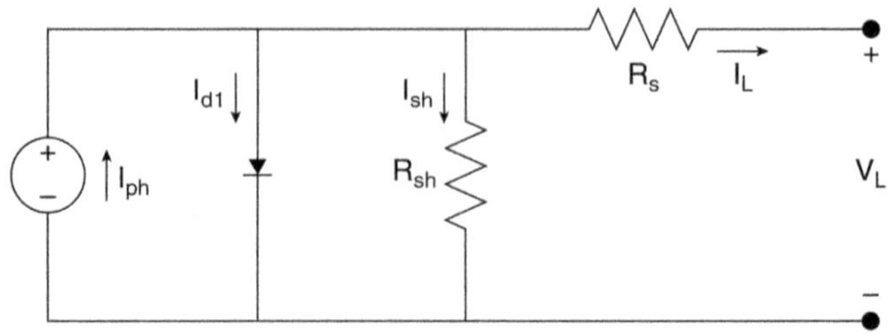

Figure 6.1 SDM equivalent circuit.

Source: Senthilkumar et al. (2023).

6.2.1 Single-diode model

SDM is quite simple to implement and develop. The SDM circuit is depicted in Figure 6.1. The diode ideality factor (n), the light-induced current (Iph), the diode dark saturation current (Id1), the series resistance (Rs), the shunt resistance (Rsh), and the light induced current (Iph) are the six uncalculated variables in the shown SDM model that should be calculated (Senthilkumar et al., 2023).

SDM is frequently used to obtain the solar cell's current-voltage (I-V) curve.

6.2.2 Double-diode model

Figure 6.2 displays the corresponding system of DDM.

DDM accounts for losses due to recombination current within the depletion area, which improves accuracy compared to SDM, which does not account for these losses. Dark saturation current in diodes (I01), dark saturation current in diodes (I02), quality of a diode (a1), factor for diodes (a2), series opposition (Rs), and Rsh are the six unknown parameters that need to be determined in DDM. DDM and SDM are contrasted in this study to improve the

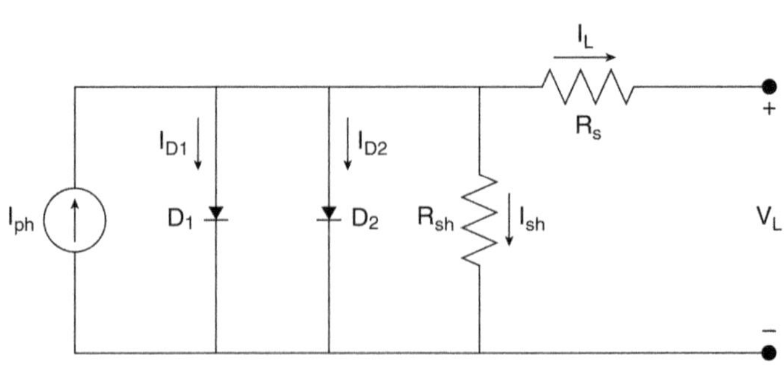

Figure 6.2 DDM equivalent circuit.

Source: Senthilkumar et al. (2023).

efficacy of solar PV systems. In contrasting SDM with DDM, SDM is simpler to design and execute but has lesser accuracy, whereas DDM is more exact and will thus increase the effectiveness of PV systems overall. It is possible to use many methods to establish the conditions of various solar PV models such as SDM, DDM, and three-diode model (TDM), among others (Senthilkumar et al., 2023).

6.2.3 Solar-powered energy

The sun sends 120,000 TW to the planet each hour, and it is thought that the world needs around 13 TW of energy annually (Lewis and Nocera, 2006). The yearly worldwide demand may therefore be met and exceeded with just 0.01% of the sun's energy that the planet gains per hour, while yet leaving opportunity for future growth and development to satisfy needs. Two basic categories may be used to roughly classify the solar energy harvesting methods. These include solar photovoltaics (SPVs) and CSP (Wang et al., 2010). In its most basic form, concentrating solar power employs mirrors to direct solar thermal energy into a heat transfer fluid. In order to create steam, which is ultimately utilised to power turbines and create electricity, the heat created is recovered. However, with an SPV, electromagnetic energy is immediately transformed into electricity. Electrons in donor molecules are excited by photons from the sun and are then transported to electrodes to produce electricity. One of the most alluring and sophisticated possibilities for supplying energy in underdeveloped nations is SPV technology. For Africa, this is particularly true, where people are dispersed across huge distances and where distinct groups have varying and changing demands for power in various rural locations. Additionally, as a result of economic, environmental, technological, or an amalgamation of a few or all of these problems, providing and maintaining traditional energy systems would be very difficult.

6.2.4 Inorganic versus organic solar cells

An estimated 90% of the worldwide demand share for commercial solar cell devices is now held by silicon-based cells. This tendency may be attributed to a number of variables, including economic (common, easily accessible commercial methods for silicon device fabrication) as well as technological (in-depth knowledge of silicon processing, dependable technology, for example). Current research has shown that multijunction concentrator devices, on the other hand, have total efficiencies exceeding 30%, and then crystalline silicon (25%) came in second (Huang et al., 2013). However, a quick and simple approach includes using unprocessed quartz to create metallurgical silicon by reducing it with carbon, which is produced at a reasonable cost but requires a lot of energy. Then, to purify the metallurgical silicon, it is changed into chlorosilanes (with HCl), and they are later changed, by utilising a chemical vapor deposition (CVD) method, into silicon for solar power (for example, the widely used Siemens method). Other more contemporary methods make use of plasma or direct metallurgical methods, although these modern methods take a lot of energy and may be bad for the environment (Pizzini, 2010; Braga et al., 2008). The adoption of silicon-based PV production is constrained by these major limitations in emerging economies (several countries in the tropics, like South Africa). Organic solar cell (OSC) technologies are advantageous in light of these difficulties and conditions. In a nutshell, OSCs may be summed up as polymer-based PV units, very small organic compounds, or a mix of both, either alone or in conjunction. Using a nanomaterial of some kind, these devices may be constructed using well-known methods for processing polymers since they are often made of polymeric materials. This implies that they may be easily created in a variety of sizes and forms(Figure 6.3; Abdulrazzaq et al., 2013) They can be readily customised and created for specialised purposes, and they can be made inexpensively. They may also be constructed on different substrates (Jeon and Lee, 2012). An OSC's fundamental structure consists of two layers of transport that convey electrons or holes; they are separated by an active layer and two electrodes surrounding the entire device. When light excites the active layer, it generates pairs of electrons and holes. These service providers are broken down at the location where the active layer and the transport layers meet, where they are subsequently the electrodes' collection and used to provide energy to the target system (Weickert et al., 2011; Deibel et al., 2010). Material on both sides of the connection is doped that improves the charge carrier's separation that happens over a solar cell based on silicon p-n junction. The situation with other inorganic-based solar cells is similar.

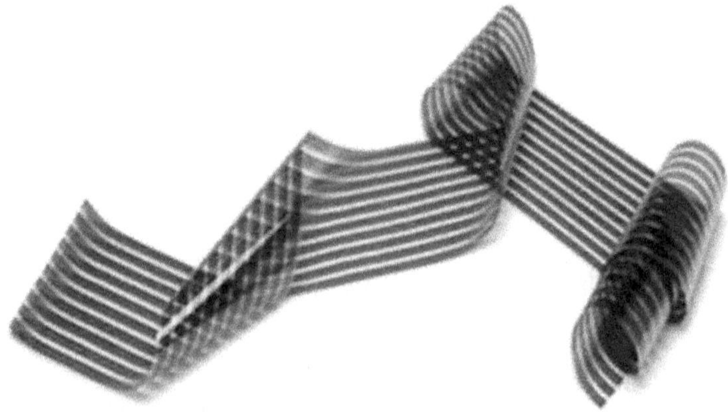

Figure 6.3 A flexible substrate was used to cast OSC. It is flexible and may be rolled, curved or stretched.

Source: Abdulrazzaq et al. (2013).

6.2.5 Carrier generation and photoexcitation

The process that leads to split between the electron-hole pair in OSCs is an extended procedure that is still a subject of intense research, although the heterojunction is one of the most important aspects of their activity. The process that produces and separates excitons eventually provides a useful image. Recent literature describes how current works (Janssen and Nelson, 2013). From the highest occupied molecular orbital (HOMO) to the lowest unoccupied molecular orbital (LUMO), photons with the required energy stimulate donor molecule electrons, according to study results thus far. Following that, the complex of photoexcited charge transfer changes to donor-acceptor (D-A) interface (either by migration or diffusion). Similar to how there might be a lot complexes of energy transition states involving the donor, or a combination of the two at the D-A interaction, the transfer of energy through complexes that transfer charges may entail one or further changes inside the donor molecule. The donor has somewhat greater levels of HOMO and LUMO energy than the recipient. This is required to speed up energy transfer for the receiver and minimise recombination inside the giver. When the charge is in the photocurrent state, it breaks into charge carriers for electrons and holes, notwithstanding the possibility of thermal or electrical losses that are close to or where the D-A interface is (Weickert et al., 2011; Deibel et al., 2010).

6.2.5.1 Calculating efficiencies in cells

The calculation of the cell's existing voltage, which is accomplished by computing either the power conversion efficiency (PCE) or the external quantum efficiency (EQE), controls the PV characteristics of OSCs and skeletal stem cells (SSCs). The EQE is outlined as the number of electrons gathered per photon incident (Abdulrazzaq et al., 2013). PCE of a cell is the proportion of solar energy that is received by a cell that is transformed into electricity. It also refers to the cell's effectiveness when illuminated by a conventional energy source. A high PCE percentage indicates that cell production will be high, and things might be improved by developing new materials or gadget designs (Janssen and Nelson, 2013).

6.2.5.2 Effects of physicochemical variables on the effectiveness of organic cells

In contrast to SSCs, OSCs have a lower PCE and a shorter lifespan, which restricts their practical use (Jeon and Lee, 2012). The active layers and electrodes' limited lifespan are a result of their physical and chemical deterioration. Chemical degradation is the process of organic materials or electrodes going through chemical changes as a result of oxygen and moisture (Figure 6.4; Jørgensen et al., 2008). Charge buildup, morphological alterations, and metal migration into the active layer are the typical causes of physical deterioration

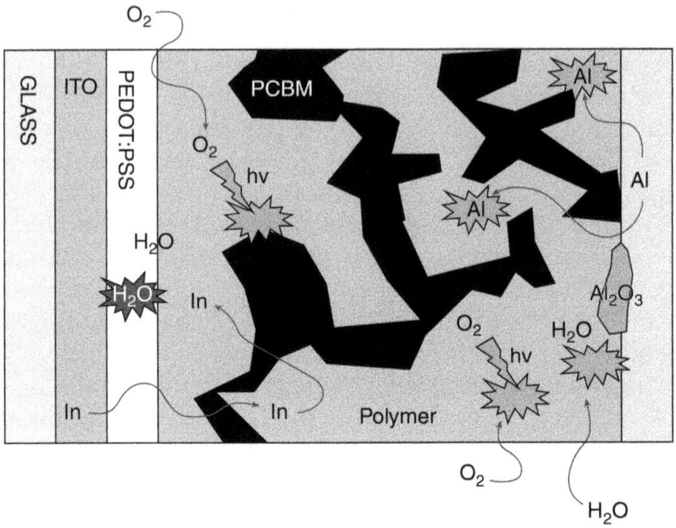

Figure 6.4 Oxygen and water contamination, as well as the penetration of metal electrode particles, causes the active layer to degrade. When making the gadget, oxygen and water are introduced. PCBM = Phenyl-C61-butyric acid methyl ester; ITO = indium tin oxide.

Source: Jørgensen et al. (2008).

(Jeon and Lee, 2012). Due to oxidation, delamination, dedoping, and organometallic interfacial chemistry, electrodes deteriorate. Energetic layers deteriorate as a result of morphological changes, impurity inclusion, photochemical reactions, and thermochemical reactions (Peters et al., 2012). Deposition of a thin layer for electron extraction, the hole extraction layer, and the electrodes can reduce degradation caused by the poor photostability at the electrode-active layer interface (Williams et al., 2013). An energetic layer containing side groups with chemical activity has reportedly been shown to be more prone to deterioration than one without. The establishment of trap between the gap and photochemical processes in the active layer are responsible for this significant loss in functionality (Peters et al., 2012). Other factors, including as lower hole mobility, trap-mediated recombinations, and charge buildup in various trap states, can result in decreased performance. The recombination processes at the D-A interface are one of the major obstacles to getting high PCE in OSCs. Improving device shape is one strategy used to reduce recombination (Wang et al., 2011), modifying electrode materials (Cox et al., 2011), employing changed D-A materials (Price et al., 2011), and even increasing optical absorption (Park et al., 2009). Zhang et al.(2012) have found a considerable decrease in losses from recombination, increasing the PCE of the device by almost 3.40%. Another factor that affects OSC efficiency is

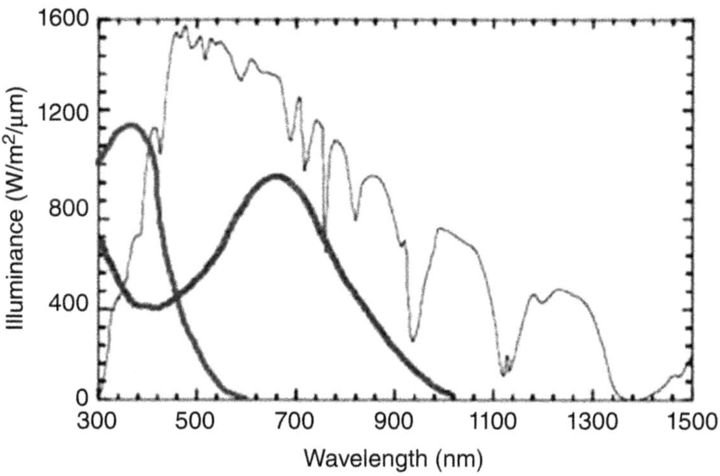

Figure 6.5 A polymer with various band gaps.

Source: Ye and Chi (2011).

the conjugated polymers' band gap. Only a tiny portion of the incident solar energy is absorbed when the semiconducting conjugated polymer's band gap is substantial. For instance, only 25% of the solar energy might be taken in by a polymer with a 2.0 eV band gap. Figure 6.5 illustrates how absorption increases to between 70% and 80% when the band is 1.2 eV less (Ye and Chi, 2011). The band gap should be as minimal as possible for maximal absorption. Utilising materials having a band gap that have energy lower than 1.5 eV is of interest in order to achieve neutral state absorption at 600 nm or above. CNTs can be added to the bulk heterogeneity (BHJ) to increase OSC efficiency. For instance, Somani and colleagues observed that adding double-walled CNTs (DWCNTs) significantly improved cell function (Somani et al., 2006). In contrast to Fullerenes which are only useful as electron acceptors, CNTs have an increased affinity and the process of delivering electrons to electrodes (Ferguson et al., 2013). The reader is advised to consult references because this is the subject of another review (Aitola et al., 2013; Xu et al., 2012). Arena and coworkers found that the CNTs' dispersion in the conjugated layer of polymer boosted hybrid cells which saw a 0.49% improvement in efficiency (Arena et al., 2008). By combining CNTs doped with heteroatoms like Ne or boron, OSC efficiency may be further increased. When nitrogen or boron is added to CNTs to create defects like N-CNTs or B-CNTs, it changed the materials' chemical and structural properties and electrical characteristics (Keru et al., 2013). By adding N-CNTs to OSCs, Lee and colleagues obtained a high efficiency of 5.29%. Prior to the addition of N-CNTs, the cell had an efficiency of 4.68% (Lee et al., 2011).

6.3 CARBON NANOTUBE INTEGRATION INTO ORGANIC SOLAR CELLS

Conductive nanostructures, such as CNTs, facilitate the transmission of carriers to the electrodes by improving charge separation in the BHJ. D-A type solar cells may be created by joining CNTs using the conjugated polymers' electrons (Somani et al., 2008). By creating a stronger at the CNT/polymer interface's electric field, which prevents the photogenerated charges recombining, the CNTs in this situation not only serve as electron acceptors but also promote the separability of excitons. More than any other semiconductor, CNTs in a polymer matrix's BHJ provide high electron mobility. The crystallinity that has been induced in Poly(3-octylthiophene) (P3OT) or other also improves hole transport. Higher hole mobility is made possible by interchain transport, thanks to the conducting polymer's highly ordered supramolecular organisation (Cataldo et al., 2012). Before they are spin-coated onto the substrate, CNTs are disseminated in the donor polymer solution. Indium tin oxide (ITO) is an example of an opaque conducting oxide, which is often used as the substrate. First, a poly (3,4-ethylenedioxylthiophine) is used as a hole conducting layer: As shown in Figure 6.6, the use of poly(styrenesulfonate) aids in the insertion of certain holes into the ITO electrode. The D-A is subsequently placed as an active layer. Because conjugated polymers may be casting semicrystalline sheets onto a surface, such as

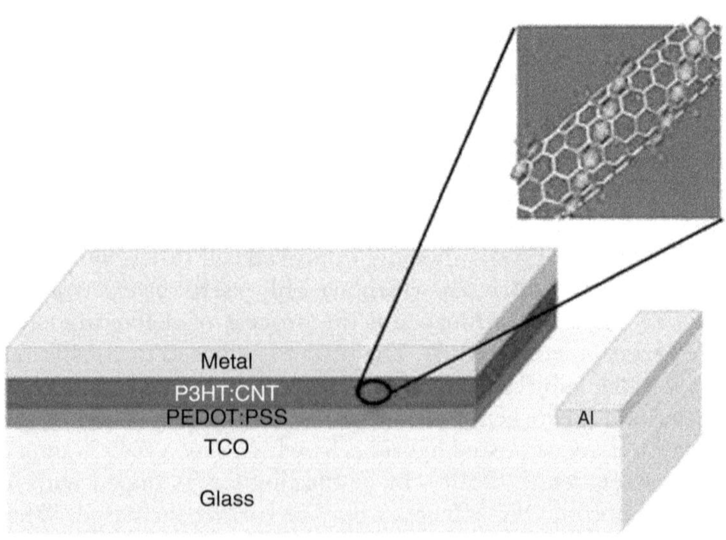

Figure 6.6 An example of an OSC schematic illustrates the sandwiched active layer with CNTs acting as acceptors between the electrodes. High-magnification image of the active layer inset demonstrates how polymer wraps around CNTs. CNT = Carbon nanotube.

Source: Abdulrazzaq et al. (2013).

P3OT, they are frequently utilised as donated parts; it may also occasionally have a very thin coating of lithium fluoride below (Günes et al., 2007). To enhance a cell's PV qualities, a third element, such as a dye, may occasionally be added to the active layer. Bhattacharyya et al. (2004) fashioned a cell that doubled the effectiveness of the previous one. They accomplished this by adding maleimide (PM), a dye. Selective Ne or boron doping of CNTs improves charge dispersion of the polymer while also tuning the work function (Hwang et al., 2012). The work function of CNTs increases as B doping decreases their Fermi-level energy. The work function of pure CNTs was 4.6 eV; after B doping, it increased to 5.2 eV, which is extremely near to the donor HOMO (Lee et al., 2011). Due to the close proximity of the PCBM's LUMO to the N-CNTs' work function (0.2 eV), an electron may pass between them with ease, and the electrode will then receive the transmitted electrons (Jana et al., 2013).

6.3.1 Carbon nanotubes' beneficial features for organic solar cells

Various techniques are used to create carbon nanotubes, depending on the resources at hand and the goal of the research. The majority of techniques, however, vary from the arc discharge, laser ablation (Iijima and Ichihashi, 1993), or CVD approach (Danafar et al., 2011). Because it is simple to use and scale up, the CVD approach is the most preferred technology (Kumar and Ando, 2011). The final characteristics of the CNTs are directly influenced by the synthesis method. SWCNTs or MWCNTs are preferred by the laser ablation and arc discharge procedures. MWCNTs are typically made using CVD techniques; however, SWCNTs and DWCNTs can also be generated. The physical-chemical characteristics of the ensuing nanocomposites, which are created when CNTs are added to a polymer matrix, change dramatically from those of the original polymer substance. The relevance of including CNTs in OSCs and the significance of CNTs' physical-chemical characteristics in the polymer matrix are discussed in the part after.

6.3.2 CNT physicochemical characteristics

The conductivity of CNTs is thousand times more powerful than Cu, making them effective electrical conductors (Kumar and Ando, 2011), and as a result, they can offer a good and reliable network that creates a percolating channel for the movement of charge carriers when included in that for OSCs. Upon electromagnetic spectrum absorption in the near-infrared range, semiconducting CNTs also produce excitons (Sakellariou et al., 2013). When included into OSCs, this has the ability to enhance the quantity of excitons produced within a device and leads to more electron and hole disintegrations. As CNTs' are large, more than 1000:1, just a little quantity is needed to create composite materials via physical characteristics that are considerably distinct from the parent polymer's properties. They have special properties that create these things valuable as supplements to create very strong composites. They also have very tiny dimensions (Kim et al., 2010) (Jorio and Dresselhaus, 2008).

6.3.2.1 Carbon nanotubes' electronic characteristics

Depending on how the graphene sheets are rolled to create each individual tube, single-walled CNTs may exhibit metallic or semiconducting characteristics (Oosthuizen and Nyamori, 2011). There is overlap of the orbitals in each of the various six-membered rings in a fully aligned SWCNT. The SWCNTs are regarded as conductors because this overlap creates a delocalized region for the passage of electrons. When the orientation of the six-membered rings is twisted, the orbital overlap changes, and the conductivity can turn semiconducting. A pair of numbers (n, m) that define the kind of nanotube is used to determine the direction of the graphene sheet plane and the nanotube diameter (Harris, 2004). The numbers are equal in the armchair configuration (n = m), one of them is zero (m or n = 0) in the zigzag orientation, and the pipe is when the integers are nonzero as chiral and unequal. All other orientations are semiconducting, whereas armchair-style tubes are metallic (Jones et al., 2010). Using semiclassical models developed for graphite, it is possible to characterise the electrical conductivity of SWCNTs at high temperatures, and they exhibit 2D quantum transport properties when operating at low temperatures (Mohlala et al., 2006). Recent analyses have emphasised the restrictions faced while adding SWCNT structures into OSCs in terms of electrical characteristics (Guldi and Costa, 2013). The inhomogeneity of a sample of SWCNTs is still the most important factor to take into account. The samples typically comprise a combination of semiconducting and highly conducting tubes after SWCNTs have been synthesised and purified. As a result, some of the tubes may improve the transmission of charge carriers to the electrodes, while others would serve as locations for charge carrier recombination. With advancements in separation techniques or chirality control in SWCNT synthesis, these restrictions may soon be removed. The resultant nanocomposites can have electrical characteristics that are different from their parent materials when CNTs, whether multi-walled or single-walled, are introduced into a polymer matrix. Whether the loading or weight percentage of CNTs is above or below the percolation threshold will determine this. When the concentration of CNTs is at or above the percolation threshold, conductivity dramatically increases. The percolation thresholds for CNT/polymer composites have been examined by Bauhofer and Kovacs (2009). The dispersion, aspect ratio, purity, and alignment of the CNTs are known to have an impact on the percolation threshold (Yang et al., 2005). Additionally, it has been hypothesised that homogenous dispersion makes it simpler for well-dispersed CNTs to construct an electrical channel. Increased photocurrents result from higher electrical conductivity, which raises the cell device's total efficiency (Jun et al., 2012). SWCNTs and MWCNTs can be used as transparent electrodes or as electron or hole carriers within the donor or acceptor polymer matrix because their electrical properties can be tuned to a certain extent so that the HOMO and LUMO match the donor and acceptor polymers appropriately (Li et al., 2007).

6.3.2.2 Carbon nanotubes' mechanical characteristics

Because they are formed of continuous networks of sp2 carbon-carbon bonds and have a tubular structure, carbon nanotubes are among the strongest and most robust substances that have ever been discovered. The tensile strength of CNTs has been measured to be greater than 100s of gigapascals, and they have a theoretical Young modulus on the order of terapascals. To put such figures into context, CNTs are frequently compared to steel, with basic descriptions saying that they are six times lighter and 100 times stronger than steel (Laird and Li, 2013; Shokrieh and Rafiee, 2010). Furthermore, CNTs are highly flexible and can repeatedly bend over 90 degrees without breaking (Iijima et al., 1996). At high pressures (>1.5 GPa), they experience structural changes that are permanent, whereas deformation is often completely elastic at these pressures. CNTs also have a substantially lower Young modulus in the radial direction of 10s of gigapascals, and their radial tensile strength is lower with values of 1 GPa. The CNTs' mechanical strength boosts the composite's strength (Shokrieh and Rafiee, 2010) and may be of interest to OSCs. It opens up new design and implementation options for solar cells in general when the mechanical properties of the CNTs are transferred to the polymer matrix or when the mechanical properties of the parent materials are changed to those of the new composite. Larger-surface-area PV systems may be installed on structures that can't sustain the weight of some inorganic systems, thanks to the adoption of lighter and more durable materials. More design and implementation alternatives would also be available, thanks to the flexibility, such as OSCs on a bridge column or any other highly curved surface. However, the CNTs must be functionalized in order to debundle them and make it easier for them to disperse in a solution or polymer matrix in order to produce superior mechanical characteristics with polymer/CNT nanocomposites.

6.3.2.3 Carbon nanotubes' thermal characteristics

In comparison to other carbonaceous substances, CNTs are oxidation-resistant at high temperatures. SWCNTs and MWCNTs exhibit various temperature behaviours. SWCNTs have a more defined structure and experience fewer deformations than MWCNTs, making them more thermally stable (Ruoff et al., 2003). CNTs have a greater axial thermoconductivity while being an insulator in the radial direction. At room temperature, the thermoconductivity is predicted to be 7000 W/m-K (Padgett and Brenner, 2004). As a result, the incorporation of CNTs in the polymer matrix creates a thermally conductive percolating network, increasing thermal conductivity (Han and Fina, 2011). That is analogous to that of diamond. This can increase the thermal conductivity of OSCs and, as a result, help with thermal deterioration issues.

6.3.2.4 Carbon nanotubes' chemical characteristics

Although CNTs are usually chemically neutral, their surface curvature increases their reactivity. Reactivity improvements result from the orbitals in the CNT walls being out of alignment. Smaller-diameter CNTs compared to larger-diameter ones are more reactive (Lin et al., 2003). Owing to their low responsiveness, CNTs may have their surfaces modified by adding functionalities that contain oxygen through acid oxidation (Ozin et al., 2005). This improves their compatibility with several polymeric matrices as well as their solubility in polar liquids. Covalent chemistry allows for further modification; examples include fluorination, ozonolysis, and acrylation (Harris, 2004). Functionalization, a term used to describe these alterations to CNTs, is covered in subsection 6.3.3.

6.3.3 CNT functionalization

In order to remove contaminants such amorphous carbon, metal particles, and graphitic nanoparticles, all of which interfere with the intended qualities or final result, raw CNTs are often not utilised to create CNT/polymer composites (Aqel et al., 2012). Gas-phase oxidation can be used to eliminate these contaminants (Rong et al., 2010), the use of acid (Edwards et al., 2011), thermal treatment and annealing (Andrews et al., 2001), ultrasonication (Aqel et al., 2012), magnetic division (Aqel et al., 2012), microfiltration (Hou et al., 2008), or a mix of a few of the aforementioned techniques (Serp et al., 2003). As was already indicated, adding CNTs to a conjugated polymer can enhance the polymer's mechanical, thermal, and electrical characteristics, which could be useful in the production of OSCs. However, surface modification is better for surface bonding. Covalent or noncovalent surface enhancement is possible (Ma et al., 2010).

6.3.3.1 Noncovalent functionalization

Through relationship between the delocalized electrical components of conjugated chains of polymers and CNTs, noncovalent modification entails adsorption on the CNTs' walls (Liu, 2005). Due to the dispersion pressures between individual tubes, CNTs typically cluster to form bundles. Because of a large aspect ratio, the packages may exceedingly be challenging to separate, necessitating via means of ultrasonic exploratory methods in the processing period. Individual CNTs and the chain of polymers interact to change the polymer's conformation, which can result around the CNT in helical winding (Spitalsky et al., 2010).

6.3.3.2 *Covalent functionalization*

The introduction of covalent functional groupings bound to the CNTs' skeleton structure by covalent modification aids in their improved dispersion inside of the polymer. However, covalent functionalization may unfavourably change the electrical characteristics of CNTs, which may negatively impact the final OSCs (Spitalsky et al., 2010).

6.4 PARAMETERS ESTIMATION ALGORITHM FOR SOLAR CELLS

For the simulation and computation of the primary features of the PV system, parameter stretching of such cells is crucial. Typically, these variables include optical current, optimal, series, and shunt resistance. The cell's model of its circuit is nonlinear. For the purpose of estimation, monitoring sun cell quality and assessing their performance are essential. According to Figure 6.7, these techniques may be divided into three groups: analytical, restorative, and evolutionary computational approaches (Subudhi and Pradhan, 2011).

Many of the methodological approaches used to categorize the methods used are described in the following subsections.

6.4.1 Methods for extraction of analytical parameters

Common techniques for estimating parameters using mathematical equations are analytical approaches. When it comes to an analytical method's key characteristics, the computation is quick and the outcomes are generally correct. Due to the fact that just one iteration is necessary for this operation, analytical parameter extraction methods are highly straightforward and take substantially less computing time. An easy analytical approach was suggested by De Blas et al. It has to do with how the cell model is built without taking the size of Rsh into account. Therefore, for the four-parameter model, only y is appropriate (De Blas et al., 2002). It used the Lambert W function in a variety

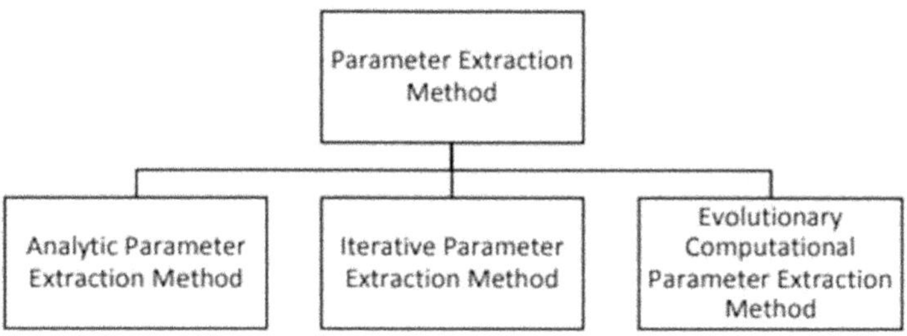

Figure 6.7 Classification of different methods.

Source: (Subudhi and Pradhan, 2011).

of ways to retrieve parameters from solar cells (Jain and Kapoor, 2004). The present equation, however, cannot be directly solved using simple functions. This approach was used to examine the I-V performance of a business sun panel at various irradiance and temperature conditions (Cubas et al., 2014). The suggested approach is based on function. The model parameters were then derived after the Co-content function had initially been determined using exact and clear analytical formulas (Ortiz-Conde et al., 2006). A straightforward approach based on the Lambert W function was provided by Zhang et al. to estimate all the solar cell's parameters. Numerous solar devices, including sun cells, sun modules, stand-alone organic cells, and Dye-Sensitized Solar Cells (DSSCs), have all been analysed using this technique (Zhang et al., 2011). Laudani et al. provided a technique that made use of model simplified forms. The convergence, computational cost, and execution time benefits are therefore substantial (Laudani et al., 2014). A pattern search (PS) optimisation approach for solar cell estimate was suggested by AlRashidi and colleagues. PS was a derivative-free approach that allowed objective functions that weren't always continuous, distinct, or differentiable some freedom. His optimisation algorithm's benefits were conceptual, easy to use, and effectiveness (AlRashidi et al., 2012). Jain et al. (2004) contrasted the earlier use of a new analytical approach technique that reads the eigenfunction theory (special trans function theory [STFT]) to determine a solar cell's characteristics. The whole set of I-V characteristics and the output results may be used using the STFT approach without any approximation (Singh et al., 2013). The model for the solar cell was an integrated circuit I-V characteristic model. A linear function that connected the current of a cell to certain unknown characteristic was used to approximate this model (Al-Hamadi, 2014). The relaxation parameters of solar cells may be calculated analytically using a single exponential model; according to research by Q. X. Jia et al. under this method, the output characteristic under illumination circumstances is taken into account as a parameter that determines the optimal coefficient of diode connection (Jia et al., 1995). A quick least-squares approximation approach for figuring out a PV Rp model's electrical properties was suggested by J. Ma and K. L. Ma. The electrical properties of an n-Rp model may be assessed using this approach at operational temperatures and high radiation levels. According to the experimental findings, the proposed solution was the computational cost of a single PV model (Ma et al., 2014). Standard test conditions' (STCs) analytical approaches are effective for some models, but they are probably not adequate for solar models or a wide range of suspended weather situations. Thus, the PV panel mathematical model that has been described is plain to see (Subudhi and Pradhan, 2014).

6.4.2 Iterative methods for parameter extraction

During the recent years, multiple iterative techniques have been accessible. To estimate solar cell parameters, R. Khezzar et al. examined three alternative approaches: an operational approach using a homeometric calculation.

However, due to its high value's minimal contribution to the model, Rsh was disregarded in favour of the other four factors (Khezzar et al., 2009). It was suggested to compare the Vilalo iterative approach with the trust-area-dogleg method. Villalva's iterative technique only extracted the two slower Rs and Rsh, but the trust-area-dogleg method retrieved all five parameters (Mohapatra et al., 2013). Finding approaches is the foundation of the Newton-Roy Raphson method (NRM). As a result, NRM was the foundation for the majority of works on progressive parameter extraction, including Hejri et al. (2014) and Ishaque et al. (2011). But its accuracy and convergence are impacted by the erroneous choice of the beginning circumstances. A technique developed by Ghani and colleagues used the Newton, which is on the W function, to numerically determine the values of sets and resistors. As a result, current may be explicitly represented about voltage in addition to for its rather simple separation (Ghani et al., 2012). The NRM procedure also needs a Jacobian matrix to be calculated. NRM is exceedingly difficult, time-consuming, and error-prone since this Jacobian matrix has 25 singular terms and 2 derivatives (Subudhi and Pradhan, 2014). The divide-by-zero singularity problem, which might get worse if the beginning conditions of a reset are appropriately carried out, is another issue with all NRM techniques. The restrictions of the Rs, Rsh, and n parameters are not taken into account by these NRM techniques. In order to simplify the difficulty of determining the five factors that lead to a value of Rs and a high value of Rsh (Villalva et al., 2009) in accordance with the requirements of the PV module, these approaches make a lot of assumptions. As a result, the parameters that these NRMs produce are incorrect. A parameter extraction technique known as the complete parameter extraction approach has been used to overcome the singularity problem (Phang et al., 1984).

6.4.3 Evolutionary computational extraction methods

6.4.3.1 Particle swarm optimization

Ye et al. (2009) used the particle swarm optimization (PSO) method to estimate PV cell and then compared the results with those from other algorithms such as genetic algorithm (GA) for SDM and DDM. They came to the conclusion that the PSO algorithm offers excellent computational speed along with good accuracy in the estimated parameters. Regarding a silicon sun cell with a 57 mm(D), Hamid et al. (2013) suggested that, using the PSO technique, SDM parameters with five unknown values may be estimated, and they acknowledged the dependability of projected parameter correctness with other methods. Khanna et al. (2014) developed the PSO method.

6.4.3.2 Genetic algorithm

In contrast to other available estimating methods, Bastidas-Rodriguez et al. (2017) found that the estimated parameters are extremely accurate for an SDM with five creators' jogging elements for outside circumstances and the GA technique for various environmental conditions, such as heat and sun irradiation. To be able to improve the existing sun cell abstraction approaches, Jervase et al. (2001) suggested GA. As a consequence, errors at the derived parameters were reduced to 5% at module standards. According to Harrag et al., a GA-based method for the five- and seven-parameter versions should be avoided. Harrag and Messalti (2015) performed an effort to boost the likelihood of quickly and accurately discovering the global minimum. Throughout this technique, the solution area has no restrictions.

6.4.3.3 Algorithm for artificial bee colony

In contrast to BFA, PSO, GA, and HS, which are additional optimisation algorithms, the ABC formula developed by Chen et al. to determine PV parameters (Chen et al., 2018) provides a parameters value with greater accuracy. It has also been found that ABC offers benevolent numerous model goal attributes which are possible. Jamadi et al. (2016) suggested an efficient ABC technique to determine PV cell characteristics for DDM models, compared to Algorithms PS, ABSO, GA, and PS, it was shown that parameter estimates could be done quickly and accurately. Additional works on this method may be found in the literature (Pilakkat and Kanthalakshmi, 2019).

6.4.3.4 Simulated annealing

When compared to some other optimisation methods, the SA approach for parameterizing battery models described by Messaoud (2020) has extended battery life and extremely excellent efficiency. The same approach was proposed by AlRashidi et al. (2013), employing three stages to derive SDM parameters. In the first phase, parameters are determined based on their current values in real time, and compared to current algorithms, the projected method provides the solar panel with active performance. Brondani et al. (2017) proposed the SA method for estimating solar properties and discovered that it produces precise parameter estimates.

6.4.3.5 Differential evaluation

Using the DE method, Abido and Khalid (2018) evaluated seven parameters for the DDM-6 solar panels made using monocrystalline silicon, polycrystalline silicon, and thin film technologies. Seven parameters from the DE method were retrieved, and their values were compared to I-V curves. It was discovered that the projected approach is more accurate in estimating panel parameters.

6.4.3.6 Harmony search based

Askarzadeh and Rezazadeh (2012) proposed silicon solar cells with a diameter of 57 mm for SDM and DDM, and they found a lower root mean square error (RMSE) value compared to PS and SA and other methods, as well as a relatively little discrepancy between experimentally determined values from I-V curves and those obtained from the simulation. When compared to FA, Satapathy et al.'s (2017) discussion of the hybrid firefly method based on HS for use in microgrid applications shows that stability is significantly increased. It also has a rapid convergence time and compact randomisation.

6.4.3.7 Pattern search optimization

Derick et al. (2017) through modelling and testing the PS method presented to create the solar cell parameter of the SDM, also to look at how well the MATLAB PS technique works for parameter approximation of the light of varied ecological circumstances, such as varying levels of solar radiation and temperature; the PS algorithm's findings provide improved accuracy with a quicker convergent moment. AlHajri et al.'s (2012) PS technique for SDM and DDM accurately abstracts the indicated models' parameter standards. Beigi and Maroosi (2018) calculated the DDM parameters using the PS method, compared the approximations to those obtained from other optimisation techniques, and came to the conclusion that the PS approach produced superior results for solar cell modelling.

6.4.3.8 Artificial immune system

For many types of two-module systems, the AIS method performs better than other algorithms like GA or PSO. Jacob et al.'s (2015) evaluation of cell parameters DDM discovered that the AIS algorithm's performance was good.

6.4.3.9 Burning algorithm

Babu et al. (2016) developed an FA faced to calculate the parameters of a solar PV system. The developed algorithm reduces the likelihood of premature meetings and the computational difficulty, and the parameters values obtained from the developed algorithm are much more precisely close to the numbers shown on the solar panel's data sheet.

6.4.3.10 Bacterial foraging algorithm

When compared to other optimisation algorithms like GA and AIS, the BFA method presented information on three different types of solar cell models, see Rajasekar et al. (2013), and delivered improved outcomes in terms of parameters like consistency, accuracy, high precision, minimal merging time, and other metrics.

6.4.3.11 *Flower pollination algorithm*

A fast conclusion about the best course of action to reduce RMSE values, and a faster convergence time in a variety of solar irradiations and temperatures, suggested FPA for PV parameter estimations by Alam et al. (2015). The recommended approach just requires a small number of control parameters.

6.4.3.12 *Hybrid flower pollination algorithm*

Under varied solar irradiance and temperature, Ram et al. (2017) suggested hybrid flower pollination algorithm (HFPA) for solar parameter extractions regarding both SDM and DDM; the proposed method delivers a satisfactory parameter estimation of very low RMSE value at very low solar irradiation levels. In order to efficiently and precisely find for SDM or DDM, Xu and Wang (2017) built a straightforward HFPA and claimed showing the provided method shows better outcomes, such as accuracy, model stability, and convergence rates, across a range of temperatures and sun irradiations.

6.5 OPTIMIZATION CNT SOLAR

6.5.1 Optimisation characteristics of CNT

Numerous issues, including inadequate light absorption caused by high-gap polymers and fullerene materials that restrict the density of short-circuit current, affect the overall performance of organic photovoltaic (OPV). Poor open circuit voltage is caused by the donor and acceptor materials' unoptimized band structures. High carrier improved spectrum invisibility in the visible and infrared areas that are provided by band-gap CNT-based solar cells for cost-effective OPV. The optimisation of PV characteristics for the purpose of enhancing overall performance has not yet been described in CNT OPV research. Therefore, efforts should be made to choose PVs' properties [166].

6.5.1.1 *Structural characterization of CNT*

You may think of carbon nanotubes as rolled-up graphene sheets. The index air p (n, m) controls how the graphene sheet is folded. The number of unit vectors of graphene honeycomb crystal lattice in each direction is represented by the integers n and m (Chowdhury, 2015).

6.5.1.2 *Optimization of Voc of organic solar cell with CNT incorporation*

The market's OPV device efficiency is still rather poor. The open circuit voltage (Voc) of an OPV device is one of the crucial efficiency-controlling parameters; thus, it's critical to comprehend its physical causes and potential enhancement techniques. Voc is linearly dependent on the disparity in work function, according to earlier OPV research. Later research by Scharber et al. (2006) shown that Voc changes linearly when the energy level varies between HOMO and LUMO. The offset (HOMOD-LUMOA) restricts the highest Voc in OPV devices, as recently demonstrated by Rand et al. (2007). There are equivalent lab ports for both the metal/insulator/metal (MIM) model and the offset model (HOMOD-LUMOA). The two models' differences are to be merged via congruent pretence. The experimental Voc values that have been published are a little lower than the offset model (HOMOD-LUMOA) values. This study modifies a formula for maximising the Voc value.

6.5.1.3 Analysis of the organic solar cell's short circuit current density (Isc)

For future development of PV devices, short-circuit current optimisation is crucial. Given that the in-circuit current (Isc) in OPV device plays a significant role in regulating efficiency, it is crucial to comprehend its physical causes and potential enhancement strategies. Knowing the causes, a measure creation is essential for optimising the short circuit current. Photon absorption is closely connected to current in a short circuit. Due of the substantial space between the polymer and the fullerene in OSCs, the highest limit for Isc is constrained by the reemission of a photon with visible wavelengths after absorption. We require low-bandgap materials to enhance the OPV's absorption and range even more. The absorption range may be expanded by CNTs to include the visible and infrared spectrums, increasing the short-circuit current density (Chowdhury, 2015).

6.5.2 Enhancing photovoltaic properties by adding CNTs to different layers

Introduction of SWNTs and MWNTs has improved both the OSCs' active layer and charge transport layer. Changing the chirality of specific SWNTs allows for the optimisation of PV characteristics. Then, to enhance the PV capabilities of the PTB7/PC71BM organic cell, mixed SWNTs and MWNTs were utilised in the transmission layer, active (Chowdhury, 2015).

6.5.2.1 Organic solar panels that use donors like carbon nanotubes

By altering the chirality of the CNTs employed as donors, the PV properties of Voc and Isc have been optimised. The devices described here include a glass substrate covered with ITO, a PEDOT: layer 40 nm, and an electrode to get the best performance (Chowdhury, 2015).

6.5.2.2 Chosen chiral SWNTs in charge transport and photoactive layer

Recently, it has become more difficult to develop the active layer for OPVs using low-bandgap materials. Depending on their characteristics, CNTs can be employed as materials. ITO/PEDOT's photoactive layers and the SWNT metal hole transport layer both include chiral single semiconducting CNTs that have been chemically and specifically functionalized: PSS/PTB7:PC71BM/Acl. PTB7 and PC71BM have similar broad band gaps and constrained visible ranges of absorption. Therefore, the PTB7/PC71BM solar device is limited since the maximum quantity of solar energy is not employed. By incorporating CNTs into various OPV layers, this constraint can be somewhat alleviated. ITO/PEDOT: PSS/PTB7:PC71BM/Al-based devices in three sets with study or optimisation of PV qualities. SWNTs are incorporated into sets 1, 2, and 3's PTB7:PC71BM, PEDOT: PSS, and both layers, respectively. Hyral CNTs organised in PTB7 and subsequently an active layer with PC71BM were produced to improve the absorption in the infrared region (Figure 6.8) (Chowdhury, 2015).

6.5.2.3 Mixed SWCNTs in photoactive layer

In ITO/PEDOT: PSS/PTB7:PC71BM/Al bulk junction solar cell devices, both metal semiconducting SWCNTs have been studied as active layer additions. The active solution processability is shown by the random distribution of nanotubes in the active layer. The insertion of CNTs into the active layer

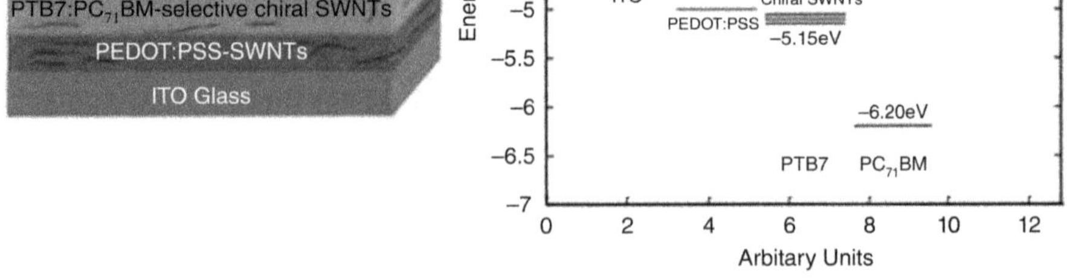

(a) (b)

Figure 6.8 (a) SWNT-equipped device construction with hole transport and photoactive layers. (b) Diagram showing the organic solar cell's energy levels with chiral SWNTs chosen specifically for the photoactive layer. SWNT = Single-walled nanotube.

Source: Chowdhury (2015).

is a straightforward and scalable technique. Performance of three different device types was examined: the regular PTB7/PC71BM device (referred to as the "reference"), the PTB7/PC71BM device with U-single-walled carbon nanotubes (USWNTs) as an additive (referred to as the "U-SWNT"), and the PTB7/PC71BM device with S-SWNTs as an additive (referred to as the "SSWNT") SWNT (semiconductor 95% pure, average D = 1.4 nm). Single-walled carbon nanotubes (SWCNTs) are often generated as mixed bundles of metal and semiconductor tubes depending on various chiralities. The negative impact of metal OSCs' active layer contains SWCNTs which cannot be overlooked despite the fact that only one-third of them are made of metal. The optical current (Jph) and incident light intensity (I) have a connection that can be expressed as Jph I, and Jph's dependency on I and power is connected to the device's biomolecular recombination. Low levels of biomolecular recombination are indicated by a number that is almost unity (Liu et al., 2009). Because there are metal SWCNTs, the value drops as the SWCNT fraction rises, rather than being entirely derived from the outside. For devices with unordered SWCNTs, the short-circuit current is calculated using the equation r Jph I for I = 100 mw/cm^2. Using Eq, Jsc for ordered devices is calculated. The construction of the PTB7/PC71BM device with mixed (ordered and unordered) SWNTs in the photoactive layer is shown in Figure 6.9a. The predicted orbital parameters for S-SWNT are n = 2.23, Rsh = 302 cm^2, and Rs = 2.35 cm^3. The environmental parameters for U-SWNTs were calculated to be Rs = 3.31 cm^2, Rsh = 489 cm^2, and n = 1.99 (Abeygunasekara et al., 2015) (Figure 6.9b). The series resistance is undoubtedly increased by the presence of m-SWNTs in disordered devices.

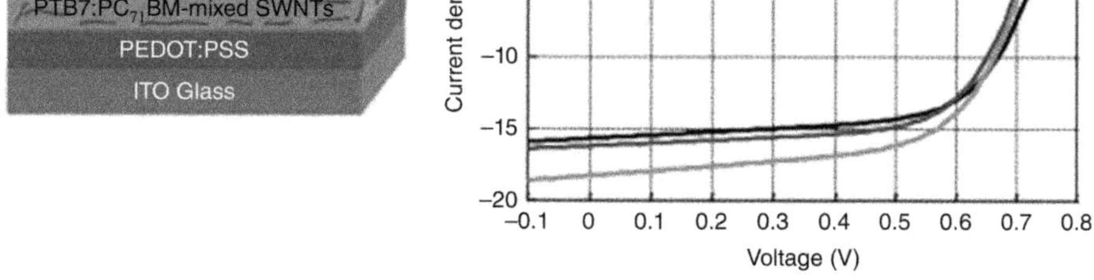

Figure 6.9 (a) Device structure schematic. (b) S-SWNT and U-SWNT light J-V parameters for SWNTs included into PTB7: PC71BM units. SWNT = Single-walled nanotube.

Source: Scharber et al. (2006).

6.5.2.4 Doped multi-walled CNTs in photoactive layer

A minor part of the metal CNTs in the photoactive area may result in undesirable electron-hole coupling in the absence of charge selectivity. Additionally, an orbital is formed by CNT aggregates with uneven dispersion. Multi-walled CNTs are often made of metal. As a result, charge selection is crucial to enhancing the functionality of multi-walled CNTs included in the photoactive layer. In BHJ solar cells, highly selective hole or electron transport enhancers such as Ne or boron multi-walled carbon nanotubes (N-CNT or B-CNT) can boost device performance. The structure of the PTB7/PC71BM device with Ne or boron in the photoactive layer is shown in Figure 6.10a, which was employed in this study to enhance the PV characteristics. The functionality of CNTs must be near to the HOMO or the LUMO of the neighbouring semiconductor materials for effective electron. In contrast, the extra-electrons from the doped nitrogen atoms caused the work function of N-CNT to be lowered to 4.4 eV (Lee et al., 2011). Figure 6.10b demonstrates how well the work functions of N-CNT and B-CNT fit the LUMO of PC71BM and the HOMO of PTB7, respectively. The selective transport of light-excited electrons and holes is easily explained by this functional variety. While the intermolecular or interchain hopping process naturally delays the carrier transport capabilities of organic semiconductors, the addition of selectively charged and doped CNTs is projected to improve carrier transport. For instance, an N-CNT might act as a b-bump to promote the transmission of b-electron between PC71BM molecules. Similar to this, adding B-CNT to PTB7 can improve the conveyance of holes in PTB7 films. Eq was used to calculate the EQE of solar cells (Lu et al., 2013).

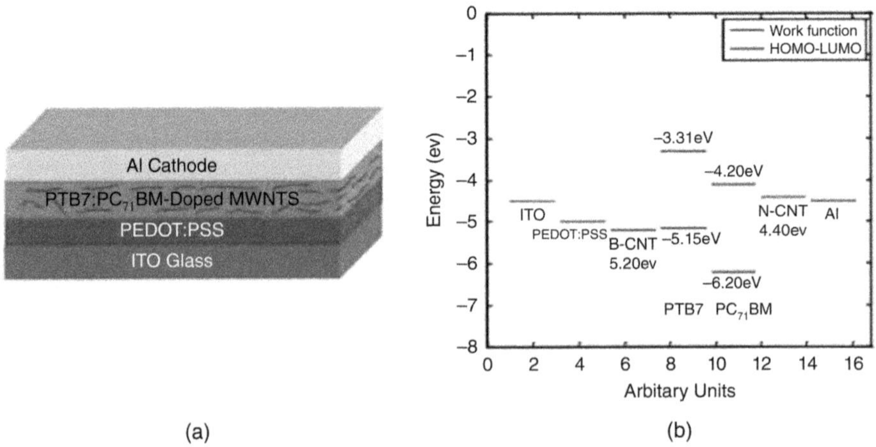

Figure 6.10 (a) Device structure schematic. (b) Diagrams illustrating the OSC's energy levels with B-CNT and N-CNT in the photoactive layer. MWNT = Multi-walled nanotube.

Source: Lu et al. (2013).

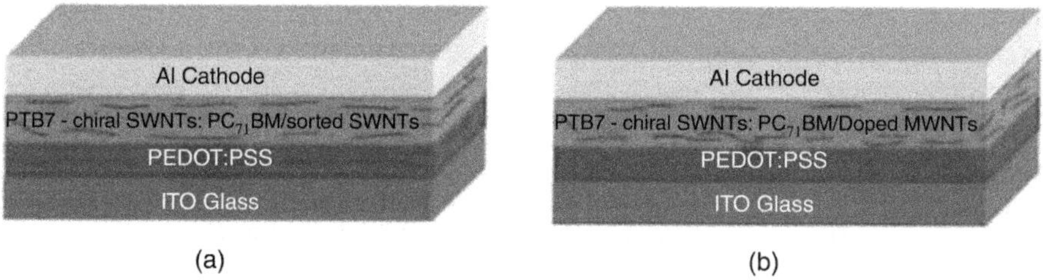

Figure 6.11 Device structure schematic. (a) Semiconducting nanotubes that are sorted and chiral in a photoactive layer. (b) A layer of photoactive materials with doped MWNTs and chiral SWNTs. SWNT = Single-walled nanotube, MWNT = multi-walled nanotube.

Source: Chowdhury (2015).

6.5.2.5 Using selected and mixed chiral CNTs to maximize PCE in the photoactive layer

The majority of the time, CNTs are found in mixed or bundled forms that include both metallic and semiconducting CNTs. Utilising metallic or semiconducting CNTs in OSCs is exceedingly challenging. Other unique chiral semiconductor SWNTs are employed in PTB7:PC71BM with mixed SWNTs and MWNTs to improve the effectiveness of power conversion. In the photoactive layer of solar cells, CNTs are employed as effective selective charge transport materials. The absorption rate is enhanced by employing particular chirality s-SWNTs (7,3), (6,5), (8,1), (7,5), and (10,2) in donor PTB7. Range of wavelengths is 960–1130 nm. Because these particular SWNTs' HOMO level is near to PTB7 and as a result exhibit gamma band absorption, p-Hatton absorption, and infrared enhancement, their bandgap ranges of 1.19–1.27 eV can be advantageous. The PTB7:PC71BM structure employs two sets of tools by evaluating or enhancing the PV properties. Set 1 placed chiral-selective s-SWNTs mixed with PTB7 donor in the photoactive layer. Figure 6.11a depicts the device structure's schematic. Set chiral-selective s-SWNTs and PTB7 donor with two doped MWNTs in the photoactive layer. Figure 6.11b depicts a schematic of the device construction.

6.6 CONCLUSIONS AND FUTURE SCOPE

The incorporation of carbon nanotubes in OSCs shows great potential for improving the efficiency and performance of solar energy systems. The advantageous properties of carbon nanotubes, such as their electronic, mechanical, thermal, and chemical properties, make them ideal candidates for enhancing the functionality of OSCs. Various parameter estimation algorithms, including analytical methods and evolutionary computational methods like

PSO, GA, and artificial bee colony algorithm, can be used to optimize the parameters of solar cells and improve their efficiency. Optimizing the PV characteristics of carbon nanotube-incorporated OSCs at different layers can further enhance their performance. By utilizing carbon nanotubes as donors in OSCs or incorporating selective chiral SWNTs in photoactive and charge transport layers, it is possible to maximize PCE. Future scope for research in this area includes exploring new functionalization techniques for carbon nanotubes, investigating novel ways to optimize the parameters of solar cells using advanced algorithms, and further studying the impact of incorporating carbon nanotubes at different layers on the overall performance of OSCs. Additionally, there is potential for developing hybrid systems that combine different types of carbon nanotubes to achieve even higher efficiencies in solar energy conversion.

REFERENCES

Abdulrazzaq, O. A., Saini, V., Bourdo, S., Dervishi, E., & Biris, A. S. (2013). Organic solar cells: A review of materials, limitations, and possibilities for improvement. Particulate Science and Technology, 31(5), 427–442.

Abeygunasekara, W. L., Hiralal, P., Samaranayake, L., Chien, C. T., Kumar, A., Flewitt, A. J., ...Amaratunga, G. A. (2015). Incorporating semiconducting single-walled carbon nanotubes as efficient charge extractors in organic solar cells. Applied Physics Letters, 106(12), 123305.

Abido, M. A., & Khalid, M. S. (2018). Seven-parameter PV model estimation using differential evolution. Electrical Engineering, 100, 971–981.

Aitola, K., Halme, J., Feldt, S., Lohse, P., Borghei, M., Kaskela, A., ...Hagfeldt, A. (2013). Highly catalytic carbon nanotube counter electrode on plastic for dye solar cells utilizing cobalt-based redox mediator. Electrochimica Acta, 111, 206–209.

Alam, D. F., Yousri, D. A., & Eteiba, M. B. (2015). Flower pollination algorithm based solar PV parameter estimation. Energy Conversion and Management, 101, 410–422.

AlHajri, M. F., El-Naggar, K. M., AlRashidi, M. R., & Al-Othman, A. K. (2012). Optimal extraction of solar cell parameters using pattern search. Renewable Energy, 44, 238–245.

Al-Hamadi, H. A. (2014). Estimation of photovoltaic cell parameters using piecewise linear approximation. International Journal of Computer Science and Electronics Engineering, 2(1), 2320–4028.

AlRashidi, M. R., El-Naggar, K. M., & AlHajri, M. F. (2012). Heuristic approach for estimating the solar cell parameters. Recent Researches in Applied Information Science, 2, 80–83.

AlRashidi, M. R., El-Naggar, K. M., & AlHajri, M. F. (2013). Solar cell parameters estimation using simulated annealing algorithm. World Academy of Science Engineering and Technology, 7, 149–152.

Andrews, R., Jacques, D., Qian, D., & Dickey, E. C. (2001). Purification and structural annealing of multiwalled carbon nanotubes at graphitization temperatures. Carbon, 39(11), 1681–1687.

Aqel, A., Abou El-Nour, K. M., Ammar, R. A., & Al-Warthan, A. (2012). Carbon nanotubes, science and technology part (I) structure, synthesis and characterisation. Arabian Journal of Chemistry, 5(1), 1–23.

Araújo, K., Boucher, J. L., & Aphale, O. (2019). A clean energy assessment of early adopters in electric vehicle and solar photovoltaic technology: Geospatial, political and socio-demographic trends in New York. Journal of Cleaner Production, 216, 99–116.
Arena, A., Donato, N., Saitta, G., Galvagno, S., Milone, C., & Pistone, A. (2008). Photovoltaic properties of multi-walled carbon nanotubes deposited on n-doped silicon. Microelectronics Journal, 39(12), 1659–1662.
Askarzadeh, A., & Rezazadeh, A. (2012). Parameter identification for solar cell models using harmony search-based algorithms. Solar Energy, 86(11), 3241–3249.
Babu, T. S., Ram, J. P., Sangeetha, K., Laudani, A., & Rajasekar, N. (2016). Parameter extraction of two diode solar PV model using fireworks algorithm. Solar Energy, 140, 265–276.
Bastidas-Rodriguez, J. D., Petrone, G., Ramos-Paja, C. A., & Spagnuolo, G. (2017). A genetic algorithm for identifying the single diode model parameters of a photovoltaic panel. Mathematics and Computers in Simulation, 131, 38–54.
Bauhofer, W., & Kovacs, J. Z. (2009). A review and analysis of electrical percolation in carbon nanotube polymer composites. Composites Science and Technology, 69(10), 1486–1498.
Beigi, A. M., & Maroosi, A. (2018). Parameter identification for solar cells and module using a hybrid firefly and pattern search algorithms. Solar Energy, 171, 435–446.
Benzekkour, N., Ghomrani, F. Z., Gabouze, N., Kermadi, S., Boumaour, M., & Ferdjani, K. (2009, May). Formation of rough TiO2 thin films on glass and porous silicon by sol-gel method. In Materials Science Forum (Vol. 609, pp. 139–143). Trans Tech Publications Ltd.
Bhattacharyya, S., Kymakis, E., & Amaratunga, G. A. J. (2004). Photovoltaic properties of dye functionalized single-wall carbon nanotube/conjugated polymer devices. Chemistry of Materials, 16(23), 4819–4823.
Braga, A. F. B., Moreira, S. P., Zampieri, P. R., Bacchin, J. M. G., & Mei, P. R. (2008). New processes for the production of solar-grade polycrystalline silicon: A review. Solar Energy Materials and Solar Cells, 92(4), 418–424.
Brondani, M. F., Sausen, A. T. Z. R., Sausen, P. S., & Binelo, M. O. (2017). Battery model parameters estimation using simulated annealing. TEMA (São Carlos), 18, 127–137.
Cataldo, S., Salice, P., Menna, E., & Pignataro, B. (2012). Carbon nanotubes and organic solar cells. Energy and Environmental Science, 5(3), 5919–5940.
Chen, X., Xu, B., Mei, C., Ding, Y., & Li, K. (2018). Teaching–learning–based artificial bee colony for solar photovoltaic parameter estimation. Applied Energy, 212, 1578–1588.
Choudhary, P., & Srivastava, R. K. (2019). Sustainability perspectives-a review for solar photovoltaic trends and growth opportunities. Journal of Cleaner Production, 227, 589–612.
Chowdhury, A. H. (2015). Optimization of photovoltaic characteristics of organic solar cell by incorporating carbon nanotube. http://lib.buet.ac.bd:8080/xmlui/handle/123456789/3522
Cox, M., Gorodetsky, A., Kim, B., Kim, K. S., Jia, Z., Kim, P., ...Kymissis, I. (2011). Single-layer graphene cathodes for organic photovoltaics. Applied Physics Letters, 98(12), 123303.
Cubas, J., Pindado, S., & De Manuel, C. (2014). Explicit expressions for solar panel equivalent circuit parameters based on analytical formulation and the Lambert W-function. Energies, 7(7), 4098–4115.
Danafar, F., Fakhru'l-Razi, A., Salleh, M. A. M., & Biak, D. R. A. (2011). Influence of catalytic particle size on the performance of fluidized-bed chemical vapor deposition synthesis of carbon nanotubes. Chemical Engineering Research and Design, 89(2), 214–223.

De Blas, M. A., Torres, J. L., Prieto, E., & Garcıa, A. (2002). Selecting a suitable model for characterizing photovoltaic devices. Renewable Energy, 25(3), 371–380.
De Groote, O., & Verboven, F. (2019). Subsidies and time discounting in new technology adoption: Evidence from solar photovoltaic systems. American Economic Review, 109(6), 2137–2172.
Deibel, C., Strobel, T., & Dyakonov, V. (2010). Role of the charge transfer state in organic donor–acceptor solar cells. Advanced Materials, 22(37), 4097–4111.
Derick, M., Rani, C., Rajesh, M., Busawon, K., & Binns, R. (2017). Estimation of solar photovoltaic parameters using pattern search algorithm. In Emerging Trends in Electrical, Electronic and Communications Engineering: Proceedings of the First International Conference on Electrical, Electronic and Communications Engineering (ELECOM 2016), Bagatelle, Mauritius, November 25–27, 2016 (pp. 184–191). Springer International Publishing.
Dincer, I. (2000). Renewable energy and sustainable development: A crucial review. Renewable and Sustainable Energy Reviews, 4(2), 157–175.
Edwards, E. R., Antunes, E. F., Botelho, E. C., Baldan, M. R., & Corat, E. J. (2011). Evaluation of residual iron in carbon nanotubes purified by acid treatments. Applied Surface Science, 258(2), 641–648.
Faustini, M., Nicole, L., Boissiere, C., Innocenzi, P., Sanchez, C., & Grosso, D. (2010). Hydrophobic, antireflective, self-cleaning, and antifogging sol–gel coatings: An example of multifunctional nanostructured materials for photovoltaic cells. Chemistry of Materials, 22(15), 4406–4413.
Ferguson, A. J., Blackburn, J. L., & Kopidakis, N. (2013). Fullerenes and carbon nanotubes as acceptor materials in organic photovoltaics. Materials Letters, 90, 115–125.
Garnett, E., & Yang, P. (2010). Light trapping in silicon nanowire solar cells. Nano Letters, 10(3), 1082–1087.
Ghani, F., Duke, M., & Carson, J. (2012, December). Extraction of solar cell modeling parameters using the Lambert W-function. In Proceedings of the 50th Annual Conference, Australian Solar Energy Society (Australian Solar Council) Melbourne.
Guldi, D. M., & Costa, R. D. (2013). Nanocarbon hybrids: The paradigm of nanoscale self-ordering/self-assembling by means of charge transfer/doping interactions. The Journal of Physical Chemistry Letters, 4(9), 1489–1501.
Günes, S., Neugebauer, H., & Sariciftci, N. S. (2007). Conjugated polymer-based organic solar cells. Chemical Reviews, 107(4), 1324–1338.
Gwon, H. J., Park, Y., Moon, C. W., Nahm, S., Yoon, S. J., Kim, S. Y., & Jang, H. W. (2014). Superhydrophobic and antireflective nanograss-coated glass for high performance solar cells. Nano Research, 7, 670–678.
Hamid, N. F. A., Rahim, N. A., & Selvaraj, J. (2013, November). Solar cell parameters extraction using particle swarm optimization algorithm. In 2013 IEEE conference on clean energy and technology (CEAT) (pp. 461–465). IEEE.
Han, Z., & Fina, A. (2011). Thermal conductivity of carbon nanotubes and their polymer nanocomposites: A review. Progress in Polymer Science, 36(7), 914–944.
Harrag, A., & Messalti, S. (2015, December). Extraction of solar cell parameters using genetic algorithm. In 2015 4th International Conference on Electrical Engineering (ICEE) (pp. 1–5). IEEE.
Harris, P. J. (2004). Carbon nanotube composites. International Materials Reviews, 49(1), 31–43.
Hejri, M., Mokhtari, H., Azizian, M. R., Ghandhari, M., & Söder, L. (2014). On the parameter extraction of a five-parameter double-diode model of photovoltaic cells and modules. IEEE Journal of Photovoltaics, 4(3), 915–923.
Hou, P. X., Liu, C., & Cheng, H. M. (2008). Purification of carbon nanotubes. Carbon, 46(15), 2003–2025.

Huang, X., Han, S., Huang, W., & Liu, X. (2013). Enhancing solar cell efficiency: The search for luminescent materials as spectral converters. Chemical Society Reviews, 42(1), 173–201.
Hwang, S. K., Lee, J. M., Kim, S., Park, J. S., Park, H. I., Ahn, C. W., ... Kim, S. O. (2012). Flexible multilevel resistive memory with controlled charge trap B-and N-doped carbon nanotubes. Nano Letters, 12(5), 2217–2221.
Iijima, S., Brabec, C., Maiti, A., & Bernholc, J. (1996). Structural flexibility of carbon nanotubes. The Journal of Chemical Physics, 104(5), 2089–2092.
Iijima, S., & Ichihashi, T. (1993). Single-shell carbon nanotubes of 1-nm diameter. Nature, 363(6430), 603–605.
Ishaque, K., Salam, Z., & Taheri, H. (2011). Accurate MATLAB simulink PV system simulator based on a two-diode model. Journal of Power Electronics, 11(2), 179–187.
Jacob, B., Balasubramanian, K., Babu, T. S., & Rajasekar, N. (2015, February). Parameter extraction of solar PV double diode model using artificial immune system. In 2015 IEEE international conference on signal processing, informatics, communication and energy systems (SPICES) (pp. 1–5). IEEE.
Jain, A., & Kapoor, A. (2004). Exact analytical solutions of the parameters of real solar cells using Lambert W-function. Solar Energy Materials and Solar Cells, 81(2), 269–277.
Jamadi, M., Merrikh-Bayat, F., & Bigdeli, M. (2016). Very accurate parameter estimation of single-and double-diode solar cell models using a modified artificial bee colony algorithm. International Journal of Energy and Environmental Engineering, 7, 13–25.
Jana, D., Sun, C. L., Chen, L. C., & Chen, K. H. (2013). Effect of chemical doping of boron and nitrogen on the electronic, optical, and electrochemical properties of carbon nanotubes. Progress in Materials Science, 58(5), 565–635.
Janssen, R. A., & Nelson, J. (2013). Factors limiting device efficiency in organic photovoltaics. Advanced Materials, 25(13), 1847–1858.
Jeon, S. O., & Lee, J. Y. (2012). Improved lifetime in organic solar cells using a bilayer cathode of organic interlayer/Al. Solar Energy Materials and Solar Cells, 101, 160–165.
Jervase, J. A., Bourdoucen, H., & Al-Lawati, A. (2001). Solar cell parameter extraction using genetic algorithms. Measurement Science and Technology, 12(11), 1922.
Jia, Q. X., Ebihara, K., & Ikegami, T. (1995). Analytical solution for solar cell model parameters from illuminated current-voltage characteristics. Philosophical Magazine B, 72(4), 375–382.
Jones, W. E. Jr, Chiguma, J., Johnson, E., Pachamuthu, A., & Santos, D. (2010). Electrically and thermally conducting nanocomposites for electronic applications. Materials, 3(2), 1478–1496.
Jørgensen, M., Norrman, K., & Krebs, F. C. (2008). Stability/degradation of polymer solar cells. Solar Energy Materials and Solar Cells, 92(7), 686–714.
Jorio, A., & Dresselhaus, G. (2008). Carbon Nanotubes: Advanced Topics in the Synthesis, Structure, Properties and Applications (Vol. 111). M. S. Dresselhaus (Ed.). Berlin: springer.
Jun, G. H., Jin, S. H., Park, S. H., Jeon, S., & Hong, S. H. (2012). Highly dispersed carbon nanotubes in organic media for polymer: Fullerene photovoltaic devices. Carbon, 50(1), 40–46.
Keru, G., Ndungu, P. G., & Nyamori, V. O. (2013). Nitrogen-doped carbon nanotubes synthesised by pyrolysis of (4-{[(pyridine-4-yl) methylidene] amino} phenyl) ferrocene. Journal of Nanomaterials, 2013, 2–2.
Khanna, V., Das, B. K., Bisht, D., Vandana, & Singh, P. K. (2014). Estimation of photovoltaic cells model parameters using particle swarm optimization. In Physics of Semiconductor Devices: 17th International Workshop on the Physics of Semiconductor Devices 2013 (pp. 391–394). Springer International Publishing.

Khezzar, R., Zereg, M., & Khezzar, A. (2009, November). Comparative study of mathematical methods for parameters calculation of current-voltage characteristic of photovoltaic module. In 2009 International Conference on Electrical and Electronics Engineering-ELECO 2009 (pp. I–24). IEEE.

Kim, D. Y., Yun, Y. S., Bak, H., Cho, S. Y., & Jin, H. J. (2010). Aspect ratio control of acid modified multiwalled carbon nanotubes. Current Applied Physics, 10(4), 1046–1052.

Kumar, M., & Ando, Y. (2011). Carbon nanotube synthesis and growth mechanism. In Carbon nanotubes-synthesis, characterization, applications, Intech Open, 147–170.

Laird, E. D., & Li, C. Y. (2013). Structure and morphology control in crystalline polymer–carbon nanotube nanocomposites. Macromolecules, 46(8), 2877–2891.

Laudani, A., Fulginei, F. R., & Salvini, A. (2014). High performing extraction procedure for the one-diode model of a photovoltaic panel from experimental I–V curves by using reduced forms. Solar Energy, 103, 316–326.

Lee, J. M., Park, J. S., Lee, S. H., Kim, H., Yoo, S., & Kim, S. O. (2011). Selective electron-or hole-transport enhancement in bulk-heterojunction organic solar cells with N-or B-doped carbon nanotubes. Advanced Materials, 23(5), 629–633.

Lewis, N. S., & Nocera, D. G. (2006). Powering the planet: Chemical challenges in solar energy utilization. Proceedings of the National Academy of Sciences, 103(43), 15729–15735.

Li, C., Chen, Y., Wang, Y., Iqbal, Z., Chhowalla, M., & Mitra, S. (2007). A fullerene–single wall carbon nanotube complex for polymer bulk heterojunction photovoltaic cells. Journal of Materials Chemistry, 17(23), 2406–2411.

Lin, T., Bajpai, V., Ji, T., & Dai, L. (2003). Chemistry of carbon nanotubes. Australian Journal of Chemistry, 56(7), 635–651.

Liu, L., Stanchina, W. E., & Li, G. (2009). Effects of semiconducting and metallic single-walled carbon nanotubes on performance of bulk heterojunction organic solar cells. Applied Physics Letters, 94(23), 233309.

Liu, P. (2005). Modifications of carbon nanotubes with polymers. European Polymer Journal, 41(11), 2693–2703.

Lu, L., Xu, T., Chen, W., Lee, J. M., Luo, Z., Jung, I. H., ... Yu, L. (2013). The role of N-doped multiwall carbon nanotubes in achieving highly efficient polymer bulk heterojunction solar cells. Nano Letters, 13(6), 2365–2369.

Ma, J., Man, K. L., Ting, T. O., Zhang, N., Lim, E. G., Guan, S. U., ... Lei, C. U. (2014). Simple computational method of predicting electrical characteristics in solar cells. Elektronika ir Elektrotechnika, 20(1), 41–44.

Ma, P. C., Siddiqui, N. A., Marom, G., & Kim, J. K. (2010). Dispersion and functionalization of carbon nanotubes for polymer-based nanocomposites: A review. Composites Part A: Applied Science and Manufacturing, 41(10), 1345–1367.

Mathew, S., Yella, A., Gao, P., Humphry-Baker, R., Curchod, B. F., Ashari-Astani, N., ... Grätzel, M. (2014). Dye-sensitized solar cells with 13% efficiency achieved through the molecular engineering of porphyrin sensitizers. Nature Chemistry, 6(3), 242–247.

Messaoud, R. B. (2020). Extraction of uncertain parameters of single-diode model of a photovoltaic panel using simulated annealing optimization. Energy Reports, 6, 350–357.

Mohan, V., & Senthilkumar, S. (2022). IoT based fault identification in solar photovoltaic systems using an extreme learning machine technique. Journal of Intelligent & Fuzzy Systems, 43(3), 3087–3100.

Mohapatra, A., Nayak, B. K., & Mohanty, K. B. (2013, February). Comparative study on single diode photovoltaic module parameter extraction methods. In 2013 International Conference on Power, Energy and Control (ICPEC) (pp. 30–34). IEEE.

Mohlala, M. S., Liu, X. Y., & Coville, N. J. (2006). Synthesis of multi-walled carbon nanotubes catalyzed by substituted ferrocenes. Journal of Organometallic Chemistry, 691(22), 4768–4772.
Oosthuizen, R. S., & Nyamori, V. O. (2011). Carbon nanotubes as supports for palladium and bimetallic catalysts for use in hydrogenation reactions. Platinum Metals Review, 55(3), 154–169.
Ortiz-Conde, A., Sánchez, F. J. G., & Muci, J. (2006). New method to extract the model parameters of solar cells from the explicit analytic solutions of their illuminated I–V characteristics. Solar Energy Materials and Solar Cells, 90(3), 352–361.
Ozin, G. A., Arsenault, A. C., & Cademartiti, L. (2005). Nanochemistry: A Chemical Approach to Nanomaterials. RSC, London.
Padgett, C. W., & Brenner, D. W. (2004). Influence of chemisorption on the thermal conductivity of single-wall carbon nanotubes. Nano Letters, 4(6), 1051–1053.
Park, S. H., Roy, A., Beaupré, S., Cho, S., Coates, N., Moon, J. S., ... Heeger, A. J. (2009). Bulk heterojunction solar cells with internal quantum efficiency approaching 100%. Nature Photonics, 3(5), 297–302.
Peters, C. H., Sachs-Quintana, I. T., Mateker, W. R., Heumueller, T., Rivnay, J., Noriega, R., ... McGehee, M. D. (2012). The mechanism of burn-in loss in a high efficiency polymer solar cell. Advanced Materials, 24(5), 663–668.
Phang, J. C. H., Chan, D. S. H., & Phillips, J. R. (1984). Accurate analytical method for the extraction of solar cell model parameters. Electronics Letters, 20(10), 406–408.
Pilakkat, D., & Kanthalakshmi, S. (2019). An improved P&O algorithm integrated with artificial bee colony for photovoltaic systems under partial shading conditions. Solar Energy, 178, 37–47.
Pizzini, S. (2010). Towards solar grade silicon: Challenges and benefits for low cost photovoltaics. Solar Energy Materials and Solar Cells, 94(9), 1528–1533.
Price, S. C., Stuart, A. C., Yang, L., Zhou, H., & You, W. (2011). Fluorine substituted conjugated polymer of medium band gap yields 7% efficiency in polymer–fullerene solar cells. Journal of the American Chemical Society, 133(12), 4625–4631.
Rajasekar, N., Kumar, N. K., & Venugopalan, R. (2013). Bacterial foraging algorithm based solar PV parameter estimation. Solar Energy, 97, 255–265.
Ram, J. P., Babu, T. S., Dragicevic, T., & Rajasekar, N. (2017). A new hybrid bee pollinator flower pollination algorithm for solar PV parameter estimation. Energy Conversion and Management, 135, 463–476.
Rand, B. P., Burk, D. P., & Forrest, S. R. (2007). Offset energies at organic semiconductor heterojunctions and their influence on the open-circuit voltage of thin-film solar cells. Physical Review B, 75(11), 115327.
Raut, H. K., Ganesh, V. A., Nair, A. S., & Ramakrishna, S. (2011). Anti-reflective coatings: A critical, in-depth review. Energy & Environmental Science, 4(10), 3779–3804.
Rong, H., Liu, Z., Wu, Q., & Lee, Y. H. (2010). A facile and efficient gas phase process for purifying single-walled carbon nanotubes. Current Applied Physics, 10(4), 1231–1235.
Ruoff, R. S., Qian, D., & Liu, W. K. (2003). Mechanical properties of carbon nanotubes: Theoretical predictions and experimental measurements. Comptes Rendus Physique, 4(9), 993–1008.
Sakellariou, G., Priftis, D., & Baskaran, D. (2013). Surface-initiated polymerization from carbon nanotubes: Strategies and perspectives. Chemical Society Reviews, 42(2), 677–704.
Satapathy, P., Dhar, S., & Dash, P. K. (2017). Stability improvement of PV-BESS diesel generator-based microgrid with a new modified harmony search-based hybrid firefly algorithm. IET Renewable Power Generation, 11(5), 566–577.

Scharber, M. C., Mühlbacher, D., Koppe, M., Denk, P., Waldauf, C., Heeger, A. J., & Brabec, C. J. (2006). Design rules for donors in bulk-heterojunction solar cells—Towards 10% energy-conversion efficiency. Advanced Materials, 18(6), 789–794.

Senthilkumar, S., Haidari, M., Devi, G., Britto, A. S. F., Gorthi, R., & Sivaramkrishnan, M. (2022a, October). Wireless bidirectional power transfer for E-vehicle charging system. In 2022 International Conference on Edge Computing and Applications (ICECAA) (pp. 705–710). IEEE.

Senthilkumar, S., Kumar, T. S., Ramachandran, L., Devarajan, D., Mohan, V., & Chitrakala, G. (2022b). Solar powered pesticide sprayer with Mobile charger and LED light. Fuel, 26(2), 205–210.

Senthilkumar, S., Mohan, D. V., & Chitrakala, G. (2020). Evolutionary algorithms for solar photovoltaic parameters estimation–A review. International Journal of Future Generation Communication and Networking, 13(2), 348–360.

Senthilkumar, S., Mohan, V., & Krithiga, G. (2023). Brief review on solar photovoltaic parameter estimation of single and double diode model using evolutionary algorithms. International Journal of Engineering Technologies and Management Research, 10(1), 64–78.

Serp, P., Corrias, M., & Kalck, P. (2003). Carbon nanotubes and nanofibers in catalysis. Applied Catalysis A: General, 253(2), 337–358.

Shi, E., Zhang, L., Li, Z., Li, P., Shang, Y., Jia, Y., ... Cao, A. (2012). TiO2-coated carbon nanotube-silicon solar cells with efficiency of 15%. Scientific Reports, 2(1), 884.

Shokrieh, M. M., & Rafiee, R. (2010). A review of the mechanical properties of isolated carbon nanotubes and carbon nanotube composites. Mechanics of Composite Materials, 46, 155–172.

Singh, N. S., Jain, A., & Kapoor, A. (2013). An exact analytical method for calculating the parameters of a real solar cell using special trans function theory (STFT). International Journal of Renewable Energy Research, 3(1), 201–206.

Somani, S. P., Somani, P. R., & Umeno, M. (2008). Carbon nanotube incorporation: A new route to improve the performance of organic–inorganic heterojunction solar cells. Diamond and Related Materials, 17(4-5), 585–588.

Somani, S. P., Somani, P. R., Umeno, M., & Flahaut, E. (2006). Improving photovoltaic response of poly (3-hexylthiophene)/n-Si heterojunction by incorporating double walled carbon nanotubes. Applied Physics Letters, 89(22).

Spitalsky, Z., Tasis, D., Papagelis, K., & Galiotis, C. (2010). Carbon nanotube–polymer composites: Chemistry, processing, mechanical and electrical properties. Progress in Polymer Science, 35(3), 357–401.

Subudhi, B., & Pradhan, R. (2011, December). Characteristics evaluation and parameter extraction of a solar array based on experimental analysis. In 2011 IEEE Ninth International Conference on Power Electronics and Drive Systems (pp. 340–344). IEEE.

Subudhi, B., & Pradhan, R. (2014). Development of new parameter extraction scheme and maximum power point tracking system. PHD thesis, National Institute of Technology, Rourkela.

Villalva, M. G., Gazoli, J. R., & Ruppert Filho, E. (2009). Comprehensive approach to modeling and simulation of photovoltaic arrays. IEEE Transactions on Power Electronics, 24(5), 1198–1208.

Wang, C., Abdul-Rahman, H., & Rao, S. P. (2010). A new design of luminescent solar concentrator and its trial run. International Journal of Energy Research, 34(15), 1372–1385.

Wang, D. H., Moon, J. S., Seifter, J., Jo, J., Park, J. H., Park, O. O., & Heeger, A. J. (2011). Sequential processing: Control of nanomorphology in bulk heterojunction solar cells. Nano Letters, 11(8), 3163–3168.

Weickert, J., Dunbar, R. B., Hesse, H. C., Wiedemann, W., & Schmidt-Mende, L. (2011). Nanostructured organic and hybrid solar cells. Advanced Materials, 23(16), 1810–1828.
Williams, G., Wang, Q., & Aziz, H. (2013). The photo-stability of polymer solar cells: Contact photo-degradation and the benefits of interfacial layers. Advanced Functional Materials, 23(18), 2239–2247.
Xu, F., Zhu, W. Q., Yan, L., Xu, H., Xiong, L. H., & Li, J. H. (2012). Single walled carbon nanotube anodes based high performance organic light-emitting diodes with enhanced contrast ratio. Organic Electronics, 13(2), 302–308.
Xu, S., & Wang, Y. (2017). Parameter estimation of photovoltaic modules using a hybrid flower pollination algorithm. Energy Conversion and Management, 144, 53–68.
Yang, M., Koutsos, V., & Zaiser, M. (2005). Interactions between polymers and carbon nanotubes: A molecular dynamics study. The Journal of Physical Chemistry B, 109(20), 10009–10014.
Ye, L., Zhang, Y., Zhang, X., Hu, T., Ji, R., Ding, B., & Jiang, B. (2013). Sol–gel preparation of SiO2/TiO2/SiO2–TiO2 broadband antireflective coating for solar cell cover glass. Solar Energy Materials and Solar Cells, 111, 160–164.
Ye, M., Wang, X., & Xu, Y. (2009). Parameter extraction of solar cells using particle swarm optimization. Journal of Applied Physics, 105(9), 094502.
Ye, Q., & Chi, C. (2011). Conjugated polymers for organic solar cells. In Solar Cells–New Aspects and Solutions (pp.453–474). InTech.
Zhang, C., Zhang, J., Hao, Y., Lin, Z., & Zhu, C. (2011). A simple and efficient solar cell parameter extraction method from a single current-voltage curve. Journal of Applied Physics, 110(6).
Zhang, L., & Resasco, D. E. (2009). Single-walled carbon nanotube pillars: A superhydrophobic surface. Langmuir, 25(8), 4792–4798.
Zhang, Y., Basel, T. P., Gautam, B. R., Yang, X., Mascaro, D. J., Liu, F., & Vardeny, Z. V. (2012). Spin-enhanced organic bulk heterojunction photovoltaic solar cells. Nature Communications, 3(1), 1043.
Zhu, H., Wei, J., Wang, K., & Wu, D. (2009). Applications of carbon materials in photovoltaic solar cells. Solar Energy Materials and Solar Cells, 93(9), 1461–1470.

Chapter 7

Enhancement of Casson SWCNT-MWCNT-graphene/ethanol tri-hybrid nanofluid's heat transfer over a spinning disk

Bhagyashri Patgiri, Neelav Sarma, and Ashish Paul

7.1 INTRODUCTION

Ternary hybrid nanofluids have marvelous applications such as solar water heating, space vehicles, nuclear reactors, and chillers, owing to their superior thermal conductivity. Elnaqeeb *et al.* (2021) described water-based THNFs' flow with altering shapes and densities of the considered nanoparticles. Mahmood *et al.* (2022) studied the flow of Cu+ Fe_3O_4 + SiO_2-polymer THNF across one heated cylinder with shape influences. Elsebaee *et al.* (2023) reported MgO + TiO_2 + Fe_3O_4/H_2O THNF flow over one slender stretching sheet. Afterward, many researchers investigated THNF flows across various geometries; one may see the allied studies in References (Ahmed *et al.*, 2021; Animasaun *et al.*, 2022; Cao *et al.*, (2022), Sarada *et al.*, 2022; Zahan *et al.*, 2022; Patil *et al.*, 2023; Paul *et al.*, 2023a, b; Paul *et al.*, 2024a, b, c, d; Patgiri and Paul, 2024; Paul and Patgiri, 2024).

Magnetohydrodynamics (MHD) has recently grabbed noticeable recognition in several industries because of its usefulness in medical industries (identifying diseases), nuclear reactors, cooling industries, petroleum industries, etc. Some inspections on MHD fluid flow are inserted References (Raptis *et al.*, 2004; Balsara, 2010; Chen, 2010; Grigoriadis *et al.*, 2010; Ibáñez and Cuevas, 2010; Lok *et al.*, 2011; Subhas Abel *et al.*, (2012); Zheng *et al.*, 2012; Mahian *et al.*, 2013; Mishra *et al.*, 2013; Shateyi, 2013; Abolbashari *et al.*, 2014; Pekmen and Tezer-Sezgin, 2014; Daniel and Daniel, 2015; Smolentsev *et al.*, 2015; Malik *et al.*, 2016; Naramgari and Sulochana, 2016; Daniel *et al.*, 2017; Arifuzzaman *et al.*, 2018; Soid *et al.*, 2018; Aly and Pop, 2019; Lund *et al.*, 2019; Anuar *et al.*, 2020; Manjunatha *et al.*, 2020; Asjad *et al.*, 2021; Krishna *et al.*, 2021; Mogharrebi *et al.*, 2021; Zainal *et al.*, 2021a, b; Adnan and Ashraf, 2022; Ali *et al.*, 2022; Chabani *et al.*, 2022; Ghorbani *et al.*, 2022; Guedri *et al.*, 2023; Kodi *et al.*, 2022; Prameela *et al.*, 2022; Majeed *et al.*, 2023; Patel *et al.*, 2023; Umavathi *et al.*, 2023)).

The flow caused due to Casson fluid is of worthy industrial applications because of superb implications in drilling processes, biological treatments, lubrication, etc. Some investigations on Casson fluid flow are included in

 DOI: 10.1201/9781003527251-7

References (Sandeep *et al.*, 2016; Babu *et al.*, 2017; Tamoor *et al.*, 2017; Hamid *et al.*, 2019; Sivaraj *et al.*, 2019; Abdelmalek *et al.*, 2020; Oyekunle *et al.*, 2021; Alzahrani *et al.*, 2022; Bejawada *et al.*, 2022; Khan *et al.*, 2022; Sahoo and Nandkeolyar, 2023).

Flow prompted by spinning disks is of noteworthy importance because of their many industrial uses such as spinning machinery, computer storage devices, air cleaning machinery, gas turbine rotors, and medical equipment. One may perceive the linked studies in References (Thomas and Davies, 2010; Griffiths, 2015; Appelquist *et al.*, 2016; Imtiaz *et al.*, 2016; Yin *et al.*, 2017; Joshi *et al.*, 2018; Doh *et al.*, 2019; Ram *et al.*, 2019; Zangooee *et al.*, 2019; Nazia *et al.*, 2022; Islam *et al.*, 2023; Kumar and Sharma, 2023; Park *et al.*, 2023; Sarangi *et al.*, 2023; Zhao *et al.*, 2023; Samantaray *et al.*, 2024).

The main motive of this chapter is to analyze heat and mass dissipation in the 3D flow of Casson SWCNT-MWCNT-graphene/ethanol THNF past a spinning disk under slip impact by employing bvp4c procedure. As far as the authors' consciousness, no such analysis has been presented yet. Graphical and tabular demonstrations for important flow variables are analyzed in detail.

7.2 MATHEMATICAL FORMULATION

In this analysis, authors consider the following assumptions:

- An incompressible 3D flow.
- MWCNT, graphene, and SWCNT are the opted nanoparticles.
- The opted carrier fluid is ethanol.
- A spinning disk is chosen as our flow geometry.
- $z = 0$ is assumed as the axis of rotation.
- The spinning disk revolves with a fixed angular speed Ω.
- Ambient temperature is presumed to be T.
- Surface temperature is presumed to be T_w.
- Ambient concentration is presumed to be C.
- Surface concentration is presumed to be C_w.
- The velocity constituent in r direction is speculated as u.
- The velocity constituent in ϕ direction is speculated as v.
- The velocity constituent in z-direction is speculated as w.
- B_1 represents magnetic strength.
- The depiction of our flow problem is evinced in Figure 7.1.

Now, the momentum, energy, and mass equations of the flow system can be typified as follows (Tassaddiq *et al.*, 2020; Jawad *et al.*, 2021):

$$\frac{\partial u}{\partial r}+\frac{u}{r}+\frac{\partial w}{\partial z}=0, \tag{7.1}$$

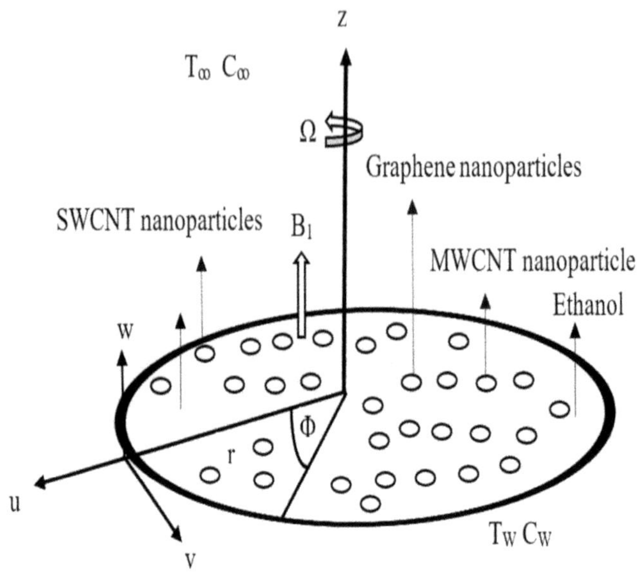

Figure 7.1 Flow geometry. *Abbreviations*: SWCNT, single-walled carbon nanotube; MWCNT, multi-walled carbon nanotube.

$$u\frac{\partial u}{\partial r}-\frac{v^2}{r}+w\frac{\partial u}{\partial z}=\frac{\mu_{thnf}}{\rho_{thnf}}\left(1+\frac{1}{\lambda}\right)\left(-\frac{u}{r^2}+\frac{\partial^2 u}{\partial z^2}+\frac{\partial^2 u}{\partial r^2}+\frac{1}{r}\frac{\partial u}{\partial r}\right)-\frac{\sigma_{thnf}}{\rho_{thnf}}\cdot u\cdot B_1^2. \quad (7.2)$$

$$u\frac{\partial v}{\partial r}-\frac{uv}{r}+w\frac{\partial v}{\partial z}=\frac{\mu_{thnf}}{\rho_{thnf}}\left(1+\frac{1}{\lambda}\right)\left(-\frac{v}{r^2}+\frac{\partial^2 v}{\partial z^2}+\frac{\partial^2 v}{\partial r^2}+\frac{1}{r}\frac{\partial v}{\partial r}\right)-\frac{\sigma_{thnf}}{\rho_{thnf}}\cdot v\cdot B_1^2, \quad (7.3)$$

$$u\frac{\partial w}{\partial r}+w\frac{\partial w}{\partial z}=\frac{\mu_{thnf}}{\rho_{thnf}}\left(1+\frac{1}{\lambda}\right)\left(\frac{\partial^2 w}{\partial r^2}+\frac{1}{r}\frac{\partial w}{\partial r}+\frac{\partial^2 w}{\partial z^2}\right), \quad (7.4)$$

$$u\frac{\partial T}{\partial r}+w\frac{\partial T}{\partial z}=\frac{k_{thnf}}{(\rho c_p)_{thnf}}\left(\frac{\partial^2 T}{\partial r^2}+\frac{1}{r}\frac{\partial T}{\partial r}+\frac{\partial^2 T}{\partial z^2}\right), \quad (7.5)$$

$$u\frac{\partial C}{\partial r}+w\frac{\partial C}{\partial z}=D_{thnf}\left(\frac{\partial^2 C}{\partial z^2}+\frac{1}{r}\frac{\partial C}{\partial r}+\frac{\partial^2 C}{\partial r^2}\right). \quad (7.6)$$

Stipulations of the boundary are as follows:

$$\left.\begin{array}{l} u=a\left(\frac{\partial u}{\partial z}\right),\ v=r\Omega+a\left(\frac{\partial v}{\partial z}\right),\ w=0,\ T=T_W \text{ and } C=C_w \text{ at } z=0 \\ u\to 0,\ v\to 0,\ w\to 0,\ T\to T_\infty \text{ and } C\to C_\infty \text{ as } z\to\infty \text{ at } z\to\infty \end{array}\right\}. \quad (7.7)$$

The symbols f, nf, hnf, and thnf represent the fluid in operation, an amalgamation of nanoparticles and fluid, a blend of nanoparticles with different

Table 7.1 Physical properties of graphene, MWCNT, SWCNT and ethanol

Name	*ρ (Kg/m³)*	*k (W/mK)*	*c_p (J/KgK)*
SWCNT	2600	6600	425
MWCNT	1600	3000	796
Graphene	2250	2500	2100
Ethanol	789	0.17	2460

Source: Haider et al. (2022), Rashid et al. (2020), and Sajid et al. (2023).

Abbreviations: SWCNT, single-walled carbon nanotube; MWCNT, multi-walled carbon nanotube.

compositions, and a combination of three different nanoparticle-fluid mixtures. Also, the symbols λ, B, k, μ, σ, D, a, r, Ω, and ρ are described in the nomenclature portion. Also, the physical attributes of graphene, MWCNT, SWCNT, and ethanol are expressed in Table 7.1.

Also, the mathematical bond between the water, graphene/ethanol, MWCNT-graphene/ethanol, and SWCNT-MWCNT-graphene/ethanol is as follows:

$$\mu_{nf} = \mu_f(1-\phi_1)^{-2.5},\ \mu_{hnf} = \mu_f(1-\phi_1)^{-2.5}(1-\phi_2)^{-2.5},$$
$$\mu_{thnf} = \mu_f(1-\phi_1)^{-2.5}(1-\phi_2)^{-2.5}(1-\phi_3)^{-2.5},$$

$$\rho_{nf} = (1-\phi_1)\rho_f + \phi_1\rho_{s1},\ \rho_{hnf} = (1-\phi_2)\{(1-\phi_1)\rho_f + \phi_1\rho_{s1}\} + \phi_2\rho_{s2},$$

$$\rho_{thnf} = (1-\phi_3)\left[(1-\phi_2)\{(1-\phi_1)\rho_f + \phi_1\rho_{s1}\} + \phi_2\rho_{s2}\right] + \phi_3\rho_{s3},$$

$$(\rho_{c_p})_{nf} = \phi_1(\rho_{c_p})_{s1} + (1-\phi_1)(\rho_{c_p})_f,$$
$$(\rho_{c_p})_{hnf} = \left\{(1-\phi_1)(\rho_{c_p})_f + \phi_1(\rho_{c_p})_{s1}\right\}(1-\phi_2) + \phi_2(\rho_{c_p})_{s2},$$

$$(\rho_{c_p})_{thnf} = \left[(1-\phi_2)\left\{(1-\phi_1)(\rho_{c_p})_f + \phi_1(\rho_{c_p})_{s1}\right\} + \phi_2(\rho_{c_p})_{s2}\right](1-\phi_3) + \phi_3(\rho_{c_p})_{s3},$$

$$k_{nf} = k_f\left\{\frac{(1-\phi_1) - 2\cdot\phi_1\cdot\left(\frac{k_{s1}}{k_{s1}-k_f}\right)\ln\cdot\left(\frac{k_{s1}+k_f}{2k_f}\right)}{(1-\phi_1) - 2\cdot\phi_1\cdot\left(\frac{k_f}{k_{s1}-k_f}\right)\ln\cdot\left(\frac{k_{s1}+k_f}{2k_f}\right)}\right\}, k_{hnf}$$

$$= k_f\left\{\frac{(1-\phi_2) - 2\cdot\phi_2\cdot\left(\frac{k_{s2}}{k_{s2}-k_f}\right)\ln\cdot\left(\frac{k_{s2}+k_f}{2k_f}\right)}{(1-\phi_2) - 2\cdot\phi_2\cdot\left(\frac{k_f}{k_{s2}-k_f}\right)\ln\cdot\left(\frac{k_{s2}+k_f}{2k_f}\right)}\right\},$$

$$k_{thnf} = k_f \left\{ \frac{(1-\phi_3) - 2\cdot\phi_3\cdot\left(\frac{k_{s3}}{k_{s3}-k_f}\right)\ln\cdot\left(\frac{k_{s3}+k_f}{2k_f}\right)}{(1-\phi_3) - 2\cdot\phi_3\cdot\left(\frac{k_f}{k_{s3}-k_f}\right)\ln\cdot\left(\frac{k_{s3}+k_f}{2k_f}\right)} \right\}.$$

Here Φ_1, Φ_2, and Φ_3 are mean volume fractions of graphene, MWCNT, and SWCNT nanoparticles and suffixes s1, s2, and s3 are used for graphene, MWCNT, and SWCNT nanoparticles, respectively.

To alter dimensional Eqs. (7.1)–(7.7) into a dimensionless form, we choose the following similarity reformation (Rehman et al., 2018):

$$\left.\begin{array}{l} u = r\Omega A'(\eta),\ v = r\Omega B(\eta),\ w = -\sqrt{2\Omega\vartheta}\ A(\eta), \\ \eta = \left(\sqrt{\frac{2\Omega}{\vartheta}}\right)z,\ \theta(\eta) = \frac{(T-T_\infty)}{(T_w-T_\infty)},\ \phi(\eta) = \frac{(C-C_\infty)}{(C_w-C_\infty)} \end{array}\right\}. \tag{7.8}$$

By using Eq. (7.8) in Eqs. (7.1)–(7.7), we get

$$\frac{2\left(1+\frac{1}{\lambda}\right)}{D_1D_2}A''' - \left[(A')^2 - 2AA'' - B^2\right] - \frac{2D_5}{D_2}\cdot M\cdot A' = 0, \tag{7.9}$$

$$\frac{2\left(1+\frac{1}{\lambda}\right)}{D_1D_2}B'' - [\ 2A'B - 2AB'] - \frac{2D_5}{D_2}\cdot M\cdot B = 0, \tag{7.10}$$

$$\theta''\cdot D_4 + D_3\cdot Pr\cdot A\cdot\theta' = 0, \tag{7.11}$$

$$\phi'' + D_6\cdot Sc\cdot A\cdot\phi' = 0, \tag{7.12}$$

and

$$\left.\begin{array}{r} A(0)=0,\ A'(0)=s\cdot A''(0),\ B(0)=1+s\cdot B'(0),\ \theta(0)=1,\ \phi(0)=1\ \text{at}\ \eta=0 \\ A(\infty)\to 0,\ A'(\infty)\to 0,\ B(\infty)\to 0,\ \theta(\infty)\to 0,\ \phi(\infty)\to 0\ \text{as}\ \eta\to\infty \end{array}\right\}. \tag{7.13}$$

Here,

$$D_1 = \frac{\mu_{thnf}}{\mu_f},\ D_2 = \frac{\rho_{thnf}}{\rho_f},\ D_3 = \frac{(\rho_{c_p})_{thnf}}{(\rho_{c_p})_f},\ D_4 = \frac{K_{thnf}}{K_f},\ D_5 = \frac{\sigma_{thnf}}{\sigma_f},\ D_6 = \frac{D_{thnf}}{D_f}.$$

Also,

$$\vartheta_f = \frac{\mu_f}{\rho_f}(\text{Kinematic Viscosity}),\ M = \frac{\sigma_f B_1^2}{2\Omega\rho_f}(\text{Magnetic Parameter}),$$

$$\mathrm{Re}=\frac{\Omega r^2}{\vartheta_f}(\text{Reynold's Number}),\ \mathrm{Pr}=\frac{\vartheta_f\left(\rho c_p\right)_f}{k_f}(\text{Prandtl number}),$$

$$s=a\left(\sqrt{\frac{2\Omega}{\vartheta_f}}\right)(\text{Velocity slip parameter}),\ \mathrm{Sc}=\frac{\vartheta_f}{D_B}(\text{Schimdt number}).$$

The dimensionless engineering quantities are as follows:

Skin friction coefficient:

$$C_1\mathrm{Re}^{\frac{1}{2}}=\left(1+\frac{1}{\lambda}\right)\cdot\frac{\sqrt{\left(A''(0)\right)^2+\left(B(0)\right)^2}}{\left(1-\phi_1\right)^{2.5}\left(1-\phi_2\right)^{2.5}\left(1-\phi_3\right)^{2.5}}.$$

Local Nusselt number:

$$N_1\mathrm{Re}^{-\frac{1}{2}}=-\left(\frac{k_{thnf}}{k_f}\right)\theta'(0)=-D_4\cdot\theta'(0).$$

Local Sherwood number:

$$S_1\mathrm{Re}^{-\frac{1}{2}}=-\phi'(0).$$

7.3 METHODOLOGY AND VALIDATION

We use the famous shooting-RK-4 method to solve the acquired equations (7.9)– (7.13). The thickness of boundary layer and mesh size are respectively assumed to be 20 and 0.01 and convergence criteria are achieved up to 10^{-10}. Also, we validate our outcomes with Mustafa (2017) for limiting conditions as displayed in Table 7.2.

7.4 RESULTS AND DISCUSSIONS

Here we focus on the implications of several elegant parameters on radial velocity $A'(\eta)$, tangential velocity $B(\eta)$, temperature $\theta(\eta)$, concentration $\phi(\eta)$ dispersions and skin friction $C_1\sqrt{Re}$, local Nusselt $N_1\sqrt{Re}$, and Sherwood number $S_1\sqrt{Re}$.

Figures 7.2 and 7.3 elaborate that the SWCNT nanoparticle's volume fraction ϕ_3 aggrandizes the radial velocity $A'(\eta)$ but curtails the temperature $\theta(\eta)$.

Table 7.2 Comparison of $A''(0)$ for different M's values with Mustafa (2017)

M	*S*	*A''(0) (Mustafa, 2017)*	*A''(0) (our output)*
0	0.25	0.259534	0.064833
0.8	0.25	0.096762	0.096762

Source: Mustafa (2017).

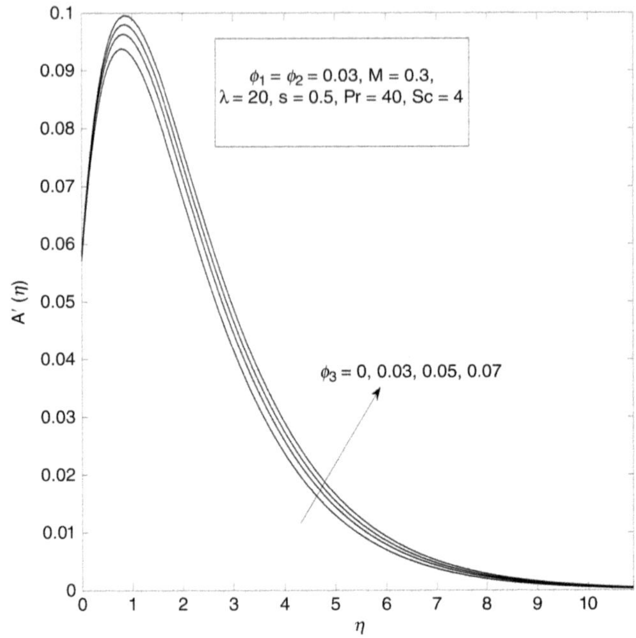

Figure 7.2 $A'(\eta)$ via ϕ_3.

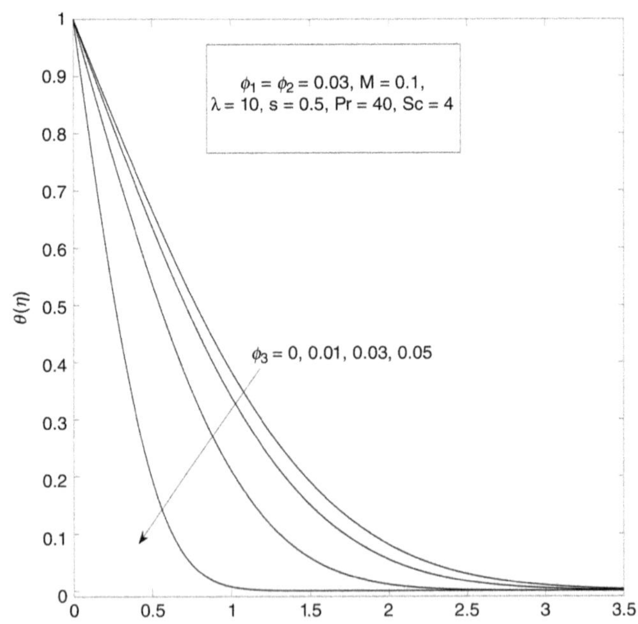

Figure 7.3 $\theta(\eta)$ via ϕ_3.

When the volume percent of graphene nanoparticles is enhanced, the fluid displays decreased viscosity, improving the fluid's radial velocity $A'(\eta)$. Also, SWCNT nanoparticles have superb thermal conductance, thereby mixing these nanoparticles, and the heat conductance of the THNF elevates, which quickens the heat transport process and, consequently, decreases temperature. Therefore, SWCNT nanoparticles may be used as a legitimate coolant in many medical and technological fields.

Figures 7.4, 7.5, and 7.6 depict that with escalating the magnetic parameter, both the velocities $A'(\eta)$ and $B(\eta)$ decelerate, but the concentration $\phi(\eta)$ accelerates. When M augments, the prompted Lorentz force augments which repels the fluid motion, and, subsequently, velocities $A'(\eta)$ and $B(\eta)$ decrease. However, an augmentation in M causes an amplification in the solutal boundary layer, and eventually the concentration $\phi(\eta)$ upsurges too.

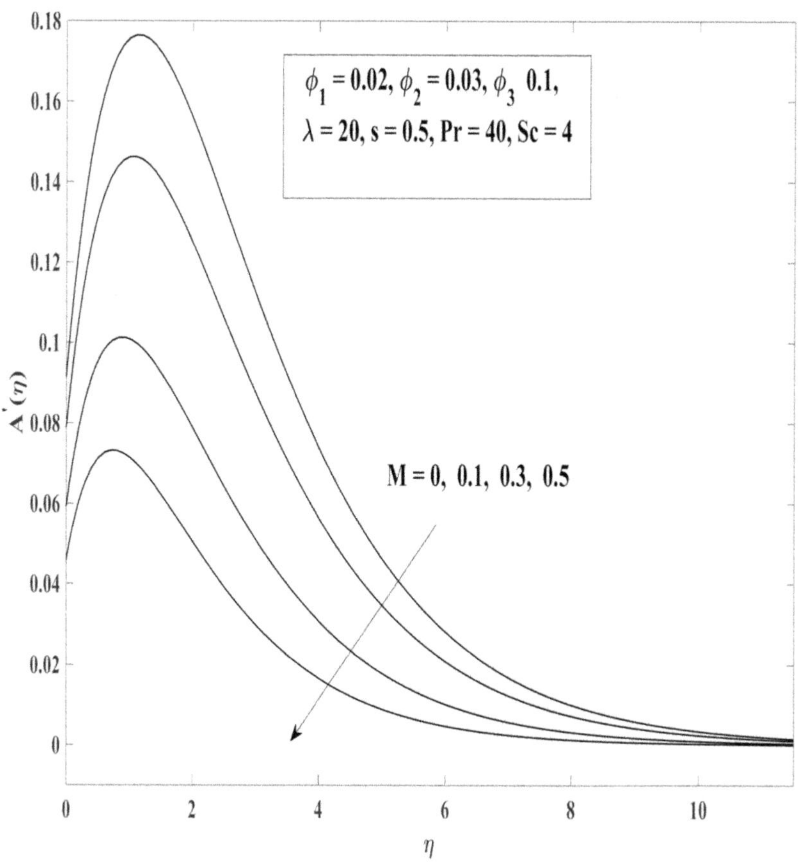

Figure 7.4 $A'(\eta)$ via M.

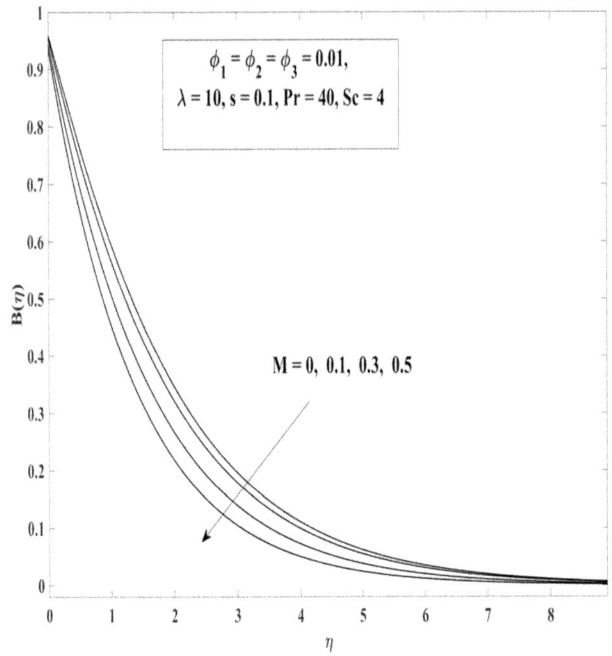

Figure 7.5 $B(\eta)$ via M.

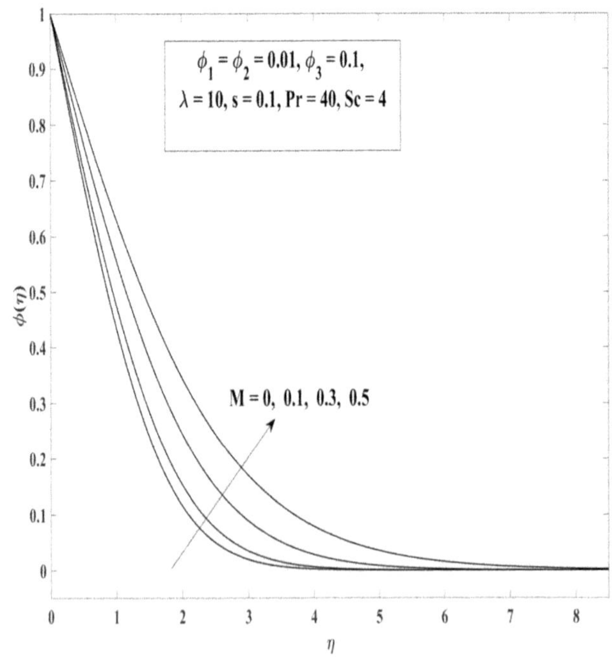

Figure 7.6 $\phi(\eta)$ via M.

Figure 7.7 shows that tangential velocity $B(\eta)$ faces retardation for elevating values of the slip parameter (s). As the s's values amplify, the flow resistance also uplifts, so $B(\eta)$ decelerates.

Figure 7.8 describes that bigger values of the λ scale down the tangential velocity $B(\eta)$. Casson parameter λ opposes the fluid flow by reducing the yield stress and therefore $B(\eta)$ decreases as λ increases.

Figure 7.9 demonstrates that the temperature $\theta(\eta)$ declines for the uplifting amplitudes of the Prandtl number (Pr). Fluid's thermal diffusive attribute depletes for mounting values of Pr which further shrinks the thickness of the linked boundary layer and correspondingly temperature scales down.

Figure 7.10 explains that the concentration $\phi(\eta)$ diminishes for the rising values of the Schmidt number (Sc). As the Sc's values uplift, the molecular diffusivity subsides, so $\phi(\eta)$ decelerates.

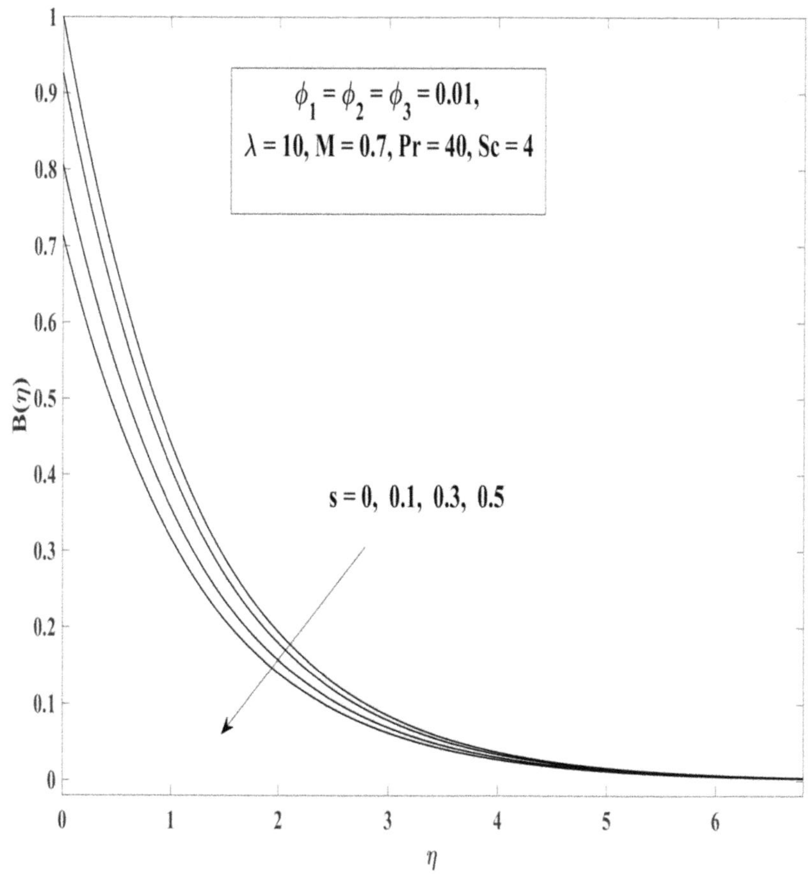

Figure 7.7 $B(\eta)$ via s.

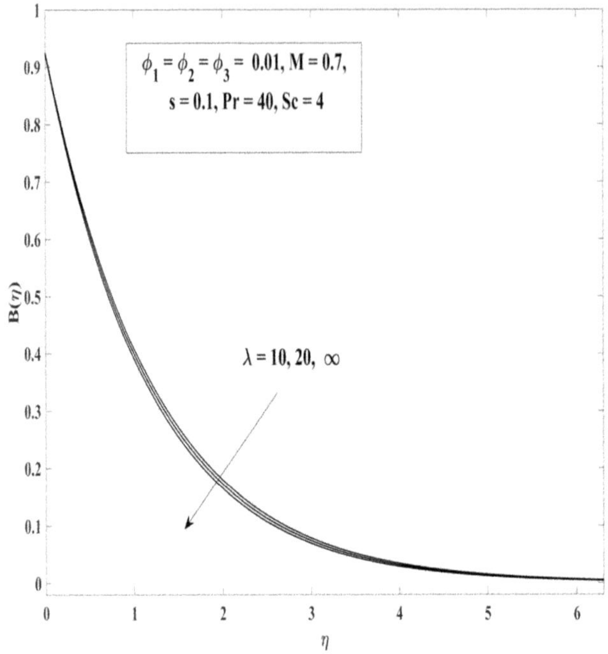

Figure 7.8 $\mathrm{B}(\eta)$ via λ.

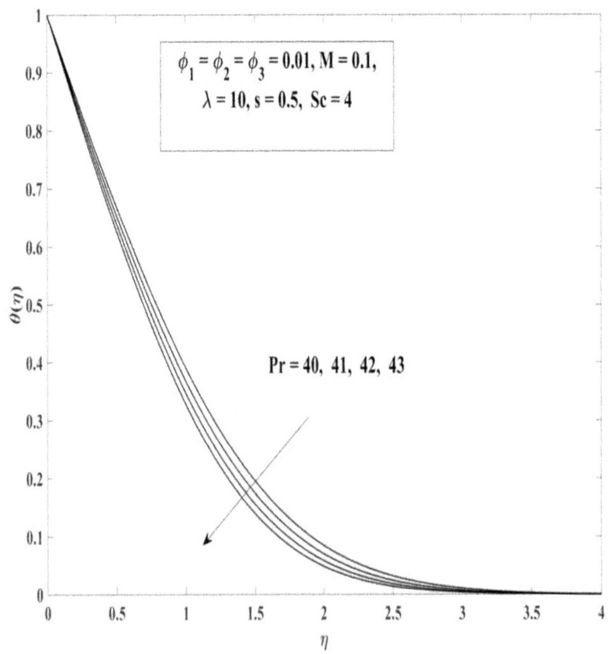

Figure 7.9 $\theta(\eta)$ via Pr.

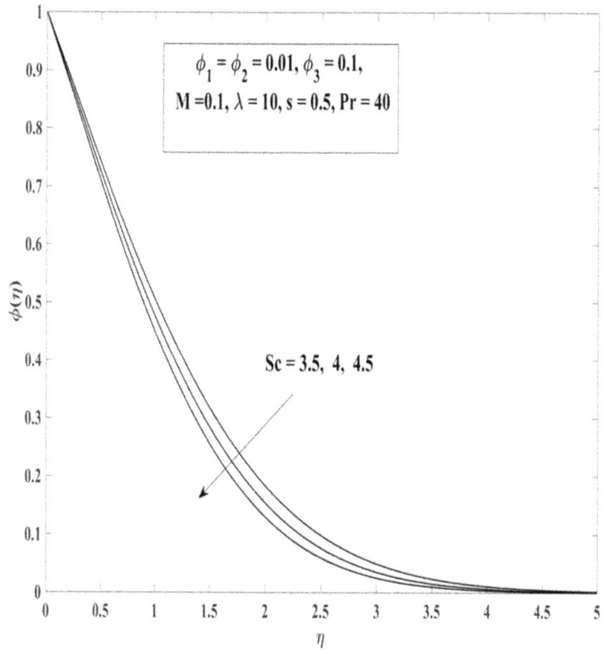

Figure 7.10 $\phi(\eta)$ via Sc.

Table 7.3 describes the effect of the Casson parameter (λ) and volume fractions ϕ_1, ϕ_2, ϕ_3 on $C_1\sqrt{Re}$ (skin friction), $N_1\sqrt{Re}$ (heat transferral rate) and $S_1\sqrt{Re}$ (mass transferral rate). This table explains that ϕ_1, ϕ_2, ϕ_3 increases $C_1Re^{1/2}$, $N_1Re^{-1/2}$, and $S_1Re^{-1/2}$. Furthermore, λ reduces $C_1Re^{1/2}$ but increases both $N_1Re^{-1/2}$ and $S_1Re^{-1/2}$. Also, for Casson SWCNT-MWCNT-graphene/ethanol ternary hybrid nanofluid, heat transmission

Table 7.3 Computation of $C_lRe^{1/2}$, $N_lRe^{-1/2}$, and $S_lRe^{-1/2}$ by taking Pr = 40, Sc = 4, s = 0.1

λ	ϕ_1	ϕ_2	ϕ_3	$C_lRe^{1/2}$	$N_lRe^{-1/2}$	$S_lRe^{-1/2}$
10	0	0	0	0.405761	0.501639	0.452362
	0.1			0.433215	0.529678	0.472154
	0.075	0.075		0.450021	0.548167	0.533618
	0.05	0.05	0.05	0.478924	0.585639	0.523416
15				0.503991	0.618658	0.553578
20				0.494423	0.619314	0.553645
∞				0.489589	0.619638	0.553661

is enhanced by 16.77%, 10.57%, and 6.83%, respectively, when compared to base fluid ethanol, nanofluid graphene/ethanol, hybrid nanofluid MWCNT-graphene/ethanol.

7.5 CONCLUSION

Here 3D flow of Casson SWCNT-MWCNT-graphene/ethanol THNFs past a spinning disk under slip condition is scrutinized. The key outcomes of this surveillance are as follows:

- Radial velocity faces retardation for growing values of M but a reverse trend is observed for ϕ_3.
- Tangential velocity deteriorated for improving values of M, s, and λ.
- Temperature deteriorated for rising values of Pr and ϕ_3.
- Concentration accelerates for higher values of M, whereas a contrary trend is noticed for Sc.
- Mass and heat relocation rates are increased for bigger values of λ, ϕ_1, ϕ_2, and ϕ_3.
- Skin friction scales down for bigger values of λ.

NOMENCLATURE

σ	electrical conductivity
ρ	density
ϑ	kinematic viscosity
μ	dynamic viscosity
λ	Casson parameter
a	slip coefficient
B_1	magnetic coefficient
C	concentration
C_∞	ambient concentration
c_p	heat capacity
C_w	surface concentration
k	thermal conductivity
M	magnetic parameter
Pr	Prandtl number
Re	Reynold number
Sc	Schmidt number
T	temperature
T_∞	ambient temperature
THNF	tri-hybrid nanofluid
T_w	surface temperature

REFERENCES

Abdelmalek, Z., Mahanthesh, B., Basir, M. F. M., Imtiaz, M., Mackolil, J., Khan, N. S., ... Tlili, I. (2020). Mixed radiated magneto Casson fluid flow with Arrhenius activation energy and Newtonian heating effects: Flow and sensitivity analysis. Alexandria Engineering Journal, 59(5), 3991–4011.

Abolbashari, M. H., Freidoonimehr, N., Nazari, F., & Rashidi, M. M. (2014). Entropy analysis for an unsteady MHD flow past a stretching permeable surface in nano-fluid. Powder Technology, 267, 256–267.

Adnan, A., & Ashraf, W. (2022). Thermal efficiency in hybrid (Al2O3-CuO/H2O) and ternary hybrid nanofluids (Al2O3-CuO-Cu/H2O) by considering the novel effects of imposed magnetic field and convective heat condition. Waves in Random and Complex Media, 1–16.

Ahmed, W., Kazi, S. N., Chowdhury, Z. Z., Johan, M. R. B., Mehmood, S., Soudagar, M. E. M., ... Ahmad, M. S. (2021). Heat transfer growth of sonochemically synthesized novel mixed metal oxide ZnO+ Al2O3+ TiO2/ DW based ternary hybrid nanofluids in a square flow conduit. Renewable and Sustainable Energy Reviews, 145, 111025. https://doi.org/10.1016/j.rser.2021.111025

Ali, L., Ali, B., & Ghori, M. B. (2022). Melting effect on Cattaneo–Christov and thermal radiation features for aligned MHD nanofluid flow comprising microorganisms to leading edge: FEM approach. Computers & Mathematics with Applications, 109, 260–269.

Aly, E. H., & Pop, I. (2019). MHD flow and heat transfer over a permeable stretching/shrinking sheet in a hybrid nanofluid with a convective boundary condition. International Journal of Numerical Methods for Heat & Fluid Flow, 29(9), 3012–3038.

Alzahrani, H. A., Alsaiari, A., Madhukesh, J. K., Naveen Kumar, R., & Prasanna, B. M. (2022). Effect of thermal radiation on heat transfer in plane wall jet flow of Casson nanofluid with suction subject to a slip boundary condition. Waves in Random and Complex Media, 1–18. https://doi.org/10.1080/17455030.2022.2030502

Animasaun, I. L., Yook, S. J., Muhammad, T., & Mathew, A. (2022). Dynamics of ternary-hybrid nanofluid subject to magnetic flux density and heat source or sink on a convectively heated surface. Surfaces and Interfaces, 28, 101654. https://doi.org/10.1016/j.surfin.2021.101654

Anuar, N. S., Bachok, N., Turkyilmazoglu, M., Arifin, N. M., & Rosali, H. (2020). Analytical and stability analysis of MHD flow past a nonlinearly deforming vertical surface in carbon nanotubes. Alexandria Engineering Journal, 59(1), 497–507.

Appelquist, E., Schlatter, P., Alfredsson, P. H., & Lingwood, R. J. (2016). On the global nonlinear instability of the rotating-disk flow over a finite domain. Journal of Fluid Mechanics, 803, 332–355.

Arifuzzaman, S. M., Khan, M. S., Mehedi, M. F. U., Rana, B. M. J., & Ahmmed, S. F. (2018). Chemically reactive and naturally convective high speed MHD fluid flow through an oscillatory vertical porous plate with heat and radiation absorption effect. Engineering Science and Technology, an International Journal, 21(2), 215–228.

Asjad, M. I., Sarwar, N., Ali, B., Hussain, S., Sitthiwirattham, T., & Reunsumrit, J. (2021). Impact of bioconvection and chemical reaction on MHD nanofluid flow due to exponential stretching sheet. Symmetry, 13(12), 2334.

Babu, M. J., Sandeep, N., Ali, M. E., & Nuhait, A. O. (2017). Magnetohydrodynamic dissipative flow across the slendering stretching sheet with temperature dependent variable viscosity. Results in Physics, 7, 1801–1807. https://doi.org/10.1016/j.rinp.2017.05.018

Balsara, D. S. (2010). Multidimensional HLLE Riemann solver: Application to Euler and magnetohydrodynamic flows. Journal of Computational Physics, 229(6), 1970–1993.

Bejawada, S. G., Reddy, Y. D., Jamshed, W., Nisar, K. S., Alharbi, A. N., & Chouikh, R. (2022). Radiation effect on MHD Casson fluid flow over an inclined nonlinear surface with chemical reaction in a Forchheimer porous medium. Alexandria Engineering Journal, 61(10), 8207–8220. https://doi.org/10.1016/j.aej.2022.01.043

Cao, W., Animasaun, I. L., Yook, S. J., Oladipupo, V. A., & Ji, X. (2022). Simulation of the dynamics of colloidal mixture of water with various nanoparticles at different levels of partial slip: Ternary-hybrid nanofluid. International Communications in Heat and Mass Transfer, 135, 106069.

Chabani, I., Mebarek-Oudina, F., & Ismail, A. A. I. (2022). MHD flow of a hybrid nano-fluid in a triangular enclosure with zigzags and an elliptic obstacle. Micromachines, 13(2), 224. https://doi.org/10.3390/mi13020224

Chen, C. H. (2010). On the analytic solution of MHD flow and heat transfer for two types of viscoelastic fluid over a stretching sheet with energy dissipation, internal heat source and thermal radiation. International Journal of Heat and Mass Transfer, 53(19–20), 4264–4273.

Daniel, Y. S., Aziz, Z. A., Ismail, Z., & Salah, F. (2017). Entropy analysis in electrical magnetohydrodynamic (MHD) flow of nanofluid with effects of thermal radiation, viscous dissipation, and chemical reaction. Theoretical and Applied Mechanics Letters, 7(4), 235–242.

Daniel, Y. S., & Daniel, S. K. (2015). Effects of buoyancy and thermal radiation on MHD flow over a stretching porous sheet using homotopy analysis method. Alexandria Engineering Journal, 54(3), 705–712.

Doh, D. H., Cho, G. R., Ramya, E., & Muthtamilselvan, M. (2019). Cattaneo-Christov heat flux model for inclined MHD micropolar fluid flow past a nonlinearly stretchable rotating disk. Case Studies in Thermal Engineering, 14, 100496.

Elnaqeeb, T., Animasaun, I. L., & Shah, N. A. (2021). Ternary-hybrid nanofluids: Significance of suction and dual-stretching on three-dimensional flow of water conveying nanoparticles with various shapes and densities. Zeitschrift für Naturforschung A, 76(3), 231–243. https://doi.org/10.1515/zna-2020-0317

Elsebaee, F. A. A., Bilal, M., Mahmoud, S. R., Balubaid, M., Shuaib, M., Asamoah, J. K., & Ali, A. (2023). Motile micro-organism based trihybrid nanofluid flow with an application of magnetic effect across a slender stretching sheet: Numerical approach. AIP Advances, 13(3). https://doi.org/10.1063/5.0144191

Ghorbani, S., Jabari Moghadam, A., Emamian, A., Ellahi, R., & Sait, S. M. (2022). Numerical simulation of the electroosmotic flow of the Carreau-Yasuda model in the rectangular microchannel. International Journal of Numerical Methods for Heat & Fluid Flow, 32(7), 2240–2259. https://doi.org/10.1108/HFF-07-2021-0495

Griffiths, P. T. (2015). Flow of a generalised Newtonian fluid due to a rotating disk. Journal of Non-Newtonian Fluid Mechanics, 221, 9–17.

Grigoriadis, D. G. E., Sarris, I. E., & Kassinos, S. C. (2010). MHD flow past a circular cylinder using the immersed boundary method. Computers & Fluids, 39(2), 345–358.

Guedri, K., Mahmood, Z., Fadhl, B. M., Makhdoum, B. M., Eldin, S. M., & Khan, U. (2023). Mathematical analysis of nonlinear thermal radiation and nanoparticle aggregation on unsteady MHD flow of micropolar nanofluid over shrinking sheet. Heliyon, 9(3). https://doi.org/10.1016/j.heliyon.2023.e14248

Haider, S. M. A., Ali, B., Wang, Q., & Zhao, C. (2022). Rotating flow and heat transfer of single-wall carbon nanotube and multi-wall carbon nanotube hybrid nanofluid with base fluid water over a stretching sheet. Energies, 15(16), 6060. https://doi.org/10.3390/en15166060

Hamid, M., Usman, M., Khan, Z. H., Ahmad, R., & Wang, W. (2019). Dual solutions and stability analysis of flow and heat transfer of Casson fluid over a stretching sheet. Physics Letters A, 383(20), 2400–2408.

Ibáñez, G., & Cuevas, S. (2010). Entropy generation minimization of a MHD (magnetohydrodynamic) flow in a microchannel. Energy, 35(10), 4149–4155.

Imtiaz, M., Hayat, T., Alsaedi, A., & Ahmad, B. (2016). Convective flow of carbon nanotubes between rotating stretchable disks with thermal radiation effects. International Journal of Heat and Mass Transfer, 101, 948–957.

Islam, N., Riasat, S., Ramzan, M., Ghazwani, H. A. S., Pasha, A. A., Kadry, S., & Eldin, S. M. (2023). Thermal efficiency appraisal of hybrid nanocomposite flow over an inclined rotating disk exposed to solar radiation with Arrhenius activation energy. Alexandria Engineering Journal, 68, 721–732.

Jawad, M., Saeed, A., Gul, T., & Bariq, A. (2021). MHD Darcy-Forchheimer flow of Casson nanofluid due to a rotating disk with thermal radiation and Arrhenius activation energy. Journal of Physics Communications, 5(2), 025008. https://doi.org/10.1088/2399-6528/abe4e0

Joshi, V. K., Ram, P., Tripathi, D., & Sharma, K. (2018). Numerical investigation of magnetic nanofluids flow over rotating disk embedded in a porous medium. Thermal Science, 22(6 Part B), 2883–2895.

Khan, K. A., Jamil, F., Ali, J., Khan, I., Ahmed, N., Andualem, M., & Rafiq, M. (2022). Analytical simulation of heat and mass transmission in Casson fluid flow across a stretching surface. Mathematical Problems in Engineering, 2022, 1–11. https://doi.org/10.1155/2022/5576194

Kodi, R., Mopuri, O., Sree, S., & Konduru, V. (2022). Investigation of MHD Casson fluid flow past a vertical porous plate under the influence of thermal diffusion and chemical reaction. Heat Transfer, 5(1), 377–394.

Krishna, M. V., Ahammad, N. A., & Chamkha, A. J. (2021). Radiative MHD flow of Casson hybrid nanofluid over an infinite exponentially accelerated vertical porous surface. Case Studies in Thermal Engineering, 27, 101229. https://doi.org/10.1016/j.csite.2021.101229

Kumar, S., & Sharma, K. (2023). Impacts of Stefan blowing on Reiner–Rivlin fluid flow over moving rotating disk with chemical reaction. Arabian Journal for Science and Engineering, 48(3), 2737–2746. https://doi.org/10.1007/s13369-022-07008-9

Lok, Y. Y., Ishak, A., & Pop, I. (2011). MHD stagnation-point flow towards a shrinking sheet. International Journal of Numerical Methods for Heat & Fluid Flow, 21(1), 61–72.

Lund, L. A., Omar, Z., & Khan, I. (2019). Analysis of dual solution for MHD flow of Williamson fluid with slippage. Heliyon, 5(3), e01345.

Mahian, O., Oztop, H., Pop, I., Mahmud, S., & Wongwises, S. (2013). Entropy generation between two vertical cylinders in the presence of MHD flow subjected to constant wall temperature. International Communications in Heat and Mass Transfer, 44, 87–92.

Mahmood, Z., Ahammad, N. A., Alhazmi, S. E., Khan, U., & Bani-Fwaz, M. Z. (2022). Ternary hybrid nanofluid near a stretching/shrinking sheet with heat generation/absorption and velocity slip on unsteady stagnation point flow.

International Journal of Modern Physics B, 36(29), 2250209. https://doi.org/10.1142/S0217979222502095

Majeed, A. H., Mahmood, R., Shahzad, H., Pasha, A. A., Raizah, Z. A., Hosham, H. A., Reddy, D. S. K., & Hafeez, M. B. (2023). Heat and mass transfer characteristics in MHD Casson fluid flow over a cylinder in a wavy channel: Higher-order FEM computations. Case Studies in Thermal Engineering, 42, 102730.

Malik, M. Y., Khan, M., Salahuddin, T., & Khan, I. (2016). Variable viscosity and MHD flow in Casson fluid with Cattaneo–Christov heat flux model: Using Keller box method. Engineering Science and Technology, an International Journal, 19(4), 1985–1992.

Manjunatha, G., Rajashekhar, C., Vaidya, H., Prasad, K. V., Makinde, O. D., & Viharika, J. U. (2020). Impact of variable transport properties and slip effects on MHD Jeffrey fluid flow through channel. Arabian Journal for Science and Engineering, 45, 417–428.

Mishra, S. R., Dash, G. C., & Acharya, M. (2013). Mass and heat transfer effect on MHD flow of a visco-elastic fluid through porous medium with oscillatory suction and heat source. International Journal of Heat and Mass Transfer, 57(2), 433–438.

Mogharrebi, A. R., D. Ganji, A. R., Hosseinzadeh, K., Roghani, S., Asadi, A., & Fazlollahtabar, A. (2021). Investigation of magnetohydrodynamic nanofluid flow contain motile oxytactic microorganisms over rotating cone. International Journal of Numerical Methods for Heat & Fluid Flow, 31(11), 3394–3412. https://doi.org/10.1108/HFF-08-2020-0493

Mustafa, M. (2017). MHD nanofluid flow over a rotating disk with partial slip effects: Buongiorno model. International Journal of Heat and Mass Transfer, 108, 1910–1916. https://doi.org/10.1016/j.ijheatmasstransfer.2017.01.064

Naramgari, S., & Sulochana, C. (2016). MHD flow over a permeable stretching/shrinking sheet of a nanofluid with suction/injection. Alexandria Engineering Journal, 55(2), 819–827.

Nazia, S., Seshaiah, B., Sreedevi, P., & Reddy, P. S. (2022). Brownian motion and thermophoresis effect on heat and mass transfer of the Reiner–Rivlin nanofluid flow over a rotating disk with thermal radiation. Waves in Random and Complex Media, 1–25.

Oyekunle, T. L., Akolade, M. T., Abubakar, J. U., & Olotu, O. T. (2021). Influence of variable properties on quadratic convective flow of Casson nanofluid past an inclined plane. Journal of Applied Sciences, Information and Computing, 2(1), 67.

Park, J. E., Kang, T. G., & Moon, H. (2023). The effect of the rotating disk geometry on the flow and flux enhancement in a dynamic filtration system. Membranes, 13(3), 291.

Patel, H., Mittal, A., & Nagar, T. (2023). Fractional order simulation for unsteady MHD nanofluid flow in porous medium with Soret and heat generation effects. Heat Transfer, 52(1), 563–584.

Patgiri, B., & Paul, A. (2024). Inspection of viscoelastic Ag+ Cu+ Fe3O4+ Al2O3/kerosene oil tetra-hybrid nanofluid flow across a stretchable rotating disk with exponentially varying viscosity. Journal of Taibah University for Science, 18(1), 2336327.

Patil, P. M., Goudar, B., & Sheremet, M. A. (2023). Tangent hyperbolic ternary hybrid nanofluid flow over a rough-yawed cylinder due to impulsive motion. Journal of Taibah University for Science, 17(1), 2199664. https://doi.org/10.1080/16583655.2023.2199664

Paul, A., & Patgiri, B. Modelling of silicone oil–based Casson hybrid nanofluid across a porous rotating disk. In Modeling and Simulation of Fluid Flow and Heat Transfer (pp. 162–179). CRC Press.

Paul, A., Patgiri, B., & Sarma, N. (2024a). Combined effect of non-linear mixed convection, and non-uniform heat source/sink on Casson ternary hybrid nanofluid across a stretched rotatory disk. Journal of Nanofluids, 13(2), 586–599. https://doi.org/10.1166/jon.2024.2136

Paul, A., Patgiri, B., & Sarma, N. (2024b). Darcy-Forchheimer flow of Ag–ZnO–CoFe2O4/H2O Casson ternary hybrid nanofluid induced by a rotatory disk with EMHD. International Journal of Ambient Energy, 45(1), 2313697. https://doi.org/10.1080/01430750.2024.2313697

Paul, A., Patgiri, B., & Sarma, N. (2024c). Transformer oil-based Casson ternary hybrid nanofluid flow configured by a porous rotating disk with hall current. ZAMM-Journal of Applied Mathematics and Mechanics/Zeitschrift für Angewandte Mathematik und Mechanik, e202300704. https://doi.org/10.1002/zamm.202300704

Paul, A., Sarma, N., & Patgiri, B. (2023a). Mixed convection of shear-thinning hybrid nanofluid flow across a radiative unsteady cone with suction and slip effect. Materials Today Communications, 37, 107522. https://doi.org/10.1016/j.mtcomm.2023.107522

Paul, A., Sarma, N., & Patgiri, B. (2023b). Thermal and mass transfer analysis of Casson-Maxwell hybrid nanofluids through an unsteady horizontal cylinder with variable thermal conductivity and Arrhenius activation energy. Numerical Heat Transfer, Part A: Applications, 1–26. https://doi.org/10.1080/10407782.2023.2297000

Paul, A., Sarma, N., & Patgiri, B. (2024d). Numerical assessment of MHD thermo-mass flow of Casson ternary hybrid nanofluid around an exponentially stretching cylinder. BioNanoScience, 1–16. https://doi.org/10.1007/s12668-024-01306-2

Pekmen, B., & Tezer-Sezgin, M. (2014). MHD flow and heat transfer in a lid-driven porous enclosure. Computers & Fluids, 89, 191–199.

Prameela, M., Gangadhar, K., & Reddy, G. J. (2022). MHD free convective non-Newtonian Casson fluid flow over an oscillating vertical plate. Partial Differential Equations in Applied Mathematics, 5, 100366.

Ram, P., Joshi, V. K., Kumar, V., & Sharma, S. (2019). Rheological effects due to oscillating field on time dependent boundary layer flow of magnetic nanofluid over a rotating disk. Proceedings of the National Academy of Sciences, India Section A: Physical Sciences, 89, 367–375.

Raptis, A., Perdikis, C., & Takhar, H. S. (2004). Effect of thermal radiation on MHD flow. Applied Mathematics and Computation, 153(3), 645–649.

Rashid, U., Baleanu, D., Liang, H., Abbas, M., Iqbal, A., & Rahman, J. U. (2020). Marangoni boundary layer flow and heat transfer of graphene–water nanofluid with particle shape effects. Processes, 8(9), 1120. https://doi.org/10.3390/pr8091120

Rehman, K. U., Malik, M. Y., Zahri, M., & Tahir, M. (2018). Numerical analysis of MHD Casson Navier's slip nanofluid flow yield by rigid rotating disk. Results in Physics, 8, 744–751. https://doi.org/10.1016/j.rinp.2018.01.017

Sahoo, A., & Nandkeolyar, R. (2023). Entropy generation in magnetohydrodynamic radiative non-Darcy slip flow of a Casson nanofluid with Hall effects and activation energy. Journal of Magnetism and Magnetic Materials, 575, 170712. https://doi.org/10.1016/j.jmmm.2023.170712

Sajid, T., Gari, A. A., Jamshed, W., Eid, M. R., Islam, N., Irshad, K., ... El Din, S. M. (2023). Case study of autocatalysis reactions on tetra hybrid binary nanofluid flow via Riga wedge: Biofuel thermal application. Case Studies in Thermal Engineering, 47, 103058. https://doi.org/10.1016/j.csite.2023.103058

Samantaray, S. S., Azam, M., Abbas, N., Misra, A., & Nayak, M. K. (2024). Impact of Cattaneo–Christov double diffusion theory and Arrhenius pre-exponential factor law on flow of radiative Reiner–Rivlin nanofluid over rotating disk. International Journal of Modern Physics B, 38(07), 2450102.

Sandeep, N., Koriko, O. K., & Animasaun, I. L. (2016). Modified kinematic viscosity model for 3D-Casson fluid flow within boundary layer formed on a surface at absolute zero. Journal of Molecular Liquids, 221, 1197–1206.

Sarada, K., Gamaoun, F., Abdulrahman, A., Paramesh, S. O., Kumar, R., Prasanna, G. D., & Gowda, R. P. (2022). Impact of exponential form of internal heat generation on water-based ternary hybrid nanofluid flow by capitalizing non-Fourier heat flux model. Case Studies in Thermal Engineering, 38, 102332. https://doi.org/10.1016/j.csite.2022.102332

Sarangi, M. K., Thatoi, D. N., Shaw, S., Azam, M., Chamkha, A. J., & Nayak, M. K. (2023). Hydrothermal behavior and irreversibility analysis of Bödewadt flow of radiative and dissipative ternary composite nanomaterial due to a stretched rotating disk. Materials Science and Engineering: B, 287, 116124. https://doi.org/10.1016/j.mseb.2022.116124

Shateyi, S. (2013). A new numerical approach to MHD flow of a Maxwell fluid past a vertical stretching sheet in the presence of thermophoresis and chemical reaction. Boundary Value Problems, 2013(1), 14.

Sivaraj, R., Benazir, A. J., Srinivas, S., & Chamkha, A. J. (2019). Investigation of cross-diffusion effects on Casson fluid flow in existence of variable fluid properties. The European Physical Journal Special Topics, 228, 35–53.

Smolentsev, S., Badia, S., Bhattacharyay, R., Bühler, L., Chen, L., Huang, Q., ... Zikanov, O. (2015). An approach to verification and validation of MHD codes for fusion applications. Fusion Engineering and Design, 100, 65–72.

Soid, S. K., Ishak, A., & Pop, I. (2018). MHD flow and heat transfer over a radially stretching/shrinking disk. Chinese Journal of Physics, 56(1), 58–66.

Subhas Abel, M., Tawade, J. V., & Nandeppanavar, M. M. (2012). MHD flow and heat transfer for the upper-convected Maxwell fluid over a stretching sheet. Meccanica, 47, 385–393.

Tamoor, M., Waqas, M., Khan, M. I., Alsaedi, A., & Hayat, T. (2017). Magnetohydrodynamic flow of Casson fluid over a stretching cylinder. Results in Physics, 7, 498–502.

Tassaddiq, A., Khan, S., Bilal, M., Gul, T., Mukhtar, S., Shah, Z., & Bonyah, E. (2020). Heat and mass transfer together with hybrid nanofluid flow over a rotating disk. AIP Advances, 10(5). https://doi.org/10.1063/5.0010181

Thomas, C., & Davies, C. (2010). The effects of mass transfer on the global stability of the rotating-disk boundary layer. Journal of Fluid Mechanics, 663, 401–433.

Umavathi, J. C., Prakasha, D. G., Alanazi, Y. M., Lashin, M. M., Al-Mubaddel, F. S., Kumar, R., & Gowda, R. P. (2023). Magnetohydrodynamic squeezing Casson nanofluid flow between parallel convectively heated disks. International Journal of Modern Physics B, 37(04), 2350031.

Yin, C., Zheng, L., Zhang, C., & Zhang, X. (2017). Flow and heat transfer of nanofluids over a rotating disk with uniform stretching rate in the radial direction. Propulsion and Power Research, 6(1), 25–30.

Zahan, I., Nasrin, R., & Khatun, S. (2022). Thermal performance of ternary-hybrid nanofluids through a convergent-divergent nozzle using distilled water-ethylene glycol mixtures. International Communications in Heat and Mass Transfer, 137, 106254. https://doi.org/10.1016/j.icheatmasstransfer.2022.106254

Zainal, N. A., Nazar, R., Naganthran, K., & Pop, I. (2021a). Heat generation/absorption effect on MHD flow of hybrid nanofluid over bidirectional exponential stretching/shrinking sheet. Chinese Journal of Physics, 69, 118–133. https://doi.org/10.1016/j.cjph.2020.12.002

Zainal, N. A., Nazar, R., Naganthran, K., & Pop, I. (2021b). MHD flow and heat transfer of hybrid nanofluid over a permeable moving surface in the presence of thermal radiation. International Journal of Numerical Methods for Heat & Fluid Flow, 31(3), 858–879.

Zangooee, M. R., Hosseinzadeh, K., & Ganji, D. D. (2019). Hydrothermal analysis of MHD nanofluid (TiO2-GO) flow between two radiative stretchable rotating disks using AGM. Case Studies in Thermal Engineering, 14, 100460.

Zhao, T. H., Khan, M. I., & Chu, Y. M. (2023). Artificial neural networking (ANN) analysis for heat and entropy generation in flow of non-Newtonian fluid between two rotating disks. Mathematical Methods in the Applied Sciences, 46(3), 3012–3030. https://doi.org/10.1002/mma.7310

Zheng, L., Liu, Y., & Zhang, X. (2012). Slip effects on MHD flow of a generalized Oldroyd-B fluid with fractional derivative. Nonlinear Analysis: Real World Applications, 13(2), 513–523.

Chapter 8

Adoption, development, and application of robotic and nanofluid technology in sustainable healthcare

A comprehensive study

Shaik Javeed Hussain and Sivasakthivel Thangavel

8.1 INTRODUCTION

By integrating robotics into healthcare, we can improve patient outcomes and transform healthcare delivery. In various medical applications, such as surgery, rehabilitation, and diagnosis, robots offer numerous benefits, including improved efficiency, precision, and cost reduction. For robotics to be implemented effectively in healthcare, more than just technical considerations must be taken into account. Integrating these technologies into healthcare practices requires consideration of human, organizational, and social factors [1]. This chapter explores how human, organizational, and social factors can impact the effectiveness and safety of robotics in healthcare.

Any future healthcare system involving robots will need to be evaluated against the existing alternative of human provision, and previous work modeling various aspects of the National Health Service (NHS) workforce suggests that this provides an application domain for demonstrating cost-effectiveness. With these factors in mind, it can be stated that there is now a more pressing and relevant set of reasons to warrant the study and innovation of robotics for healthcare. This is corroborated by recent increased interest and funding for robotics research in Europe and the formation of new specific funding initiatives for medical robotics in the US. For these reasons, the new research and technology that takes place in a space of 10–20 years could have significant implications [2].

In recent years, there have been various publications on the possible applications of robotics in medicine and some limited experimental work particularly on the use of robot assistants for physically disabled people. However, progress toward an established robotic technology for healthcare has been slow, and in some areas, there is now a real possibility of leapfrogging to a new generation of robotic and intelligent systems, bypassing purely teleoperated robot manipulators. An example of this is the projected use of mobile robots in hospitals for logistics and delivery tasks. This opens an enormous scope encompassing automation of present manual tasks, intelligent support systems for human carers, and new forms of therapy and

 DOI: 10.1201/9781003527251-8

rehabilitation for a wide range of patients, both at the hospital and home. The potential synergy with developments in artificial intelligence and smart materials is also strong.

The initial kick off of this study presents the overall picture which includes the adoption of advanced robotics, nanotechnology, and robotics into the healthcare sector. The development of a strategic level model is essential for the healthcare sector to cover up the predicted problems such as growing numbers of elderly and disabled people, staff shortages, spiraling costs, and the impact of these factors both on the quality of care and on working conditions. The methods described, if properly implemented, should entail health and social care services that are far more efficient than those of today and of a higher standard at a similar or lower cost [3].

There is great potential for improving efficiency and driving sustainable development in a variety of industries, including healthcare. Robotics can reduce waste and improve resource utilization, resulting in a more sustainable and environmentally friendly future. To ensure their effective and safe use, robotics must also consider human, organizational, and social factors. The adoption, use, and maintenance of robotics in healthcare can be influenced by these factors. The purpose of this chapter is to examine the role of human, organizational, and social factors in the sustainability of robotics in healthcare as well as how these factors can improve patient outcomes and the delivery of healthcare [4].

Research questions related to the deployment and use of automation in healthcare need to be interdisciplinary, incorporating perspectives from human factors engineering [5], information technology, organizational behavior, and sociology. This approach can be used to address the following key research questions:

What are the human factors that influence the deployment and use of automation in healthcare, such as user preferences, behavior, and performance?

What role do organizational factors play in deploying and using automation in healthcare?

In what ways do infrastructure, technology, and regulations influence automation in healthcare?

What impact do public perception, cultural norms, and ethical considerations have on healthcare automation's deployment and use?

The mixed-methods approach can be used to address these research questions, in which qualitative and quantitative methods are combined. In healthcare automation deployment and use, qualitative methods such as focus groups, interviews, and case studies can be used to gather rich, in-depth data about stakeholders' experiences, attitudes, and perceptions [6]. Large-scale data can be collected using quantitative methods, such as surveys and experiments, to support generalizability and test hypotheses.

Furthermore, human factor engineering can be used to understand the impact of automation on human performance and design technology that supports and enhances human performance, while reducing the risk of errors and hazards. Organizations can develop strategies to address the social, cultural, and political aspects of automation in healthcare by considering the interplay between technology, organizations, the environment, and society [7, 8]. This increases the chances of success and improves the technology's overall impact which are shown in Figure 8.1.

8.1.1 Human factors

Human factors refer to the psychological and physiological characteristics of individuals that influence their interactions with technology and the environment. A wide range of factors contribute to human factors in healthcare robotics, including user needs and preferences, cognitive and physical abilities, perception, and attention, as well as the design of the human-robot interface. When developing and deploying healthcare robots, these factors are critical because they can affect the technology's effectiveness, efficiency, and safety. Designing healthcare robots with human factors in mind makes them more usable, effective, and safe and meets the needs of patients, providers, and others [9, 10].

8.1.2 Organizational factors

Organizational factors refer to characteristics of healthcare organizations and societal contexts in which healthcare robots are deployed. Healthcare robots can be adopted and successfully deployed based on these factors. Several organizational factors can be considered, including the organization's culture and values, its resources (for example, funding, staff, and training), its policies and procedures, and the level of support of stakeholders (patients, healthcare providers, and regulators).

In addition to organizational factors, data privacy, liability, and patient consent can also affect the legal and ethical considerations around healthcare robots. It is crucial to understand these organizational factors to develop healthcare robots that are not only technically sound but also socially acceptable and can be effectively integrated into healthcare organizations [11]. By examining the organizational context and the factors that may impact the deployment of healthcare robots, potential challenges and opportunities can be identified, and effective and efficient solutions can be designed.

8.1.3 Societal factors

Healthcare robots are deployed in a broader social and cultural context referred to as societal factors. They can have a significant impact on the adoption and deployment of healthcare robots by shaping public perceptions

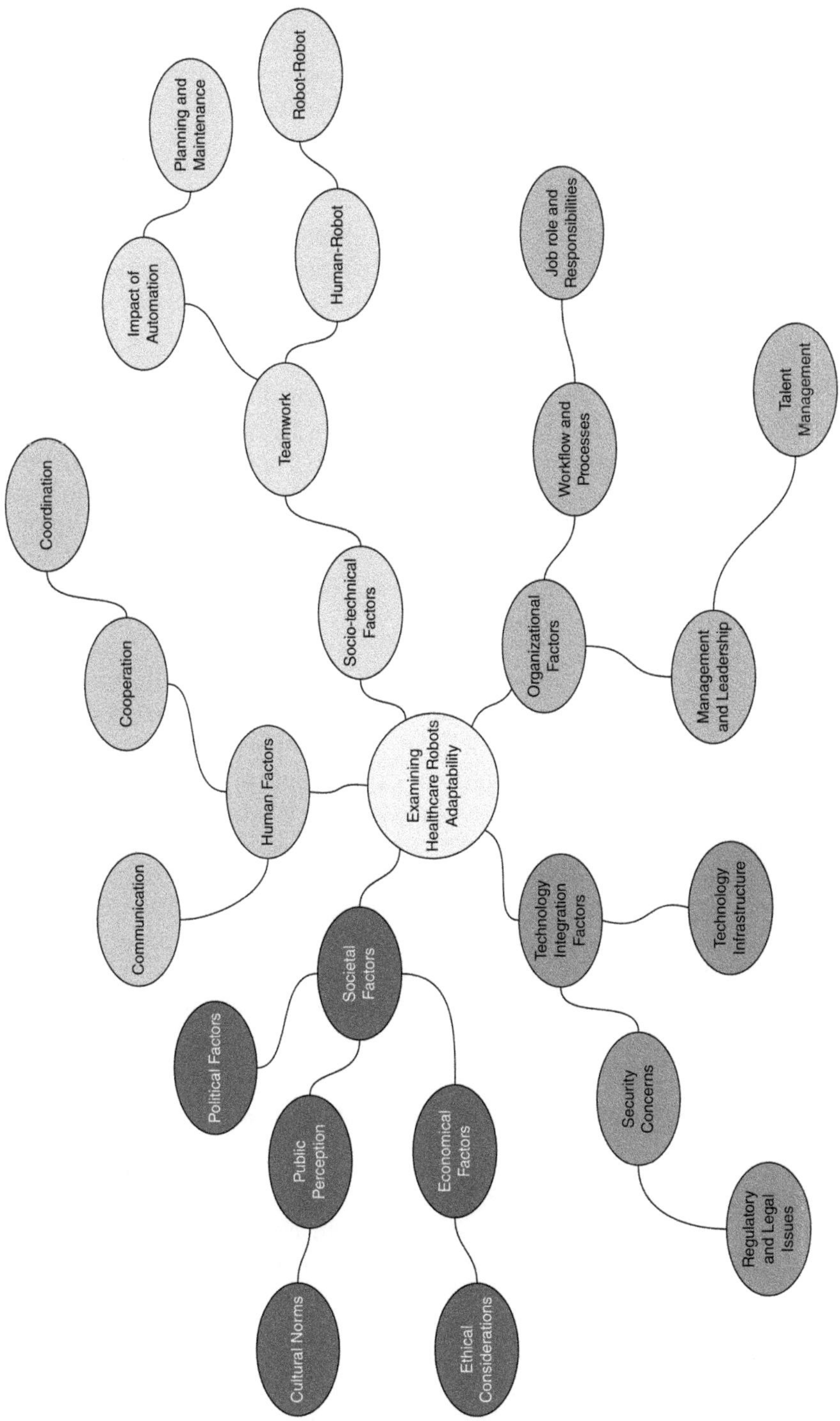

Figure 8.1 Factors affecting the adaptability of healthcare robots.

and attitudes. Among societal factors are the public perception of technology, awareness and understanding of healthcare robot capabilities and limitations, values and beliefs about healthcare, and concerns about privacy and security [12].

8.1.4 Nanofluids

Nanofluid is a new type of heat transfer fluid which uses base fluid mixed with nanometer-size solid particles. Nanofluids are stable colloidal suspensions consisting of solid particles in a liquid base, ranging in size from 1 to 100 nm. It was found that the addition of nanoparticles to the base fluid enhances the thermal conductivity of the mixture. This enhancement was found to be much more than the models predict for the addition of particles to fluid. This anomaly was the genesis of much of the research conducted in the field of nanofluids, and to date, there are no concrete theories to fully describe this phenomenon. A relatively small increase in solid volume fraction (1–5%) can lead to a substantial increase in the thermal conductivity of a fluid. Because of this, there is no need for a large solid volume fraction to take advantage of the superior thermal properties of the nanofluid, which removes the problem of sedimentation and clogging in the pipework associated with a high particle concentration.

During the last decade, there has been a noticeable rise in both development and usage of nanofluid in various engineering and medical applications all over the globe. Nanofluid can be defined as a dilute colloidal suspension containing nano-scaled particles that are called nanoparticles. One can find these particles of metal, oxide, or carbide in various base fluids like water, ethylene glycol, and engine oil. The formation or designing of nanofluid is done by dispersing the nanoparticles in base fluid with the help of vigorous stirring. The result of such process is an increase in thermal conductivity and convective heat transfer, a decrease in the specific heat, and an increase in the density of the fluid. Usually, the change in physical properties of nanofluid with respect to base fluid is determined by volume concentration and particle to fluid size parameter.

8.1.5 Importance of nanofluids in healthcare

Aretakis et al. have discussed the use of magnetic nanoparticles as a method of local drug delivery by using an external magnetic field to direct the drug to a specific site. This is advantageous as it can reduce the drug dosage needed, minimize systemic distribution, and prevent the drugs from undergoing chemical changes like they would with other conventional drug carriers [13].

The ability to use MRI contrast agents, which can allow very early diagnosis of diseases that were previously very difficult to diagnose until symptoms had progressed, is a significant advancement. One such example of these contrast agents is to detect and visualize metastatic cancer cells, which can allow immediate treatment for the patient. This contrast agent can locate the cancer cells and allow magnetic heat treatment to destroy those cells.

Nanofluids are proposed to be used in the treatment and diagnostics of diseases such as cancer, artery plaque, and parasitic infections (for example, HIV) over the next 10–15 years. The new advances in nanofluid preparation have made it much easier to manipulate and engineer surface chemistry and add functionality to the particles. This, in turn, has a much more rapid translation to clinical applications. It is a critical initiative for clinical trials of new products and the integration of new technologies for diagnosing and treating diseases.

In the neoteric era, healthcare researchers have polarized their attention toward biomedical and clinical applications of nanofluids, considering a large number of fatal diseases. Nanofluids, in general, have attracted attention toward extensive research due to their improved thermophysical properties, heat transfer, and transport characteristics. These properties, along with particle size comparable to the natural building blocks of living systems, such as DNA, have directed researchers to explore their applications in the field of life sciences. This has led to the foundation of an emerging science called nanomedicine.

8.2 BACKGROUND

Robotics is becoming increasingly common in healthcare and has the potential to greatly improve patient outcomes. To ensure the successful deployment of these technologies, several human-factors-related research questions must be addressed. These questions include understanding how to design the technology to support performance and eliminate hazards, how to incorporate effectively human-robot and robot-robot teams, and how to address the potential impact of automation on various roles within the healthcare industry. Research must also consider the societal and organizational factors that will affect the adoption and integration of these technologies [14]. Qualitative and quantitative studies can help researchers better understand the benefits, barriers, and requirements for the near-term, mid-term, and long-term adoption of healthcare robots.

For technology to be successfully deployed, several major barriers must be overcome, including the following:

Cost and technical complexity: The high cost of technology can prove to be one of the biggest barriers to its widespread adoption, particularly for small businesses and healthcare organizations with limited budgets. When it comes to technology, it may be complicated to use, which can make it difficult for organizations to integrate it into their existing systems and processes, making it challenging for them to integrate it successfully [15].

8.2.1 Socio-technical framework

In the design, development, and deployment of technology, a socio-technical framework considers social, technical, and organizational factors. Socio-technical frameworks see technology as an enabler of positive social and organizational outcomes rather than an end in itself. The framework emphasizes the need for technology to integrate with social and organizational processes and practices for it to succeed.

To integrate technology, such as robotics, into the healthcare system, a socio-technical framework can be used. The impact of technology on healthcare providers and patients, as well as on the broader healthcare system, must be examined. Additionally, deploying and utilizing technology effectively involves understanding the social and organizational factors that can affect its effectiveness, such as resistance to change, regulatory requirements, and public perceptions.

Using a socio-technical framework can help ensure that technology is deployed in a way that benefits all stakeholders, including patients, healthcare providers, and the healthcare system as a whole. By considering the long-term effects of technology use on society, the environment, and the economy, it also helps ensure that technology deployment is sustainable.

8.2.2 Users

For healthcare robots to be successful, social and organizational factors need to be considered. When healthcare robots are used with users, several potential issues can arise, including the following:

Issues related to user acceptance and technical implementation: Users can be resistant to change, particularly if they have to adapt their existing practices to accommodate new technology. Technical issues, such as hardware and software malfunctions, can result in decreased efficiency and accuracy of robots, leading to dissatisfaction among users.

Privacy, ethics, and security concerns: When it comes to deploying healthcare robots, privacy and security concerns can be a major issue, especially when it relates to sensitive information such as highly confidential medical records. It is also possible that ethical concerns arise when healthcare robots are deployed, particularly when it comes to issues such as informed consent, autonomy, and accountability when the robots are used in healthcare.

To ensure that technology is deployed in a way that supports positive outcomes for all stakeholders, including patients, healthcare providers, and the entire healthcare system, it is essential to address these factors so that technology is deployed in a way that supports positive outcomes for all stakeholders.

8.2.3 Teamwork

As the name implies, teamwork refers to the collaboration of individuals who work together as a team toward achieving a common goal. In the context of healthcare, it is essential to ensure that team members work effectively together to provide patients with the highest quality of care possible.

The concept of teamwork can be used in healthcare organizations in a wide variety of ways to provide care to patients. It can involve a wide range of professionals, including doctors, nurses, and therapists working together to provide care. The use of healthcare robots in healthcare can also be integrated into the care system, working alongside human healthcare providers in order to support the delivery of care in some cases.

To be able to work effectively as a team, clear communication, coordination, and cooperation between team members are essential. There is no doubt that the team should be able to establish a culture of trust, collaboration, and mutual respect within the team and that opportunities should be provided for members to collaborate and communicate regularly within the team.

8.2.3.1 Human-robot

In the context of healthcare delivery, human-robot teamwork refers to the collaboration between human healthcare providers and healthcare robots in order to provide high-quality care. To improve the efficiency and effectiveness of care delivery, this type of teamwork involves the integration of healthcare robots into the care team to improve efficiency and effectiveness of care delivery.

For a human-robot team to work effectively, there must be clear communication between the human and robot members, coordination, and cooperation. There is a need to create a culture of trust, collaboration, and mutual respect between the human and robotic members of the team and to provide opportunities for collaboration and communication between them.

Developing a human-robot team technique can present several challenges, including the fact that human and robotic team members often have different capabilities and limitations, as well as the requirement for clear protocols for decision-making and task distribution.

8.2.3.2 Robot-Robot

In the context of delivering healthcare, the robot-robot model is a way of thinking about the relationship and interactions between different healthcare robots. As part of this approach, consideration is given to how healthcare robots can be designed, developed, and used to work collaboratively and in concert with one another in the delivery of healthcare services.

As part of the robot-robot approach, researchers recognize that there are potential benefits associated with the collaboration between healthcare robots, such as increased efficiency, improved patient outcomes, and increased flexibility in care delivery. As part of the design and use of healthcare robots, it is also recognized that technical factors, including the capabilities and limitations of the technology, need to be taken into consideration as part of the design and use process.

For the robot-robot approach to succeed, it is important to consider ways in which healthcare robots can work together effectively, including the development of clear protocols for communication between the healthcare robots and their task allocation between the robots to ensure the success of the approach. The development of new technologies and systems to support the collaborative interaction between robots may be one way in which this can be achieved.

8.2.4 Impact of automation

To introduce new technologies in healthcare, it is important to consider the potential consequences that automation may have on these factors as well as to develop strategies to manage and mitigate any negative effects that may arise. Automation in healthcare has the potential to have a significant impact on several factors, including the following:

Processes and workflow: Automated workflows ensure that healthcare processes are streamlined and optimized, resulting in a reduction of the amount of time and effort required to complete a task.

Patient outcomes and satisfaction: Automation can improve patient outcomes by offering more accurate and timely diagnoses and treatments and by reducing the risk of human error in diagnosing and treating patients. In addition to reducing wait times and improving access to care, automation can contribute to the satisfaction of patients by providing a more personalized and efficient experience and reducing wait times.

8.2.5 Planning and maintenance

In terms of the roles involved in the planning of maintenance and inspection in healthcare, it is essential to take into account all of these roles as they all play an important role in the planning of maintenance and inspection activities. To ensure that each role is properly accounted for and involved, the following factors need to be taken into account:

Personnel responsible for maintenance and inspection of healthcare robots: These individuals are responsible for ensuring the healthcare robots are in good working order, including cleaning, testing, and repairing them as required.

Support staff: These personnel provide technical support to the healthcare robots and are responsible for troubleshooting issues as well as becoming involved in the maintenance and inspection process of the robots.

Healthcare providers: The healthcare providers, such as physicians and nurses, use healthcare robots to help deliver care and may even be involved in planning maintenance and inspection activities related to the robots.

Patients: For healthcare robots to have the greatest impact on patient care, they must be considered in the planning of maintenance and inspection activities in their input and feedback.

Management: It is the responsibility of the management, including administrators and executives, to oversee the deployment and application of healthcare robots as well as to participate in the planning of maintenance and inspection work.

The IT department: IT staff are responsible for the management and maintenance of the technology infrastructure that supports healthcare robots as well as being responsible for the planning of maintenance and inspection activities.

Regulators: Regulators such as government agencies and professional organizations are responsible for ensuring the safety and effectiveness of healthcare robots and should be involved in planning maintenance and inspection activities to ensure their safety and effectiveness.

The development of clear protocols and processes for maintenance and inspection activities can help achieve this goal through effective communication and collaboration.

8.2.6 Organizational factors

To be able to manage these changes effectively, organizations must take a proactive stance and develop strategies that will address both the challenges and the opportunities presented by automation. Many different activities can be included in this category, such as workforce planning and development, change management, and technology management. A healthcare organization that introduces automation can experience fundamental changes within the organization that impact many key areas, including the following:

Processes and workflows: Automation can streamline and optimize healthcare processes, leading to changes in how work is organized and carried out and in the way that care is delivered.

As a result of proactively addressing these changes, organizations can ensure that they are well-positioned to take advantage of the full benefits that automation in healthcare can provide.

8.2.6.1 Enablers

This chapter describes the factors that contribute to the successful deployment and use of technology in healthcare, including the use of automation. Automating healthcare can be facilitated in several ways, including the following:

Financial resources and interoperability: To develop and deploy technology in healthcare, including automation, it is fundamental to have access to sufficient financial resources. Developing interoperable systems, including hardware, software, and data standards, is critical for the successful deployment and use of technology in healthcare, including automation, as a result of the development and implementation of interoperable systems.

Automation: Expertise in the development and deployment of technology, including automation in healthcare, is essential to ensure it is developed and deployed with the highest level of reliability and efficiency. Access to technical expertise, including software development and engineering skills, is a key component of technology development and deployment. Ultimately, the success of the use of technology is going to depend on the engagement and adoption of technology, including automation, by healthcare providers, patients, and other stakeholders.

For the deployment and use of technology in healthcare, including automation, to be successful, organizations need to ensure that these enablers are in place. The organizations that invest in these enablers will be able to increase their chances of success and improve the overall impact of automation in healthcare by investing in them.

8.2.6.2 Barriers

Several obstacles can prevent the successful deployment and use of technology in the healthcare sector, including the use of automation. Healthcare automation faces many key challenges, including the following:

Cost and technical complexity: Developing and deploying technology, including automation, can be daunting for smaller organizations and those with limited resources, as the cost of doing so can be a significant barrier. For organizations without access to technical expertise or resources, the complexity of technology, including automation, can be a barrier to its successful deployment and use, particularly for those without the financial means to invest in technical expertise or resources.

Interoperability: A lack of interoperability between systems and technologies can create a barrier to the successful deployment and use of automation in healthcare due to the inefficiency of the connectivity between these systems and technologies.

Barriers: In the healthcare industry, technical barriers, such as the resistance of healthcare providers, patients, and other stakeholders, may interfere with the successful deployment and application of automation.

There are barriers to the successful deployment of and use of automation in healthcare that must be addressed by organizations to ensure the success of the project. An organization can increase its chances of success and improve the overall impact of automation in healthcare by developing strategies to overcome these barriers and acquiring the necessary skills and capabilities.

8.2.7 Technology, organizational, and environmental (TOE) framework

As a model of technology adoption and implementation in healthcare, this model includes factors such as technology, organizational, and environmental factors that affect the adoption and implementation of technology, including automation, and the interplay between these factors. To be able to deploy and use technology effectively in healthcare, an organization needs to understand the interdependencies among these factors and the factors that influence this process. The TOE framework provides a comprehensive view of the factors influencing this process.

Technology: Within the TOE framework, the technology component includes not only the technology itself but also its technical features, capabilities, and specifications, which are generally considered to be a part of the technology.

Using the TOE approach, organizations can obtain a holistic view of the factors that influence the deployment and use of technology in healthcare and can be able to better understand their interdependencies with these factors. To optimize the deployment and use of technology, including automation, in healthcare, organizations need to take into account the interplay between technology, organizational factors, and environmental factors and develop strategies to optimize the deployment and use of technology, including automation, in healthcare, and increase success chances.

8.2.8 Integration of robotics

Using robotics in healthcare has the potential to greatly improve the inspection and maintenance processes of medical equipment, resulting in more efficient delivery of healthcare services for better patient outcomes as well as a more efficient way to inspect and maintain medical equipment. When robots are used to inspect or maintain equipment, it can increase accuracy, reduce human error, and save time and resources as well as increase efficiency. To ensure the safe and effective use of robotics in this industry, human, organizational, and social factors must also be taken into consideration during the process of implementation. It is important to note that these factors can have a significant impact on the adoption and utilization of these technologies, in addition to their overall impact on the healthcare system. Our purpose in this chapter is to explore the role that human, organizational, and social factors play in the inspection and maintenance of medical equipment using robotic, as well as how they can be used together to improve the delivery of healthcare services and ensure the safety of patients. Our chapter will focus on the role of technology innovations, such as robotics, in the inspection and maintenance of medical equipment as well as how they can play a significant role in driving positive outcomes for patients, healthcare providers, and the healthcare system as a whole. There are several key concerns that must be taken into account when implementing robotics in the healthcare system, such as the following:

What are the attitudes and perceptions of healthcare providers toward the use of robotics in healthcare?
How does the physical design and interface of robotics impact their use by healthcare providers?
What are the cognitive and physical demands of operating and using robotics in healthcare?
How do training and education impact the adoption and use of robotics in healthcare?
What are the ethical and moral implications of using robotics in healthcare and how do these influence healthcare provider attitudes and behaviors?
How does patient comfort and trust impact the use of robotics in healthcare?
How do patient's age, gender, and cultural background impact their attitudes toward the use of robotics in healthcare?
How do regulations and policies impact the development and use of robotics in healthcare?
How do healthcare providers perceive the reliability and safety of robotics in healthcare and how does this impact their use?
What is the impact of human factors on the effectiveness and efficiency of using robotics in healthcare?

8.2.9 Societal

Social, cultural, and political factors are all factors that can influence the deployment and use of technology in healthcare, including the use of automation, based on their impact on the social, cultural, and political aspects of society. The important societal factors that can impact the deployment of and the use of automation in healthcare are as follows.

The public's perception of the use and deployment of technology in healthcare, and their willingness to accept and trust this technology, can be a crucial factor for the success of the deployment and usage of the technology.

The culture and values of one's society can have a profound impact on how technology is perceived and used and can also play a role in guiding how automation can be deployed and used in healthcare.

Several factors may contribute to the deployment and use of automation in healthcare, including socioeconomic factors, such as poverty, unemployment, and income inequality, which may affect technology availability and accessibility. It is important to note that political factors, such as government policies and regulations, can have a major impact on the deployment and implementation of automation in healthcare due to their influence on both the regulatory and legal setting within which technology will be used. Ethical considerations, such as privacy, security, and the use of data, can have a significant impact on the deployment and use of automation in healthcare, as they can raise serious issues concerning the protection of sensitive information.

To ensure the successful deployment and use of automation in healthcare, organizations need to consider these societal factors, and they need to develop strategies to address the social, cultural, and political implications of the technology. As an organization, you can increase your chances of success and improve the overall impact of automation in healthcare by taking into consideration the interplay between technology, organizational, and societal factors.

8.3 MATERIALS AND METHODS

8.3.1 Material

To address the research questions outlined in Section 1 more specifically, a variety of approaches were considered to address them. Some approaches were used to achieve this, including workshops, focus groups, and surveys. We chose to conduct interviews with subject matter experts in the field of healthcare robotics to optimize our use of time and resources as well as to gain a deeper understanding of the field of healthcare robotics in general.

As part of the interview process, an initial workshop was convened with key stakeholders from the healthcare sector as part of the project's stakeholder engagement process. The following were some of the key outcomes that were achieved during this workshop:

Through the workshop, we were able to establish the scope of the interviews.
To guide the interview process, an illustrative scenario was developed.
The stakeholders' extensive network of contacts was used to compile a preliminary list of interview candidates for the interviews, which would be conducted shortly.

As a result of the workshop proceedings, three sets of materials have been developed.

All participants were provided with a consent form and information sheet. As a consequence of this document, the participants had a clear understanding of the study's objectives, expectations about their involvement in the study, as well as details about their rights to anonymity and data protection as participant.

The purpose of the visual scenario (shown in Figure 8.2) was to illustrate possible applications of automation using a visual representation of the process. As part of the scenario, both short-term and long-term prospects were discussed, which helped foster a common understanding among participants about the future uses of automation, both in the short term and in the long term. There was a purpose for this process, which was to focus the interviews on human and organizational factors rather than getting bogged down in discussions about technology.

To structure the interviews and ensure that the three interviewers who participated in the project answered the same questions, a question guide was developed.

It should be noted that before commencing the study, all materials and procedures underwent a thorough review and received approval from The Global College of Engineering and Technology's Ethics Committee before proceeding with the study.

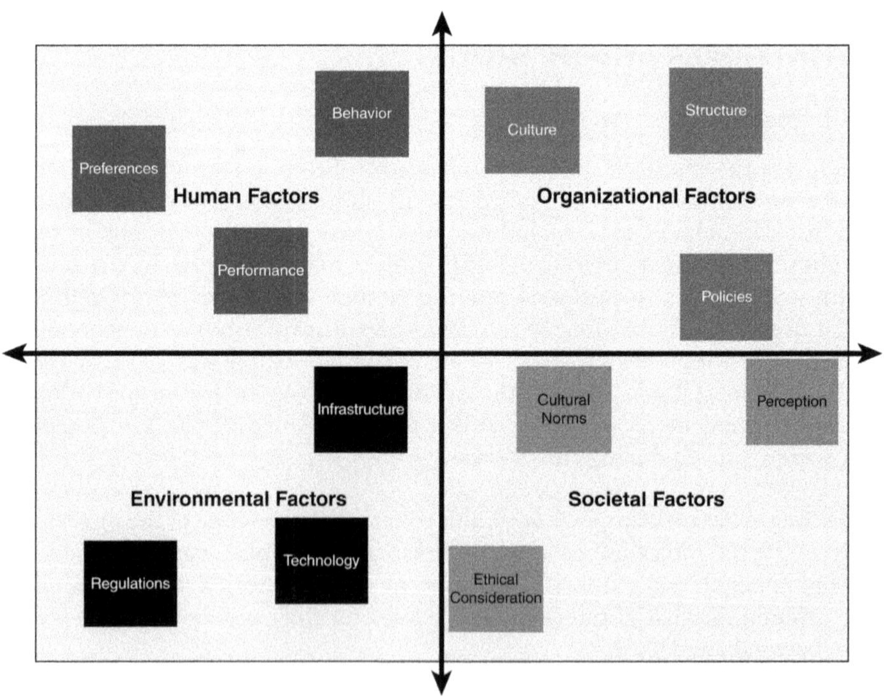

Figure 8.2 Impact of automation scenario in interview for discussion.

8.3.2 Participants

Qualitative research methods typically require the participation of stakeholders who are directly or indirectly impacted by the deployment and use of automation in healthcare. Participants can be grouped into three broad categories:

Users: People who interact directly with an automation system in the clinical setting, such as healthcare professionals, patients, and their families, can be considered to be end-users. As a result, they can provide valuable insight into the way the technology is being used in practice, the experience of the users, and if any issues arise during the process.

People responsible for making decisions: This group is made up of senior healthcare managers, administrators, and policymakers who are responsible for making decisions concerning the deployment and use of automation in healthcare. It is also possible for them to advise on the wider factors that influence the deployment and use of automation, such as funding, regulations, and policies, which can provide insight into the broader organizational and environmental factors involved.

Other stakeholders: This group includes individuals who are impacted by the deployment and use of automation in healthcare, such as technology vendors, researchers, and members of the general public, who stand to be affected by their decisions. The results of these studies can be used to provide insight into the broader societal factors that can affect the deployment and use of automation, such as public perception, cultural norms, and ethical considerations, that influence the deployment and use of automation.

To ensure that a wide range of perspectives is captured, it is important to include participants from each of these groups as part of the research study to ensure a broad range of perspectives are represented. Depending on the research questions being addressed, the scope of the study, and how many participants are available, the number of participants in each group will vary. To achieve saturation of data, the minimum number of participants to be grouped is 10–15, but there may need to be more participants to achieve saturation.

8.3.3 Procedure

In advance of the interview, the participants received a formal letter outlining the study and asking for their consent as well as an electronic copy of the scenario that was to be interviewed. A brief instruction was given to them before the meeting that they should familiarize themselves with the scenario for a few minutes before the meeting started.

The majority of the interviews were conducted via telephone, although some of them were conducted face-to-face when it was convenient for the interviewer. The first step in preparing for the study was to reiterate the study's objectives as well as follow a structured question guide. During the interview process, the interview approach was semi-structured, allowing participants to elaborate on a variety of topics of their choosing, as they deemed appropriate. There was a need for this kind of flexibility because the participants had a diverse range of expertise about different aspects of the automation framework, and they were encouraged to discuss in detail those areas where they felt most comfortable or where they had the most to offer.

The average length of each interview was approximately 45 minutes. Every time we conducted an interview, detailed contemporaneous notes were taken, especially in the case of five of the interviews that took place in public places where it was not feasible to record audio. All other interviews have been recorded and transcribed after being created as audio recordings.

To conduct a formal study, one must keep in mind that the exact steps involved will depend on the specific design, methodology, and research questions of the study. There is, however, a general framework that can be followed to carry out a robust and meaningful study on the human, organizational, and social aspects of robotic healthcare by following these steps.

8.3.4 Discussion

In line with the methodology, the coding and analysis process underwent several iterative phases. Each participant's transcripts and notes were thoroughly examined, and their audio recordings were reviewed.

To facilitate the analysis, a spreadsheet was created, with each row dedicated to a participant's record. This spreadsheet not only included anonymized participant information, such as their background, organization, and gender, but also served as a repository for the content of their respective interviews, which was systematically analyzed and documented.

The results of an analysis explored the human, organizational, and social factors in robotic healthcare. However, it will provide some potential outcomes of such an analysis as shown in Figure 8.3.

Subsequently, these identified themes were shared with the project team members for their input and feedback. Their comments were subsequently transformed into concept maps, categorizing each high-level area, and accompanied by a narrative that expounded on the themes and sub-themes, mirroring the structure of the results section described below.

In order to obtain feedback or clarification, all the participants, including those who participated in the initial workshop, were provided with copies of the report, which was then distributed to all participants for their comments or clarifications. In the process, a few minor comments were raised, which were incorporated into the final descriptions of the topics presented in this chapter.

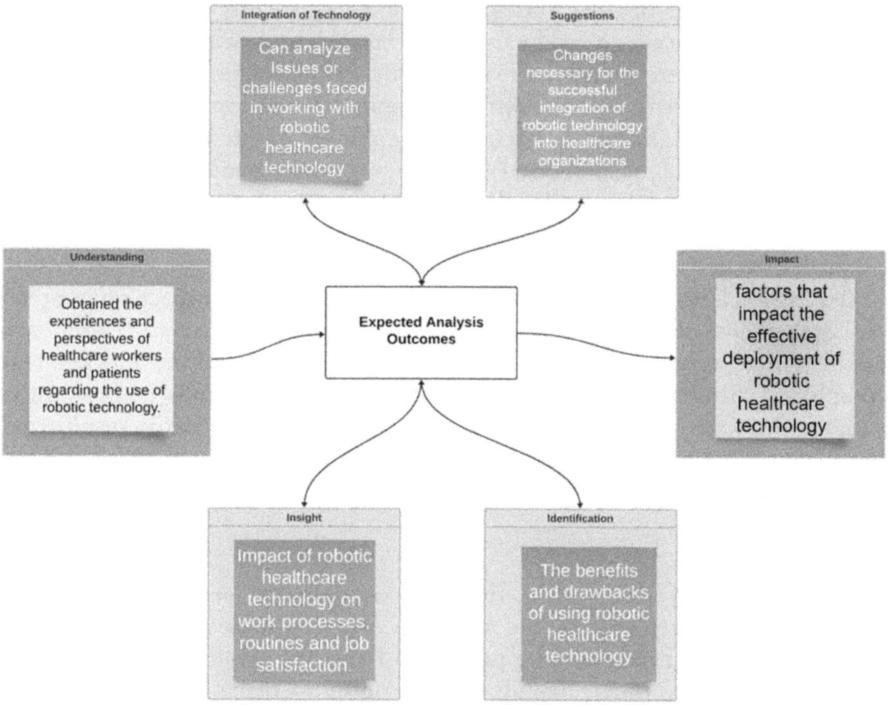

Figure 8.3 Expected analysis outcomes to gain a deep understanding of the human, organizational, and social factors in robotic healthcare.

8.4 RESULTS

8.4.1 Confidence in automation

In terms of users' perspectives, it encompasses both factors that play a role in either promoting or delaying the adoption of robotic maintenance systems for automated systems. Automation must be capable of performing tasks more efficiently and, more importantly, at a faster rate than the existing manual processes, which are the foremost requirements for automation. Furthermore, there are a number of advantages associated with the reduction of physical exertion and exposure to adverse conditions, such as inclement weather and adverse conditions. The benefits and costs associated with these maintenance tasks do not only refer to the actual maintenance tasks themselves but also to aspects like initiating the effort, completing the task, and disengaging the equipment after the job has been completed. Throughout this entire process, it is extremely important to maintain the highest level of safety.

Numerous experts within the healthcare industry expressed the notion that there is a legitimate role for automation. The paramount concern revolves around the safety, health, and sustainability of healthcare operations, in which automation plays a vital part. Nonetheless, there exists a risk that automation might be perceived negatively in terms of employment implications. To mitigate this concern, it is imperative to commit to a vision where automation results in skill enhancement, becoming a valuable addition to one's resume. Automation should be perceived as replacing only monotonous, low-value tasks, thereby allowing personnel to concentrate on more valuable activities. Moreover, there should be clear transparency regarding the redeployment of resources toward meaningful, high-priority work.

One obstacle to widespread acceptance pertains to its impact on contractual agreements and compensation. Workers remunerated on an hourly basis might exhibit reluctance to embrace automation if it leads to quicker task completion. Similarly, staff compensated based on shifts may be hesitant to adopt technology if it means that they finish their work swiftly, resulting in significant idle time, especially during night-time hours.

8.4.2 Safety

Safety consistently emerged as the foremost driving factor for healthcare experts when contemplating the adoption of autonomous maintenance systems. The prospect of removing personnel from hazardous areas around the paths in healthcare units, including challenging-to-reach locations of emergency wards and operation theaters, stood out as a compelling rationale for embracing automation. In this context, the healthcare industry could draw valuable insights from domains such as infectious wards in the healthcare unit, where robots have successfully reduced human exposure to risks.

In contrast to safety serving as merely a motivational factor, ensuring safe interactions within and around robots has emerged as an important factor that should not be overlooked. Various aspects of safety require explicit attention both during the design phase as well as during its deployment, all of which are incorporated into the multifaceted concept of safety:

Ensure that the robot is safe during operation: To prevent any possible harm to individuals in the vicinity of the robot while it is performing its functions.

To assure that the automation operates safely and efficiently: It is important to guarantee that the automated activities contribute to the safe operation of the healthcare systems.

Behavior during unsafe situations: Automation should be able to respond in a way that safeguards both itself and the other people within the environment, even when there is no fault on the part of the automation itself.

Further to what has been discussed, the introduction of novel automated work processes may necessitate the establishment of dedicated safety systems as well as the introduction of novel automated processes. Modern manufacturing systems, for instance, deploy arrangements involving light sensors and tracking mechanisms in order to trigger either hard or soft stops for robotic systems when the presence of humans is detected. While similar setups can be implemented alongside healthcare units, implementing them alongside these units would consume a lot of time and introduce an additional expense to the maintenance operations because it would involve new construction. It was determined that a significant part of the deliberations centered around determining which entity or role would be the most suitable to assume responsibility for these tasks.

8.4.3 Recommendations

The recommendations restricting from an analysis exploring the human, organizational, and social factors in robotic healthcare will depend on the findings of the report. However, some potential recommendations based on common themes include the following:

Providing training and support to healthcare workers to help them effectively use and integrate robotic technology into their work.

Addressing any issues or challenges identified in the use of robotic technology, such as technical problems or ergonomic concerns.

Developing clear protocols and guidelines for the deployment and use of robotic technology in healthcare organizations.

Engaging in ongoing evaluation and monitoring of the use of robotic technology to ensure its continued effectiveness and address any new issues that may arise.

By improving communication and collaboration among healthcare workers, patients, and technology providers, it will be possible to ensure that all stakeholders are aware of and involved in the decision-making process regarding the deployment and use of robotic technology.

As a result of the findings of a study that explored the human, organizational, and social factors associated with robotic healthcare, certain potential recommendations could be made. As a result of these recommendations, we are hoping to support the successful integration of technology and ensure that its use is safe, efficient, and beneficial for all the parties involved.

Healthcare robot's near-term adoption has the following benefits:

Improved patient outcomes: Healthcare robots can provide high-precision and consistent care to patients, which can lead to improved health outcomes.

Increased efficiency: Healthcare robots can automate routine tasks, freeing up healthcare workers to focus on more complex and higher-level tasks.

Enhanced accessibility: Healthcare robots can provide remote care and support to patients in underserved areas, improving access to care.

Reduced costs: Healthcare robots can reduce costs by reducing the need for human labor and reducing the length of hospital stays.

Near-term barriers to healthcare robot adoption include the following:

Technical challenges: There are technical challenges associated with the development and deployment of healthcare robots, including issues related to safety, reliability, and accuracy.
Cost: The cost of developing, producing, and deploying a healthcare robot is often an issue for many organizations, not just because of the high cost to develop, produce, and deploy them.
Regulatory: As a result of the use of robotics in healthcare, there are regulatory challenges associated with using these devices, including concerns about various issues related to data privacy and security.

Despite the benefits of healthcare robots, there may be some resistance to their adoption among healthcare workers, patients, and organizations due to concerns about the impact of healthcare robots on their jobs, safety, and quality of care as well as the stability of the industry.

Healthcare robots should be able to meet the following requirements in the near future:

Adoption: The goal of meeting these requirements is to ensure that the adoption of healthcare robots is safe, effective, and accessible to all stakeholders in the near term.
Ability: The ability to perform a wide range of tasks in a safe, accurate, and reliable manner. Healthcare robots must be highly reliable, accurate, and safe.
Affordable and scalable: Healthcare robots need to be affordable and scalable so that organizations can adopt them without incurring excessive costs or disrupting their existing working processes.
Compliance with regulations: Healthcare robots must comply with existing regulations and standards related to privacy, security, and safety in order to maintain their usefulness and safety.
Design of healthcare robots that are user-centered: Healthcare robots must take into account the needs and preferences of users, including healthcare workers as well as patients.

There are several benefits that may be derived from the adoption of healthcare robots in the mid-term, such as increased efficiency, accuracy and speed, better patient outcomes, a reduction of healthcare staff workload, and a reduction of human errors. A number of barriers may be encountered, including high costs for acquiring and maintaining the robots, a lack of standardization and regulatory framework, technical difficulties related to the integration and compatibility of the robots, privacy concerns, and the need for staff training and retraining. To achieve the mid-term adoption of healthcare robots, it is important to have strong and supportive leadership, a clear vision and strategy, dedicated funding and resources, interdisciplinary collaborations, and principles of ethical and inclusive design. Additionally, it is vital to ensure that the benefits of the technology are available to all populations, regardless of socioeconomic status, for it to be adopted successfully in the medium term.

Healthcare robot adoption can have a number of benefits over the long term, including the following:

Improved patient outcomes: Over time, healthcare robots will continue to improve in their ability to provide patients with high precision and consistent care, which in turn can help improve health outcomes in even greater ways.

A higher level of efficiency: As healthcare robots become more advanced and integrated into present-day healthcare systems, they will be able to automate more tasks, making it easier for healthcare workers to focus on higher-level tasks.

Enhancement of accessibility: In the future, medical robots can become more widespread and more accessible, which will enable patients in more underserved areas to receive remote care and support.

Cost reductions: As the cost of developing, producing, and deploying healthcare robots decreases over time, they will become more affordable for organizations and will lower the overall cost of healthcare as a result.

Healthcare robot adoption has been facing a number of barriers over the past few years, including the following:

Technical challenges: Although technology progresses, the development and deployment of healthcare robots will always face technical challenges, including safety, reliability, and accuracy issues.

The cost of healthcare robots: Even though over time the costs associated with these machines will shrink, they will still serve as a significant barrier to many organizations' adoption of these machines.

Legislation and regulation: As healthcare robots become more integrated into healthcare systems, there will be a plethora of regulatory challenges that will continue to arise, including those regarding the privacy and security of patient data.

Increased resistance to change: Over some time, there may be an increased resistance to the adoption of medical robots in the healthcare industry due to concerns about the impact on jobs, safety, and the quality of care, as well as the impact on workers' wages.

Healthcare robots have to meet the following requirements in order to be adopted on a long-term basis:

Commitment to technical proficiency: Healthcare robots must continue to improve their ability to perform a wide range of tasks with a high level of accuracy, reliability, and safety.
Low cost and scalable: Healthcare robots have to become more affordable and scalable as a means to enable organizations to adopt them on a large scale without incurring excessive costs or disrupting their existing work patterns.
Adherence to regulations: Healthcare robots must continue to adhere to the existing regulations and standards that relate to privacy, security, and safety as well as any new regulations that may arise in the future.
Design that is user-centered: Healthcare robots must continue to be designed with the needs and preferences of users in mind, including healthcare workers and their patients, in order to ensure their continued acceptance and use in the future.

To meet these requirements, we must ensure that the long-term adoption of healthcare robots is safe, effective, and accessible to all stakeholders, therefore providing ongoing benefits to patients, healthcare workers, and the overall healthcare system in the long run.

8.5 DISCUSSION

Ideology: A user-centered automation ideology in healthcare includes a variety of challenges and preferences that have to be met. There is always a dynamic aspect to the perspective of humans, based on the analysis, interaction design, and accessibility, since these aspects are influenced by a number of factors.
Understanding user preferences and needs: In order to develop user-friendly and effective automation systems, healthcare providers and patients must be understood.
Analysis of the tasks that will be carried out by the automation system: An analysis of the tasks and activities that will be performed by the automation system must be conducted to identify potential challenges, design solutions, and ensure that the system will work efficiently.
Human factors engineering: To eliminate hazards, improve usability, and enhance the overall user experience, human factors engineering principles should be incorporated into the design process of any project.
Designing the user's interaction with the system: The user should have an intuitive and easy-to-use interaction with the automation system, with clear instructions and feedback for them.
Safety and reliability: The automation system must be designed in a way that is both safe and reliable as well as having built-in mechanisms that detect, recover, and mitigate errors as they arise.

Data security and privacy: The automation system must be designed in such a way that it provides protection for sensitive information and ensures that data is protected.

Accessibility: An automation system should be designed in such a way that it can be accessed by anyone, regardless of their physical or cognitive abilities, and should also be inclusive of disabled individuals.

Interoperability: The automation system needs to work seamlessly with other systems and devices to ensure seamless integration and compatibility.

Support and maintenance: To ensure easy maintenance and support, the automation system should be designed with clear documentation, training materials, and support processes in place.

A user-centered automation system can be developed by examining all of these important features and addressing the needs of the end-users to enhance performance, reduce the risk of failure, and meet the needs of the end-users.

8.5.1 Limitations of study

The design, methodology, and data collection process of every study have inherent limitations that are reflected in the results of the study. An analysis of the adoption of healthcare robots is likely to have some limitations, among which are the following:

Bias in the sample: The participants in the study may not be representative of the population as a whole, and therefore, the results of the study may not generalize to other groups.

Accuracy and reliability of the data: The quality of the measurement instruments, the ability or veracity of the participants to recall information, or the accuracy and reliability of the data collected during the study may all impact the quality and reliability of the data collected during the study.

Blinded observation: The researchers who conducted the study may have a bias that affects how they interpret the results of the study.

Limited scope: It is possible that the study will only focus on a limited set of factors and might not take into account all of the variables that affect the adoption of robots in healthcare.

Period: There may be a limited amount of time in which the study will be conducted, and as a result, the results may not reflect trends or changes over a longer period.

Considerations regarding ethical considerations: There may be ethical concerns associated with the use of healthcare robots or the possibility of harm that have not been fully addressed in the study due to privacy concerns or the possibility of harm.

Researchers can enhance the validity and reliability of the results by acknowledging and addressing these limitations as well as provide a more comprehensive understanding of the challenges and opportunities associated with the adoption of healthcare robots by acknowledging and addressing these limitations.

8.6 CONCLUSIONS

In conclusion, I would like to emphasize that the integration of robotics into the healthcare sector presents both opportunities and challenges in terms of how care can be delivered in terms of using robotics. A complex interplay of technical, organizational, and societal factors has a significant influence on the development and deployment of healthcare robots. The design and implementation of healthcare robots must take into consideration a number of human factors, such as performance, teamwork, and the robot-human interface, when they are being designed and implemented. It can be concluded from the results of the study that healthcare robots can improve efficiency, accuracy, and patient outcomes, while the barriers to their adoption can be attributed to costs, safety concerns, and organizational resistance to change. For a technology to be successfully adopted, it must meet several requirements, including a user-centered design, an effective integration into clinical workflows, and a set of regulations and standards that are clear and concise.

A healthcare robot is one of the most important and sustainable technological advancements that can be made, where long-term health and sustainability are integral to providing long-term administrative role solutions through the use of healthcare robots. As a result of this, robotics has become an extremely important aspect of the healthcare system as it ensures the safe, efficient, and effective operation of the healthcare system. Developing and deploying robotics in this domain requires a comprehensive approach that takes into account the requirements of users, the collaborative dynamics among human teams working alongside robots, organizational considerations, and societal impacts as well as technological considerations.

The purpose of this chapter is to identify the human and organizational challenges that are inherent to the space of healthcare robotics through a qualitative approach. Several challenges associated with robotics were addressed by the findings of the study, including trust issues, acceptance issues, and the transformation of business processes as a result of the use of robotics. However, the research revealed that several challenges are specific to the healthcare sector as well, which were also revealed in the study. These include the imperative to contemplate the entire lifecycle of maintenance tasks, conceptualizing the teamwork between robots and human counterparts within the framework of rail maintenance sites, navigating the intricate supply chain implications that extend both upstream (robotics suppliers) and downstream (third-party maintenance contractors), and addressing public acceptance of robotics in an environment that frequently interfaces with patients and healthcare employees which is crucial for understanding and improving operations.

A key practical contribution of this research is the formulation of a series of recommendations that encompass functional and socio-technical aspects that aim to improve the acceptance and efficacy of healthcare robotics by enhancing their functional and socio-technical aspects. The scientific benefit lies in the creation of an evidence-based framework, which offers a range of insights into a range of human and organizational factors that can serve as a guide in future research in functional robotics design, human factors within robotics (for example, in robotic human-machine interfaces), as well as organizational change initiatives in the field of robotics.

ACKNOWLEDGMENTS

The authors want to acknowledge the support and facilities provided by Global college of Engineering and Technology, Muscat, Oman.

This work was supported by a grant from Ministry of Higher Education and Research, Innovation (MOHERI), Project number GCET/MOHERI/2021/01.

REFERENCES

[1] Golightly, D., Chan-Pensley, J., Dadashi, N., Jundi, S., Ryan, B., & Hall, A. (2022). Human, organisational and societal factors in robotic rail infrastructure maintenance. Sustainability, 14(4), 2123.

[2] Raje, S., Reddy, N., Jerbi, H., Randhawa, P., Tsaramirsis, G., Shrivas, N. V., ... Piromalis, D. (2021). Applications of healthcare robots in combating the COVID-19 pandemic. Applied Bionics and Biomechanics, 2021, 099510.

[3] Kyrarini, M., Lygerakis, F., Rajavenkatanarayanan, A., Sevastopoulos, C., Nambiappan, H. R., Chaitanya, K. K., ... Makedon, F. (2021). A survey of robots in healthcare. Technologies, 9(1), 8.

[4] Pradhan, B., Bharti, D., Chakravarty, S., Ray, S. S., Voinova, V. V., Bonartsev, A. P., & Pal, K. (2021). Internet of things and robotics in transforming current-day healthcare services. Journal of Healthcare Engineering, 2021, 1–15.

[5] Yang, Y., Li, X., Li, H., Zhang, C., Todo, Y., & Yang, H. (2023). Yet another effective dendritic neuron model based on the activity of excitation and inhibition. Mathematics, 11(7), 1701.

[6] Driewer, F. (2008). Teleoperation Interfaces in Human-Robot Teams. Universität Würzburg.

[7] van der Heijden, M., van Gorp, R., Sadeghi, M. A., Gostick, J., & Forner-Cuenca, A.(2022). Assessing the versatility and robustness of pore network modeling to simulate redox flow battery electrode performance. Journal of The Electrochemical Society, 169(4), 040505.

[8] Wong, K. V., &De Leon, O. (2017). Applications of Nanofluids: Current and Future, in Nanotechnology and Energy. Jenny Stanford Publishing, pp. 1056–132.

[9] Saidur, R., Leong, K. Y., & Mohammed, H. A. (2011).A review on applications and challenges of nanofluids. Renewable and Sustainable Energy Revi ews,15(3),1646–668. Doi:10.1016/j.rser.2010.11.035.

[10] Al Bassam, N., Hussain, S. A., Al Qaraghuli, A., Khan, J., Sumesh, E. P., & Lavanya, V. (2021). IoT based wearable device to monitor the signs of quarantined remote patients of COVID-19. Informatics in Medicine Unlocked, 24, 100588.
[11] Yang, X., Shang, G., & Wang, X. (2022). Biochemical, transcriptomic, gut microbiome responses and defense mechanisms of the earthworm *Eisenia fetida* to salt stress. Ecotoxicology and Environmental Safety, 239, 113684.
[12] Bowling, A., & Ebrahim, S. (2005). Handbook of Health Research Methods: Investigation, Measurement and Analysis. McGraw-Hill Education.
[13] Nieten, T., & Fishwick, P.(2007). "A Simulation-Based Approach to Decision Support for Robot-Human Team Configuration," 2007 International Symposium on Computational Intelligence in Robotics and Automation, Jacksonville, FL, USA, pp. 485–490.Doi: 10.1109/CIRA.2007.382890.
[14] Wery, J., Gaudreault, J., Thomas, A., & Marier, P. (2018). Simulation-optimisation based framework for sales and operations planning taking into account new products opportunities in a co-production context. Computers in Industry, 94, 41–51.
[15] Hussain, S. A., Al Bassam, N., Zayegh, A., & Al Ghawi, S. (2022). Prediction and evaluation of healthy and unhealthy status of COVID-19 patients using wearable device prototype data. MethodsX, 9, 101618.

Chapter 9

MHD Darcy-Forchheimer ternary *Ag-Au-Cu/H_2O* hybrid nanofluid flow over moving thin needle under the influence of Hall current and chemical reaction

Vishal Saikia and Paramananda Deka

9.1 INTRODUCTION

In various industrial and technological sectors, heat generation is a common phenomenon and ensuring its effective transport is a notable priority. Numerous industries have utilized fluids like water, air and oils for this purpose because of their widespread availability and relatively low cost. Nevertheless, deficiencies in thermal conductivity of these fluids make them inadequate to fulfill the requirements of modern industries. Introduction of nanofluids which are made by dispersing nanoparticles in conventional base fluids revolutionized the heat transfer process in various engineering and industrial fields as well as cooling applications such as in nuclear reactors and electronics. The concept of nanofluid was initially established in 1995 by Choi and Eastman (1995) using Hamilton and Crosser (1962) model to copper nanoparticles in water and also utilizing the results of Masuda *et al.*(1993) for $\gamma - Al_2O_3$ particles in water. When Lee *et al.* (1999) investigated the thermal conductivities of different metal oxides in different base fluids, they found that the size and shape of the nanoparticles impact the thermal conductivity of the nanofluid. The core applications and basic models for the study of nanofluids were described by Tiwari and Das (2007) and Buongiorno (2006). Through experimental tests, Suresh *et al.*(2011) proposed the concept of a hybrid nanofluid. Devi and Devi (2016) found that their theoretical findings mirrored those study of Suresh *et al.* (2011), proving the accuracy of their numerical model of hybrid nanofluid. In their analysis of the benefits and drawbacks of conventional and hybrid nanofluids, Sarkar *et al.* (2015) suggested that hybrid nanofluids offer several advantages over traditional nanofluids due to their favorable aspect ratio, pressure drop capability and favorable enhancement of heat transfers. Rao and Deka (2023) studied a H_2O-based hybrid nanofluid with multi-walled carbon nanotube ($MWCNT$) and alumina as nanoparticles considering the influences of chemical reactions and thermal radiation. Alharbi *et al.* (2022) examined the flow characteristics of nanofluid comprising of TiO_2, MgO and $CoFe_2O_4$ nanoparticles over extending cylinder.

Flow over thin needle has many uses in instrumentation and electronic devices. Lee (1967) investigated the flow around a slender needle

DOI: 10.1201/9781003527251-9

submerged in an incompressible Newtonian fluid. Grosan and Pop (2011) investigated the impact of varying solid volume fractions of Cu and Al_2O_3 nanoparticles in a H_2O-based hybrid nanofluid on boundary layer flow dynamics. Nazar *et al.* (2021) investigated dual solutions and carried out a stability analysis of hybrid nanofluid influenced by magnetic field over a moving needle. Ahmad *et al.* (2017) numerically investigated Buongiorno's model for flow of nanofluid over moving needle. Souayeh *et al.* (2019) performed a comparative assessment of Casson nanofluid flow over moving and fixed thin needle.

Due to their wide range of uses, the study of chemical reactions has substantially gain importance in industrial areas. Chemical reactions take place within the flow system when an external mass is introduced into the fluid. It can be categorized as homogeneous and heterogeneous. Ramzan *et al.* (2020) discussed flow over moving needle considering both homogeneous and heterogeneous reactions. Rao and Deka (2022) discussed the effects of chemical reaction in their investigation of magnetohydrodynamics (MHD) Casson nanofluid flow over porous stretching sheet using explicit finite difference method.

When a porous media is subjected to high flow rates of fluid, the Forchheimer term appears in governing equation as it incorporates the opposing inertial force. Darcy-Forchheimer flow of nanofluid comprising of SWCNT nanoparticle over unsteady expanding sheet was analyzed by Nasir *et al.* (2019). Huda *et al.*(2020) examined how the Cattaneo-Christov model affected the Darcy-Forchheimer flow through a moving needle using ethylene glycol as base fluid. For higher values of the Forchheimer number, we see a drop in the velocity profile in all these studies.

In the current study, we will investigate a H_2O-based ternary hybrid nanofluid comprising of Ag, Au and Cu as nanoparticles. We will consider a Darcy-Forchheimer porous medium and discuss the impact of Hall current, chemical reaction, needle size and some additional parameters on the flow. Viscous dissipation and Lorentz force will also be taken into consideration.

9.2 PHYSICAL DESCRIPTION AND MATHEMATICAL FORMULATION

Figure 9.1 demonstrates a horizontal thin needle configuration which is subjected to a flow of ternary hybrid nanofluid over its boundary. We consider a porous medium and modify the momentum equation to incorporate the Forchheimer term. The geometry of the problem is demonstrated in cylindrical coordinates (x, r) along axial and radial directions with velocity components (u, v). The ternary hybrid nanofluid used is water based and comprises of silver (Ag), gold (Au) and copper (Cu) as nanoparticles. Its thermophysical properties are provided in Table 9.1. The needle is assumed to move at uniform velocity u_w in the same or opposing direction of the ternary hybrid nanofluid

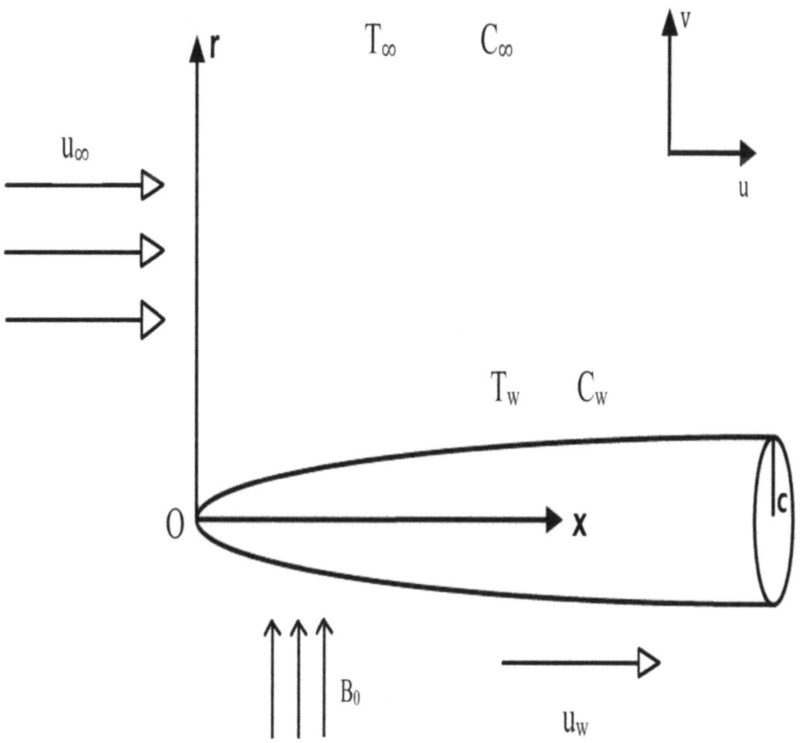

Figure 9.1 Geometric sketch of flow model.

which is flowing at constant velocity u_∞. Here, $r = R(x) = \left(\nu_f cx / U\right)^{1/2}$ represents needle's radius, where $U = u_w + u_\infty$ is the composite velocity. In the radial direction, a magnetic field of strength B_0 is applied. T_w and C_w represent the constant wall temperature and concentration at needle's surface and T_∞ and C_∞ at free stream region. The needle's thickness is assumed minimal so as not to exceed the boundary layer formed over it.

Table 9.1 Thermophysical properties of constituents

	$\rho\ (kg/m^3)$	$\kappa\ (W/mK)$	$\sigma\ (s/m)$	$C_p\ (J/kg\ K)$
H_2O	997.1	0.613	5.5×10^{-6}	4179
Ag	10500	429	6.30×10^{7}	235
Au	19300	318	4.52×10^{7}	129
Cu	8933	401	59.6×10^{6}	385

Source: Asghar *et al.* (2022) and *Rashid et al.* (2021).

Considering the aforementioned assumptions, the fluid flow problem is described by the following governing equations (Huda *et al.*, 2020; Khan *et al.*, 2021)

$$\frac{\partial}{\partial x}(ru)+\frac{\partial}{\partial r}(rv)=0, \tag{9.1}$$

$$u\frac{\partial u}{\partial x}+v\frac{\partial u}{\partial r}=\frac{\mu_{thnf}}{\rho_{thnf}}\frac{1}{r}\frac{\partial}{\partial r}\left(r\frac{\partial u}{\partial r}\right)-\frac{\mu_{thnf}}{\rho_{thnf}}\frac{u}{k}-\frac{C_b}{\sqrt{k}}u^2-\frac{\sigma_{thnf}B_0^2}{\rho_{thnf}\left(1+m^2\right)}u, \tag{9.2}$$

$$u\frac{\partial T}{\partial x}+v\frac{\partial T}{\partial r}=\frac{\kappa_{thnf}}{(\rho C_p)_{thnf}}\frac{1}{r}\frac{\partial}{\partial r}\left(r\frac{\partial T}{\partial r}\right)+\frac{\mu_{thnf}}{(\rho C_p)_{thnf}}\left(\frac{\partial u}{\partial r}\right)^2$$

$$-\frac{1}{\left(\rho C_p\right)_{thnf}}\frac{\partial q_r}{\partial r}+\frac{\sigma_{thnf}B_0^2}{(\rho C_p)_{thnf}}\frac{u^2}{\left(1+m^2\right)}+\tau\left(D_B\frac{\partial T}{\partial r}\frac{\partial C}{\partial r}+\frac{D_T}{T_\infty}\left(\frac{\partial T}{\partial r}\right)^2\right),\ \textit{and} \tag{9.3}$$

$$u\frac{\partial C}{\partial x}+v\frac{\partial C}{\partial r}=\frac{D_T}{T_\infty}\frac{1}{r}\frac{\partial}{\partial r}\left(r\frac{\partial T}{\partial r}\right)+D_B\frac{1}{r}\frac{\partial}{\partial r}\left(r\frac{\partial C}{\partial r}\right)-K_1\left(C-C_\infty\right). \tag{9.4}$$

The subjected boundary conditions are:

$$u(x,r)=u_w,\ v(x,r)=0,\ T(x,r)=.T_w,\ C(x,r)=C_w\ \ at\ r=R(x),$$

$$u(x,r)\to u_\infty\ ,\ T(x,r)\to T_\infty,\ C(x,r)\to C_\infty\ \ at\ r\to\infty. \tag{9.5}$$

As per the Rosseland approximation, the radiative heat flux term, denoted by q_r, can be represented as follows (Sparrow and Cess, 1978; Brewster, 1992; Raptis, 1998):

$$q_r=-\frac{4\sigma^*}{3k^*}\frac{\partial T^4}{\partial r}. \tag{9.6}$$

Expanding the Taylor series while ignoring higher order terms, we obtain:

$$T^4=4T_\infty^3T-3T_\infty^4.$$

Thus, Eq. (9.6) becomes

$$\frac{\partial q_r}{\partial r} = -\frac{16\sigma^*}{3k^*} T_\infty^3 \frac{\partial^2 T}{\partial r^2}. \tag{9.7}$$

Introducing following similarity transformation (Khan *et al.*, 2021)

$$\eta = \frac{Ur^2}{\nu_f x}, 0, \psi = \nu_f x f(\eta), \theta(\eta) = \frac{T - T_\infty}{T_w - T_\infty}, \phi(\eta) = \frac{C - C_\infty}{C_w - C_\infty}, \tag{9.8}$$

where ψ is the stream function such that $u = \frac{1}{r}\frac{\partial \psi}{\partial r}$, $v = -\frac{1}{r}\frac{\partial \psi}{\partial x}$ and η is the similarity variable.

Substituting Eq. (9.8) in Eqs. (9.1–9.4), we have

$$2A_1\eta f''' + 2A_1 f'' + A_2 ff'' - A_2\ Fr(f')^2 - \left(A_1\lambda + A_5\frac{M}{1+m^2}\right)f' = 0, \tag{9.9}$$

$$\begin{aligned}&(2A_4 + Rd)\theta' + 2\eta(A_4 + Rd)\theta'' + A_3 Prf\theta' + 8A_1\eta PrEc(f'')^2 \\ &+ \frac{4A_5 PrEcM}{1+m^2}(f')^2 + 2A_3 PrNb\eta\theta'\phi' + 2A_3 PrNt\eta(\theta')^2 = 0,\end{aligned} \tag{9.10}$$

$$2\phi' + 2\eta\phi'' + 2\frac{\mathrm{Nt}}{\mathrm{Nb}}\theta' + 2\frac{\mathrm{Nt}}{\mathrm{Nb}}\eta\theta'' + Lef\phi' - \frac{1}{2}LeK_c\phi = 0. \tag{9.11}$$

The boundary conditions are reduced to

$$\begin{aligned}&f(\eta) = \frac{\epsilon}{2}c,\ f'(\eta) = \frac{\epsilon}{2},\ \theta(\eta) = 1,\ \phi(\eta) = 1\ at\ \eta = c, \\ &f'(\infty) \to \frac{1-\epsilon}{2},\ \theta(\infty) \to 0,\ \phi(\infty) \to 0.\end{aligned} \tag{9.12}$$

Here,

$$A_1 = \frac{\mu_{thnf}}{\mu_{bf}},\ A_2 = \frac{\rho_{thnf}}{\rho_{bf}},\ A_3 = \frac{(\rho C_p)_{thnf}}{(\rho C_p)_{bf}},\ A_4 = \frac{\kappa_{thnf}}{\kappa_{bf}},\ A_5 = \frac{\sigma_{thnf}}{\sigma_{bf}}.$$

Table 9.2 Thermophysical properties of ternary hybrid nanofluid

Density	$\rho_{thnf} = (1-\phi_3)\left\{(1-\phi_2)\left[(1-\phi_1)\rho_{bf} + \phi_1\rho_{p1}\right] + \phi_2\rho_{p2}\right\} + \phi_3\rho_{p3}$
Dynamic viscosity	$\mu_{thnf} = \dfrac{\mu_{bf}}{\left\{(1-\phi_1)(1-\phi_2)(1-\phi_3)\right\}^{2.5}}$
Heat capacity	$(\rho C)_{thnf} = (1-\phi_3)\left\{(1-\phi_2)\left[(1-\phi_1)(\rho C_p)_{bf} + \phi_1(\rho C_p)_{p1}\right] + \phi_2(\rho C_p)_{p2}\right\} + \phi_3(\rho C_p)_{p3}$
Thermal conductivity	$\kappa_{thnf} = \dfrac{\kappa_{p3} + 2\kappa_{hnf} - 2\phi_3(\kappa_{hnf} - \kappa_{p3})}{\kappa_{p3} + 2\kappa_{hnf} + \phi_3(\kappa_{hnf} - \kappa_{p3})} \times \kappa_{hnf}$ where $\kappa_{hnf} = \dfrac{\kappa_{p2} + 2\kappa_{nf} - 2\phi_2(\kappa_{nf} - \kappa_{p2})}{\kappa_{p2} + 2\kappa_{nf} + \phi_2(\kappa_{nf} - \kappa_{p2})} \times \kappa_{nf}$ where $\kappa_{nf} = \dfrac{\kappa_{p1} + 2\kappa_{bf} - 2\phi_1(\kappa_{bf} - \kappa_{p1})}{\kappa_{p1} + 2\kappa_{bf} + \phi_1(\kappa_{bf} - \kappa_{p1})} \times \kappa_{bf}$
Electrical conductivity	$\sigma_{thnf} = \dfrac{\rho_{p3}(1+2\phi_3) + 2\sigma_{hnf}(1-\phi_3)}{\rho_{p3}(1-\phi_3) + \sigma_{hnf}(2+\phi_3)} \times \sigma_{hnf}$ where $\sigma_{hnf} = \dfrac{\rho_{p2}(1+2\phi_2) + 2\sigma_{nf}(1-\phi_2)}{\rho_{p2}(1-\phi_2) + \sigma_{nf}(2+\phi_2)} \times \sigma_{nf}$ Where $\sigma_{nf} = \dfrac{\rho_{p1}(1+2\phi_1) + 2\sigma_{bf}(1-\phi_1)}{\rho_{p1}(1-\phi_1) + \sigma_{bf}(2+\phi_1)} \times \sigma_{bf}$

Source: Souayeh and Ramesh (2023).

It is to be noted that $\epsilon = \frac{u_w}{u_w + u_\infty}$. When $\epsilon = 0$, then the ternary hybrid nanofluid is moving while needle is stationary; when $\epsilon = 1$, then the nanofluid is static while needle is stationary; when $0 < \epsilon < 1$, both the nanofluid and needle are moving in same direction.

Table 9.2 provides information about the ternary hybrid nanofluid properties, and Table 9.3 provides information about emerging parameters in Eqs. (9.9–9.11).

Skin friction coefficient (C_f), Nusselt number (Nu_x) and Sherwood number (Sh_x) are observed in this study. Its dimensional form is

$$C_f = \frac{\mu_{thnf}}{\rho_{bf} u_w^2}\left(\frac{\partial u}{\partial r}\right)_{r=R}, \quad Nu_x = -\frac{x\kappa_{thnf}}{\kappa_{bf}(T_w - T_\infty)}\left(\frac{\partial T}{\partial r}\right)_{r=R}$$

Table 9.3 Parameters information

Fr	$\frac{C_b}{\sqrt{k}}$	Forchheimer number
λ	$\frac{\nu_f}{2Uk}$	Porosity parameter
M	$\frac{\sigma_f B_0^2}{2U\rho_f}$	Magnetic parameter
Rd	$\frac{16\,\sigma^* T_\infty^3}{3\kappa k^*}$	Radiation parameter
K_c	$\frac{K_1 x}{U}$	Chemical reaction parameter
Nt	$\frac{\tau D_T (T_W - T_\infty)}{T_\infty \nu_f}$	Thermophoretic parameter
Nb	$\frac{\tau D_B (C_W - C_\infty)}{\nu_f}$	Brownian motion parameter
Le	$\frac{\nu_f}{D_B}$	Lewis number
Pr	$\frac{\nu_f (\rho C_P)}{\kappa_f}$	Prandtl number
Ec	$\frac{U^2}{(T_W - T_\infty) C_P}$	Eckert number

$$and\ Sh_x = -\frac{x}{(C_w - C_\infty)}\left(\frac{\partial C}{\partial r}\right)_{r=R}. \tag{9.13}$$

Using Eq. (8) in Eq. (13), we obtain the dimensionless form

$$C_f Re_x^{1/2} = 4c^{1/2}\frac{1}{\{(1-\phi_1)(1-\phi_2)(1-\phi_3)\}^{2.5}} f''(c),$$

$$Nu_x Re_x^{-1/2} = -2\frac{\kappa_{thnf}}{\kappa_{bf}} c^{1/2}\theta'(c)\ and\ Sh_x Re_x^{-\frac{1}{2}} = -2c^{1/2}\phi'(c), \tag{9.14}$$

where $Re_x = \frac{Ux}{\nu_{bf}}$ depicts local Reynolds number.

9.3 NUMERICAL PROCEDURE

To solve Eqs. (9.9–9.12), we use Keller box method. This approach provides a number of advantageous features, including second-order accuracy and simple programming for the solution of large number of coupled equations. Steps associated with this technique are as follows.

The set of nonlinear ordinary differential equations Eqs. (9.9–9.11), along with boundary conditions Eq. (9.12), undergo a transformation into a system of first-order equations. The transformed equations are subsequently translated into finite difference equations using central difference. To facilitate this process, we introduce new dependent variables $p(\eta), q(\eta), t(\eta), and\ v(\eta)$ as

$$f' = p,\ p' = q,\ \theta' = t,\ \phi' = v. \tag{9.15}$$

Eqs. (9.9–9.11) can be rewritten as

$$2A_1\eta q' + 2A_1 q - A_2 Frp^2 - \left(A_1\lambda + A_5\frac{M}{1+m^2} \right)p = 0, \tag{9.16}$$

$$(2A_4 + Rd)t + 2\eta(A_4 + Rd)t' + A_3\Pr(ft) + 8A_1\eta PrEcq^2$$
$$+4A_5PrEc\frac{M}{1+m^2}p^2 + 2\eta A_3PrNb(tv) + 2A_3Pr\eta Nt(t^2) = 0, \tag{9.17}$$

$$2v + 2\eta v' + 2\frac{Nt}{Nb}t + 2\eta\frac{Nt}{Nb}t' + Le(fv) - \frac{1}{2}LeK_c\phi = 0. \tag{9.18}$$

The boundary conditions given in Eq. (9.12) are described by the newly dependent variables as

$$f(c) = \frac{\epsilon}{2}c,\ p(c) = \frac{\epsilon}{2},\ \theta(c) = 1,\ \phi(c) = 1,\ p(\infty) \to \frac{1-\epsilon}{2},\ \theta(\infty) \to 0,\ \phi(\infty) \to 0. \tag{9.19}$$

Next, we consider a segment $\left[\eta_{j-1}, \eta_j\right]$ which is defined as

$$\eta_0 = c,\ \ \eta_j = \eta_{j-1} + h_j,\ \eta_J = \eta_\infty,$$

where $\Delta\eta$ spacing is denoted by h_j and $\Delta\eta_j$, j = 1, 2, ..., J denotes coordinate location.

Using central difference derivatives, the finite difference approximations Eqs. (9.15–9.18) written for mid-point $\eta_{j-\frac{1}{2}}$ are

$$f_j - f_{j-1} - \frac{h_j}{2}(p_j + p_{j-1}) = 0, \tag{9.20}$$

$$p_j - p_{j-1} - \frac{h_j}{2}\left(q_j + q_{j-1}\right) = 0, \tag{9.21}$$

$$\theta_j - \theta_{j-1} - \frac{h_j}{2}\left(t_j + t_{j-1}\right) = 0, \tag{9.22}$$

$$\phi_j - \phi_{j-1} - \frac{h_j}{2}\left(v_j + v_{j-1}\right) = 0, \tag{9.23}$$

$$\begin{aligned} &2A_1\eta\left(q_j - q_{j-1}\right) + 2A_1h_j\left(\frac{q_j + q_{j-1}}{2}\right) + A_2h_j\left(\frac{f_j + f_{j-1}}{2}\right)\left(\frac{q_j + q_{j-1}}{2}\right) \\ &- A_2Frh_j\left(\frac{p_j + p_{j-1}}{2}\right)^2 - \left(A_1\lambda + A_5\frac{M}{1+m^2}\right)h_j\left(\frac{p_j + p_{j-1}}{2}\right) = 0, \end{aligned} \tag{9.24}$$

$$\begin{aligned} &\left(2A_4 + Rd\right)h_j\left(\frac{t_j + t_{j-1}}{2}\right) + 2\eta\left(A_4 + Rd\right)\left(t_j - t_{j-1}\right) \\ &+ A_3Prh_j\left(\frac{f_j + f_{j-1}}{2}\right)\left(\frac{t_j + t_{j-1}}{2}\right) + 8A_1\eta PrEch_j\left(\frac{q_j + q_{j-1}}{2}\right)^2 \\ &+ 4A_5\,\mathrm{Pr}\,Ech_j\frac{M}{1+m^2}\left(\frac{p_j + p_{j-1}}{2}\right)^2 + 2\eta A_3PrNbh_j\left(\frac{t_j + t_{j-1}}{2}\right)\left(\frac{v_j + v_{j-1}}{2}\right) \\ &+ 2A_3Pr\eta h_jNt\left(\frac{t_j + t_{j-1}}{2}\right)^2 = 0, \end{aligned} \tag{9.25}$$

$$\begin{aligned} &2h_j\left(\frac{v_j + v_{j-1}}{2}\right) + 2\eta\left(v_j - v_{j-1}\right) + 2\frac{Nt}{Nb}h_j\left(\frac{t_j + t_{j-1}}{2}\right) + 2\frac{Nt}{Nb}\eta\left(t_j - t_{j-1}\right) \\ &+ Leh_j\left(\frac{f_j + f_{j-1}}{2}\right)\left(\frac{v_j + v_{j-1}}{2}\right) - \frac{1}{2}LeK_ch_j\left(\frac{\phi_j + \phi_{j-1}}{2}\right) = 0. \end{aligned} \tag{9.26}$$

Here, η_J is sizable enough so that it extends beyond boundary layer. Boundary conditions given in Eq. (9.19) can be rewritten as

$$f_0 = \frac{\epsilon}{2}c,\ p_0 = \frac{\epsilon}{2},\ \theta_0 = 1,\ \phi_0 = 1,\ p_J = \frac{1-\epsilon}{2},\ \theta_J = 0,\ \phi_J = 0. \tag{9.27}$$

Now we use the Newton's method and utilize following iterations (Ramya *et al.*, 2018):

$$(f)_j^{(k+1)} = (f)_j^{(k)} + \delta(f)_j^{(k)},$$

where $k = 0,1,2, \ldots$.

In order to simplify the equations, the variable k has been omitted from Eqs. (9.20–9.26), which can be written in linear form as

$$\delta f_j - \delta f_{j-1} - \frac{h_j}{2}\left(\delta p_j + \delta p_{j-1}\right) = \left(r_1\right)_{j-\frac{1}{2}}, \tag{9.28}$$

$$\delta p_j - \delta p_{j-1} - \frac{h_j}{2}\left(\delta q_j + \delta q_{j-1}\right) = \left(r_2\right)_{j-\frac{1}{2}}, \tag{9.29}$$

$$\delta \theta_j - \delta \theta_{j-1} - \frac{h_j}{2}\left(\delta t_j + \delta t_{j-1}\right) = \left(r_3\right)_{j-\frac{1}{2}}, \tag{9.30}$$

$$\delta \phi_j - \delta \phi_{j-1} - \frac{h_j}{2}\left(\delta v_j + \delta v_{j-1}\right) = \left(r_4\right)_{j-\frac{1}{2}}, \tag{9.31}$$

$$\begin{aligned}&\left(a_1\right)\delta q_j + \left(a_2\right)\delta q_{j-1} + \left(a_3\right)\delta f_j + \left(a_4\right)\delta f_{j-1} + \left(a_5\right)\delta p_j \\ &\quad + \left(a_2\right)\delta p_{j-1} = \left(r_5\right)_{j-\frac{1}{2}},\end{aligned} \tag{9.32}$$

$$\begin{aligned}&\left(b_1\right)\delta t_j + \left(b_2\right)\delta t_{j-1} + \left(b_3\right)\delta f_j + \left(b_4\right)\delta f_{j-1} + \left(b_5\right)\delta q_j + \left(b_6\right)\delta q_{j-1} \\ &\quad + \left(b_7\right)\delta p_j + \left(b_8\right)\delta p_{j-1} + \left(b_9\right)\delta v_j + \left(b_2\right)\delta t_{j-1} = \left(r_6\right)_{j-\frac{1}{2}},\end{aligned} \tag{9.33}$$

$$\begin{aligned}&\left(c_1\right)\delta v_j + \left(c_2\right)\delta v_{j-1} + \left(c_3\right)\delta t_j + \left(c_4\right)\delta t_{j-1} + \left(c_5\right)\delta f_j \\ &\quad + \left(c_6\right)\delta f_{j-1} + \left(c_7\right)\delta \phi_j + \left(c_8\right)\delta \phi_{j-1} = \left(r_7\right)_{j-\frac{1}{2}}.\end{aligned} \tag{9.34}$$

Here,

$$(a_1)_j = 2A_1\eta + A_1 h_j + \left(\frac{A_2}{2}\right)h_j f_{j-1/2},\ (a_2)_j = -2A_1\eta + A_1 h_j + \left(\frac{A_2}{2}\right)h_j f_{j-1/2},$$

$$(a_3)_j = \left(\frac{A_2}{2}\right)h_j q_{j-1/2},\ (a_4)_j = (a_3)_j,$$

$$(a_5)_j = -\left(\frac{A_2}{2}\right) Frh_j p_{j-1/2} - \left(A_1\lambda + A_5\frac{M}{1+m^2}\right)\frac{h_j}{2},\ (a_6)_j = (a_5)_j,$$

$$(b_1)_j = (2A_4 + Rd)\frac{h_j}{2} + 2\eta(A_4 + Rd) + A_3 Pr\frac{h_j}{2} f_{j-1/2} + A_3 PrNbh_j v_{j-1/2}$$
$$+ A_3 PrNt\eta h_j t_{j-1/2},$$

$$(b_2)_j = (2A_4 + Rd)\frac{h_j}{2} - 2\eta(A_4 + Rd) + A_3 Pr\frac{h_j}{2} f_{j-1/2} + A_3 PrNbh_j v_{j-1/2}$$
$$+ A_3 PrNt\eta h_j t_{j-1/2},$$

$$(b_3)_j = A_3 Pr\frac{h_j}{2} t_{j-1/2},\ (b_4)_j = (b_3)_j,\ (b_5)_j = 4A_1\eta PrEch_j q_{j-1/2},\ (b_6)_j = (b_5)_j,$$

$$(b_7)_j = 2A_5 PrEch_j\frac{M}{1+m^2} p_{j-1/2},\ (b_8)_j = (b_7)_j,\ (b_9)_j$$
$$= A_3 PrNb\eta h_j t_{j-1/2},\ (b_{10})_j = (b_9)_j,$$

$$(c_1)_j = h_j + 2\eta + Le\frac{h_j}{2} f_{j-1/2},\ (c_2)_j = h_j - 2\eta + Le\frac{h_j}{2} f_{j-1/2},$$

$$(c_3)_j = \frac{Nt}{Nb}h_j + 2\frac{Nt}{Nb}\eta,\ (c_4)_j = \frac{Nt}{Nb}h_j - 2\frac{Nt}{Nb}\eta,\ (c_5)_j = Le\frac{h_j}{2}v_{j-\frac{1}{2}},\ (c_6)_j = (c_5)_j,$$

$$(c_7)_j = -\frac{1}{4}LeK_c h_j,\ (c_8)_j = (c_7)_j,$$

$$(r_1) = f_{j-1} - f_j + h_j p_{j-1/2},\ (r_2) = p_{j-1} - p_j + h_j q_{j-1/2},$$

$$(r_3) = \theta_{j-1} - \theta_j + h_j t_{j-1/2},\ (r_4) = \phi_{j-1} - \phi_j + h_j v_{j-1/2},$$

$$(r_5) = 2A_1\eta(q_{j-1} - q_j) - 2A_1 h_j q_{j-1/2} - A_2 h_j (fq)_{j-1/2} + A_2 Frh_j p^2_{j-1/2}$$
$$+\left(A_1\lambda + A_5\frac{M}{1+m^2}\right)h_j p_{j-1/2},$$

$$(r_6) = -(2A_4 + Rd)h_j t_{j-\frac{1}{2}} + 2\eta(A_4 + Rd)(t_{j-1} - t_j)$$

$$- A_3 Prh_j (ft)_{j-\frac{1}{2}} - 8A_1\eta PrEch_j q^2_{j-\frac{1}{2}} - 4A_5 PrEch_j \frac{M}{1+m^2} p^2_{j-\frac{1}{2}}$$

$$- 2A_3 PrNb\eta h_j (tv)_{j-\frac{1}{2}} - 2A_3 PrNt\eta h_j t^2_{j-\frac{1}{2}},$$

$$(r_7) = -2h_j v_{j-\frac{1}{2}} + 2\eta(v_{j-1} - v_j) - 2\frac{Nt}{Nb} h_j t_{j-\frac{1}{2}} + 2\frac{Nt}{Nb}\eta(t_{j-1} - t_j)$$

$$- Leh_j (fv)_{j-\frac{1}{2}} + \frac{1}{2} LeK_c h_j \phi_{j-\frac{1}{2}}.$$

Eq. (9.27) becomes

$$\delta f_0 = 0,\ \delta p_0 = 0,\ \delta\theta_0 = 0,\ \delta\phi_0 = 0,\ \delta p_J = 0,\ \delta\theta_J = 0,\ \delta\phi_J = 0. \tag{9.35}$$

The linearized equations Eqs. (9.28–9.34) are solved by block elimination method which is provided in Cebeci and Bradshaw's (1988) work. Eqs. (9.28–9.34) can be expressed in matrix form as

$$A\boldsymbol{\delta} = \boldsymbol{r}, \tag{9.36}$$

where

$$A = \begin{bmatrix} [A_1] & [C_1] & & & & \\ [B_2] & [A_2] & [C_2] & & & \\ & & \ddots & & & \\ & & & \ddots & & \\ & & & & \ddots & \\ & & & [B_{j-1}] & [A_{j-1}] & [C_{j-1}] \\ & & & & [B_j] & [A_j] \end{bmatrix}, \quad \boldsymbol{\delta} = \begin{bmatrix} [\delta_1] \\ [\delta_2] \\ \vdots \\ \vdots \\ \vdots \\ [\delta_{j-1}] \\ [\delta_j] \end{bmatrix}, \quad \boldsymbol{r} = \begin{bmatrix} [r_1] \\ [r_2] \\ \vdots \\ \vdots \\ \vdots \\ [r_{j-1}] \\ [r_j] \end{bmatrix}.$$

Here, considering $E_j = \dfrac{-h_j}{2}$,

$$[A_1] = \begin{bmatrix} 0 & 0 & 0 & 1 & 0 & 0 & 0 \\ E_1 & 0 & 0 & 0 & E_1 & 0 & 0 \\ 0 & E_1 & 0 & 0 & 0 & E_1 & 0 \\ 0 & 0 & E_1 & 0 & 0 & 0 & E_1 \\ (a_2)_1 & 0 & 0 & (a_3)_1 & (a_1)_1 & 0 & 0 \\ (b_6)_1 & (b_2)_1 & (b_{10})_1 & (b_3)_1 & (b_5)_1 & (b_1)_1 & (b_9)_1 \\ 0 & (c_4)_1 & (c_2)_1 & (c_5)_1 & 0 & (c_3)_1 & (c_1)_1 \end{bmatrix}.$$

For $2 \le j \le J$,

$$[A_j] = \begin{bmatrix} E_j & 0 & 0 & 1 & 0 & 0 & 0 \\ -1 & 0 & 0 & 0 & E_j & 0 & 0 \\ 0 & -1 & 0 & 0 & 0 & E_j & 0 \\ 0 & 0 & -1 & 0 & 0 & 0 & E_j \\ (a_6)_j & 0 & 0 & (a_3)_j & (a_1)_j & 0 & 0 \\ (b_8)_j & 0 & 0 & (b_3)_j & (b_5)_j & (b_1)_j & (b_9)_j \\ 0 & 0 & (c_8)_j & (c_5)_j & 0 & (c_3)_j & (c_1)_j \end{bmatrix},$$

$$[B_j] = \begin{bmatrix} 0 & 0 & 0 & -1 & 0 & 0 & 0 \\ 0 & 0 & 0 & 0 & E_j & 0 & 0 \\ 0 & 0 & 0 & 0 & 0 & E_j & 0 \\ 0 & 0 & 0 & 0 & 0 & 0 & E_j \\ 0 & 0 & 0 & (a_4)_j & (a_2)_j & 0 & 0 \\ 0 & 0 & 0 & (b_4)_j & (b_6)_j & (b_2)_j & (b_{10})_j \\ 0 & 0 & 0 & (c_6)_j & 0 & (c_4)_j & (c_2)_j \end{bmatrix}.$$

$$For\ 1 \le j \le J-1, [C_j] = \begin{bmatrix} E_j & 0 & 0 & 0 & 0 & 0 & 0 \\ 1 & 0 & 0 & 0 & 0 & 0 & 0 \\ 0 & 1 & 0 & 0 & 0 & 0 & 0 \\ 0 & 0 & 1 & 0 & 0 & 0 & 0 \\ (a_5)_j & 0 & 0 & 0 & 0 & 0 & 0 \\ (b_7)_j & 0 & 0 & 0 & 0 & 0 & 0 \\ 0 & 0 & (c_7)_j & 0 & 0 & 0 & 0 \end{bmatrix},$$

$$[\delta_1] = \begin{bmatrix} \delta q_0 \\ \delta \theta_0 \\ \delta \phi_0 \\ \delta f_1 \\ \delta q_1 \\ \delta t_1 \\ \delta v_1 \end{bmatrix}, for\ 2 \le j \le J\ [\delta_j] = \begin{bmatrix} \delta q_{j-1} \\ \delta \theta_{j-1} \\ \delta \phi_{j-1} \\ \delta f_j \\ \delta q_j \\ \delta t_j \\ \delta v_j \end{bmatrix}, for\ 1 \le j \le J\ [r_j] = \begin{bmatrix} (r_1)_{j-\frac{1}{2}} \\ (r_2)_{j-\frac{1}{2}} \\ (r_3)_{j-\frac{1}{2}} \\ (r_4)_{j-\frac{1}{2}} \\ (r_5)_{j-\frac{1}{2}} \\ (r_6)_{j-\frac{1}{2}} \\ (r_7)_{j-\frac{1}{2}} \end{bmatrix}.$$

The block-tridiagonal form is solved using lower-upper (LU) decomposition method.

Table 9.4 Numerical values of $f''(c)$ for changing c when $\epsilon = M = m = \lambda = Fr = \phi_1 = \phi_2 = \phi_3 = 0$

c	*Salleh* et al. *(2018)*	*Aladdin* et al. *(2021)*	*Present study*
0.1	1.2888299	1.2888171	1.2615
0.15	0.9383388	0.9383388	0.9264
0.2	0.7515725	0.7515725	0.7453

The initial guesses considered in this study which satisfies the boundary conditions are as follows:

$$f_0(\eta) = 0.5\epsilon\eta + (0.5-\epsilon)\left(\eta - c - 1 + e^{c-\eta}\right),\ \theta_0(\eta) = e^{c-\eta},\ \phi_0(\eta) = e^{c-\eta}.$$

Table 9.4 compares the numerical values of $f''(c)$ with earlier literature. The outcomes obtained are seen to be in good accord with previous literature.

9.4 RESULTS AND DISCUSSION

The numerical results are presented in graphical and tabular form so that the behavior of the flow can be studied. Results are provided for Ag/H_2O, $Ag\text{-}Au/H_20$ and $Ag\text{-}Au\text{-}Cu/H_2O$ flows.

Figures 9.2 and 9.3 describe the variation in velocity and temperature for changing M. On application of external magnetic field, Lorentz force is

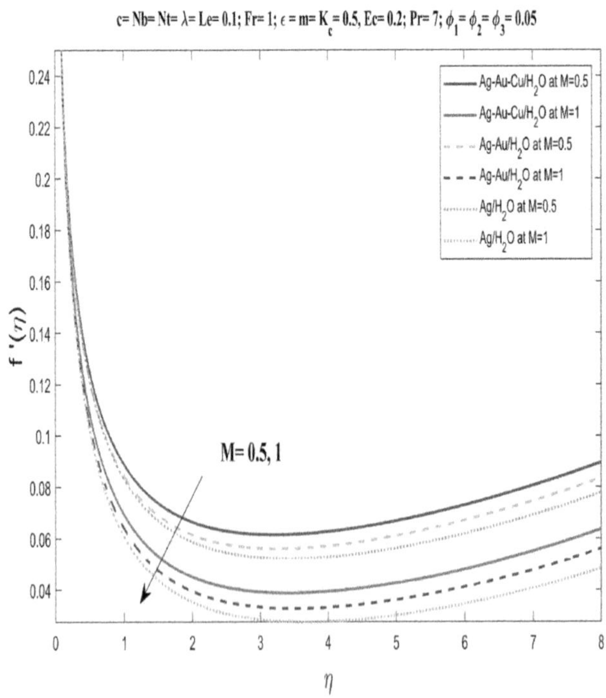

Figure 9.2 Variation in $f'(\eta)$ for changing M.

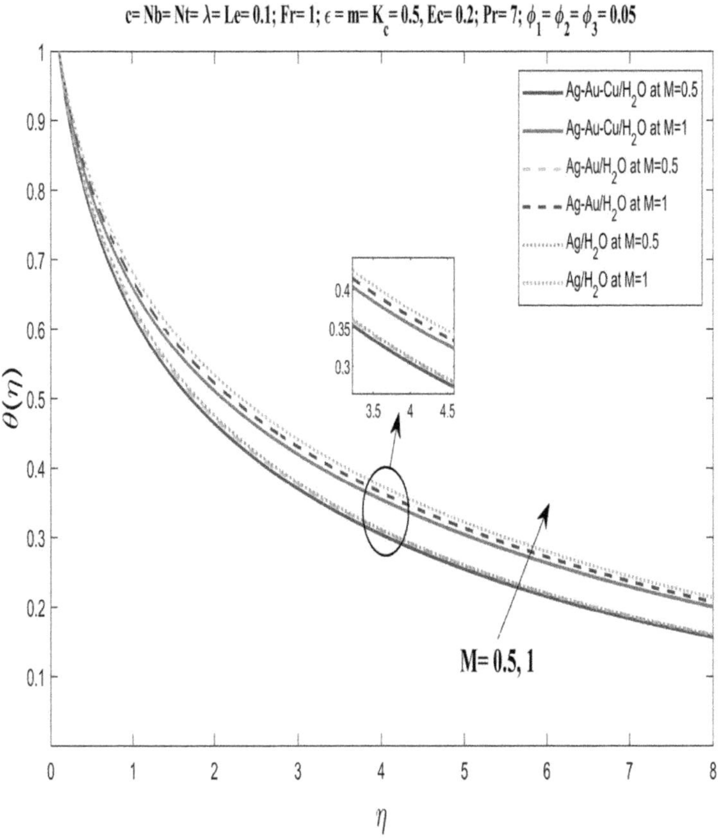

Figure 9.3 Variation in $\theta(\eta)$ for changing *M*.

generated which alters the dynamics of the flow. The fluid velocity perpendicular to the magnetic field lines is resisted by this force. Thus, the velocity distribution declines with increasing values of M. A rise in the Lorentz force, on the other hand, results in a frictional resistance to the flow, which elevates the temperature in the vicinity of the boundary layer. Thus, temperature profile increases as M increases.

Figures 9.4 and 9.5 describe the effect of c on velocity and temperature profile. The fluid velocity within the boundary layer decreases as the needle size increases because the fluid has to cover a larger cross-sectional area, which causes the velocity to decrease, whereas a thicker needle will dissipate more energy as heat due to the increase in friction between the fluid and larger surface area of the needle. This causes the temperature to rise.

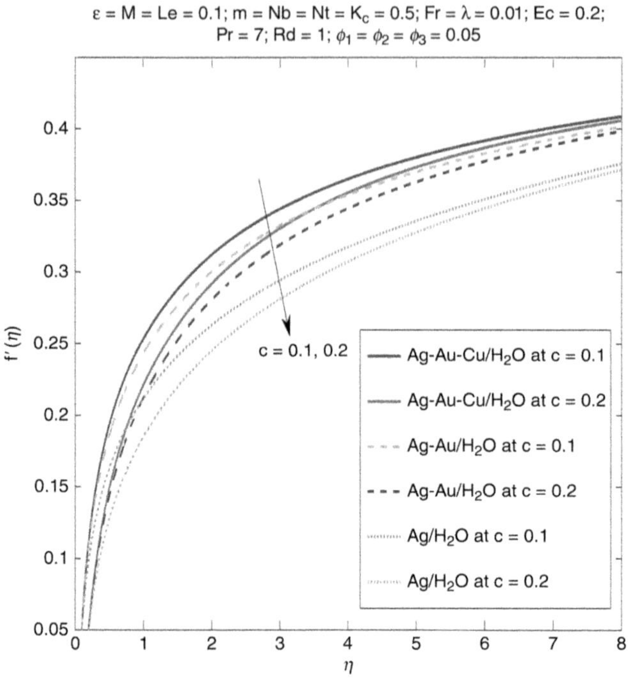

Figure 9.4 Variation in $f'(\eta)$ for changing *c*.

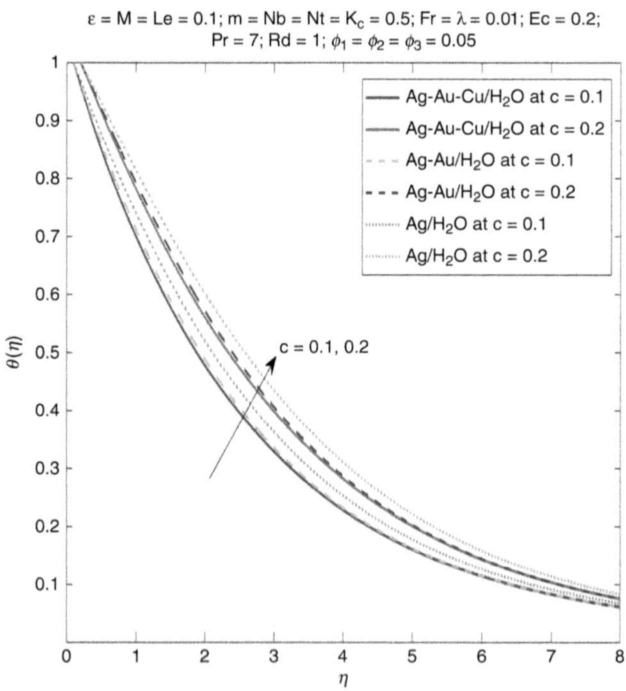

Figure 9.5 Variation in $\theta(\eta)$ for changing *c*.

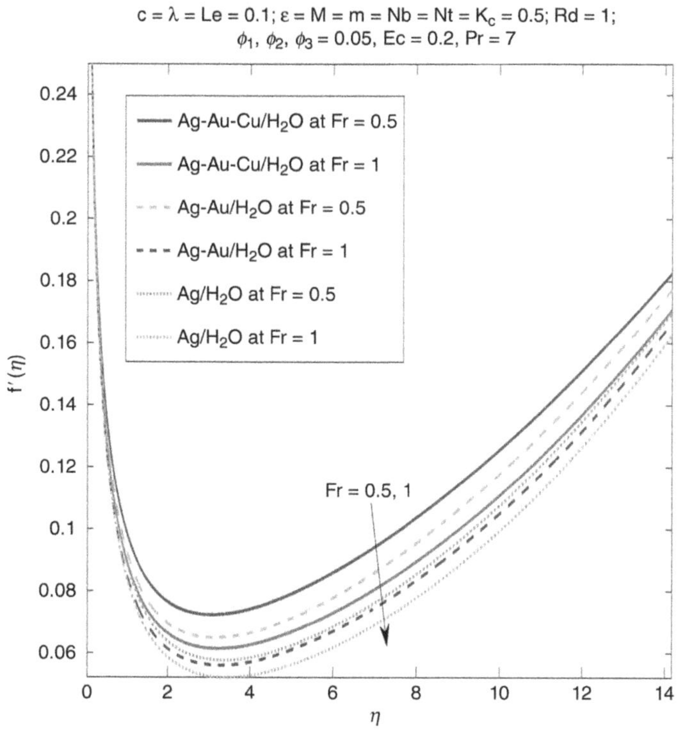

Figure 9.6 Variation in $f'(\eta)$ for changing *Fr*.

Effects of *Fr* on velocity profile are given in Figure 9.6. As the Forchheimer number rises, the inertial effects become stronger, which results in this velocity drop in the boundary layer. Figure 9.7 describes the velocity profile for increasing values of λ. Increasing λ increases the porous layer which causes the fluid to flow slower as it encounters more resistance to navigate through the boundary.

As the nanoparticles considered are metallic, the fluid exhibit electrical conductivity and we consider the impact of Hall current. Figure 9.8 illustrates the effect of m on velocity profile. The magnetic field's resistive force diminishes as m is increased. As a result, for higher values of m, the velocity profile rises. Increasing Rd increases temperature profile as shown in Figure 9.9. Increasing Rd increases the radiative heat flux which leads to more heat energy being released into the fluid. Thus, temperature rises.

The random motion of nanoparticles increases with increase in Nb, which causes an increase in collisions among the nanoparticles. Moreover, as Nb rises, the nanoparticles move closer to one another. The nanoparticles occupy less volume due to this clustering effect and thus the

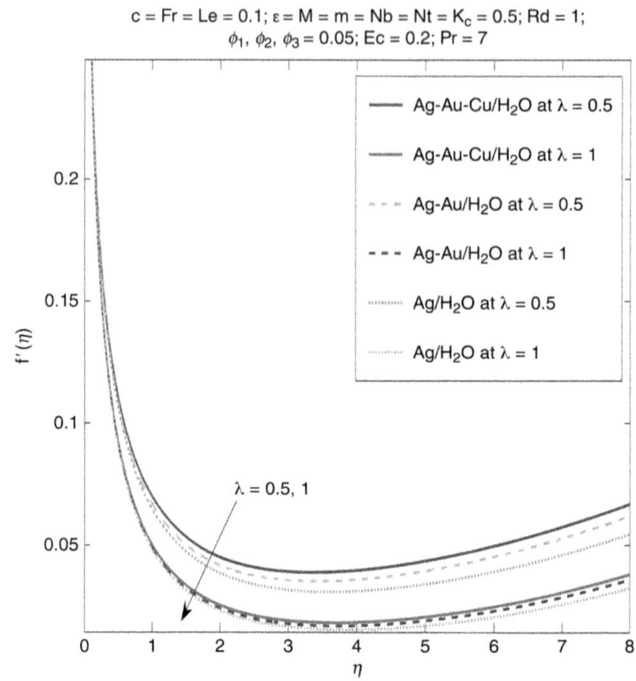

Figure 9.7 Variation in $f'(\eta)$ for changing λ.

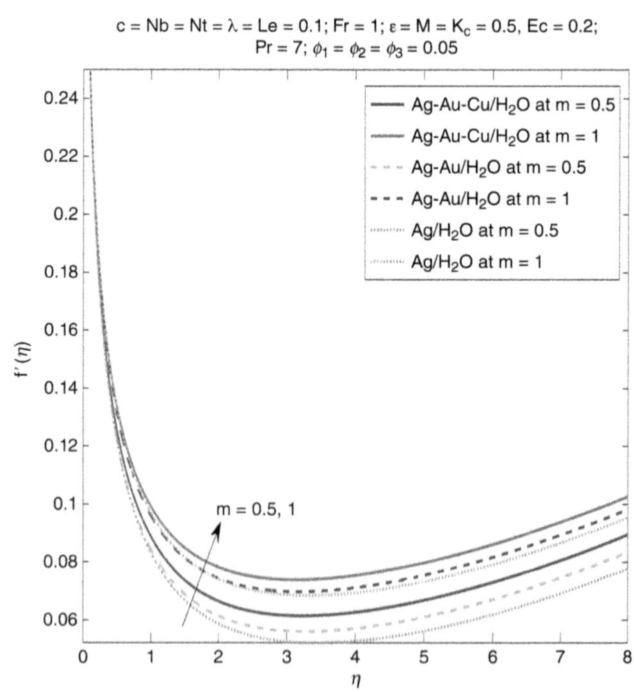

Figure 9.8 Variation in $f'(\eta)$ for changing m.

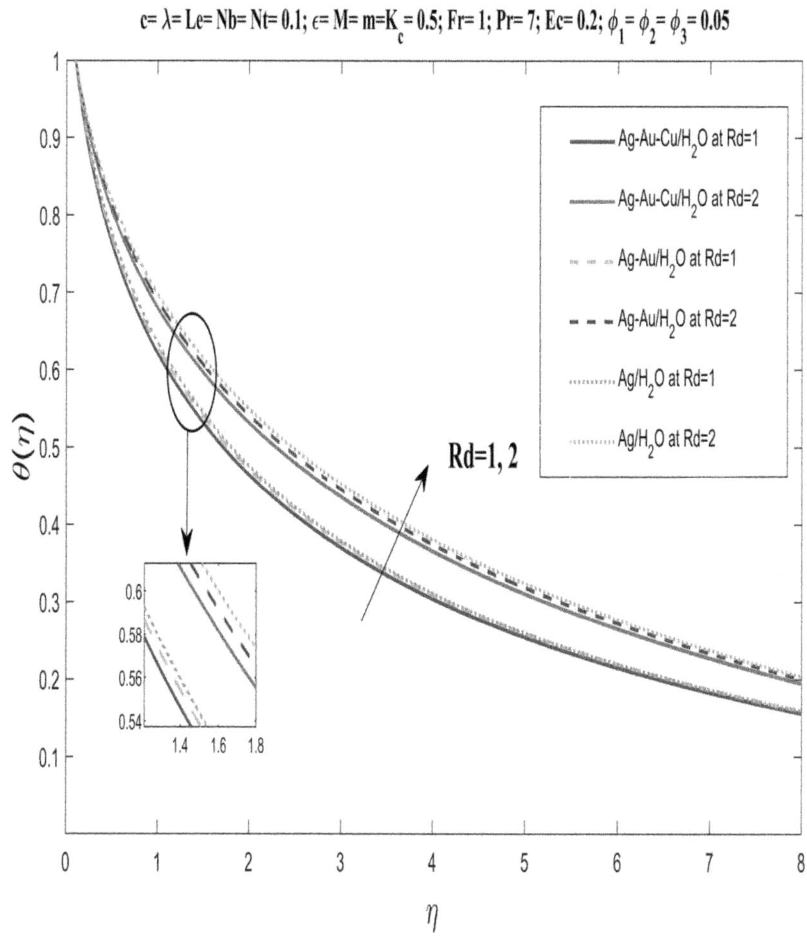

Figure 9.9 Variation in $\theta(\eta)$ for changing *Rd*.

concentration profile decreases. The effects are presented in Figure 9.10. Figure 9.11 depicts the effects of *Nt* on concentration. When *Nt* increases, the movement of particle due to thermophoresis intensifies. As the surface is warmer compared to the surrounding, there is a drop in concentration profile initially near the surface. But the movement of the nanoparticles causes an increase of nanoparticle concentration away from surface. Thus, concentration profile intensifies on increasing *Nt* when we consider flow away from the surface.

When K_c increases, more solute molecules are involved in the reaction due to which concentration profile declines as shown in Figure 9.12.

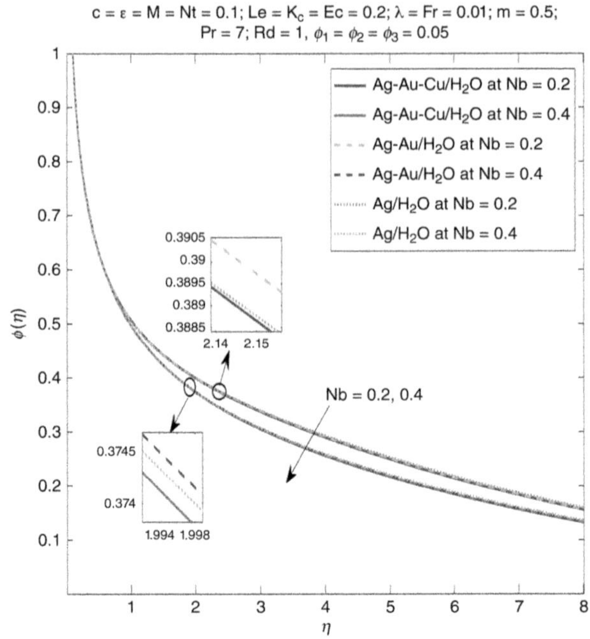

Figure 9.10 Variation in $\phi(\eta)$ for changing *Nb*.

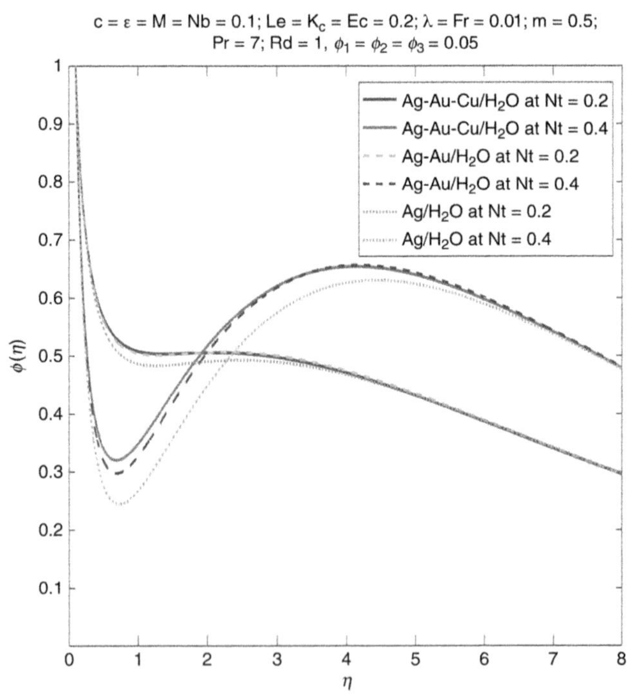

Figure 9.11 Variation in $\phi(\eta)$ for changing *Nt*.

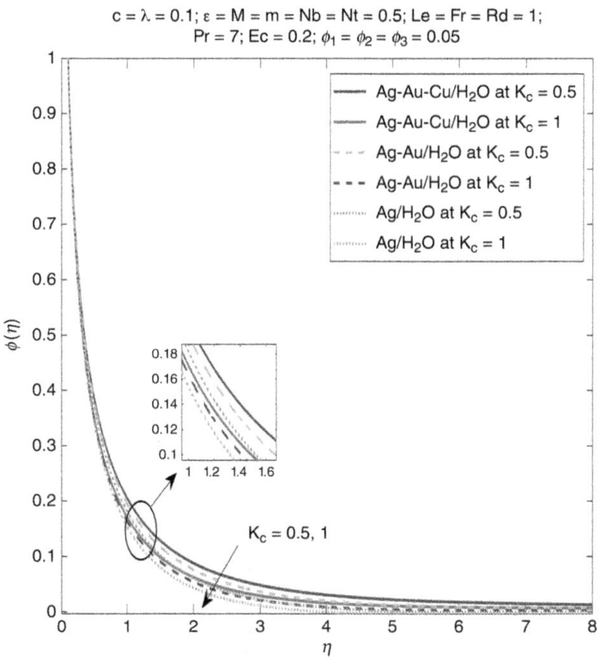

Figure 9.12 Variation in $\phi(\eta)$ for changing K_c.

Tables 9.5–9.7 aid in assessing shear stress and also the rate of heat and mass transfer for Ag-H_2O, Ag-Au/H_2O and Ag-Au-Cu-H_2O flows.

Table 9.5 Impact upon skin friction coefficient when $\phi_1 = \phi_2 = \phi_3 = 0.05$ and $Pr = 7$

c	M	m	λ	Fr	$C_f Re_x^{\frac{1}{2}}$ for Ag/H_2O	$C_f Re_x^{\frac{1}{2}}$ for Ag-Au/H_2O	$C_f Re_x^{\frac{1}{2}}$ for Ag-Au-Cu/H_2O
0.1	0.1	0.8	0.01	0.01	1.0829	1.4468	1.7165
0.2					0.9263	1.2433	1.4743
0.3					0.8527	1.1475	1.3605
	0.2				1.3673	2.0054	2.4852
	0.3				0.9651	1.5702	2.0474
	0.4				0.6537	1.1993	1.6596
		0.9			1.9292	2.5661	3.0214
		1			1.9793	2.6029	3.0662
		1.1			2.0253	2.6453	3.1072
			0.02		1.7862	2.4041	2.8557
			0.03		1.6996	2.3050	2.7411
			0.04		1.6161	2.2594	2.6291
				0.02	1.8370	2.4420	2.8965
				0.03	1.7999	2.3803	2.8222
				0.04	1.7638	2.3211	2.7514

Table 9.6 Impact upon Nusselt number when $\phi_1 = \phi_2 = \phi_3 = 0.05$ and $Pr = 7$

c	M	m	Rd	Nb	Nt	$Nu_x Re_x^{-\frac{1}{2}}$ for Ag/H_2O	$Nu_x Re_x^{-\frac{1}{2}}$ for $Ag\text{-}Au/H_2O$	$Nu_x Re_x^{-\frac{1}{2}}$ for $Ag\text{-}Au\text{-}Cu/H_2O$
0.1	0.1	0.8	1	0.1	0.1	0.9260	1.1492	1.3429
0.2						0.8726	1.0719	1.2415
0.3						0.8477	1.0349	1.1926
	0.2					0.8050	0.9980	1.1579
	0.3					0.7599	0.9582	1.1207
	0.4					0.7159	0.9172	1.0825
		0.9				0.7321	0.9327	1.0969
		1				0.7474	0.9467	1.1101
		1.1				0.7613	0.9593	1.1218
			1.2			0.7300	0.9219	1.0799
			1.4			0.7029	0.8893	1.0429
			1.6			0.6793	0.8604	1.0100
				0.2		0.5927	0.7606	0.8995
				0.3		0.5149	0.6695	0.7981
				0.4		0.4453	0.5869	0.7050
					0.2	0.5248	0.6811	0.8109
					0.3	0.4633	0.6084	0.7292
					0.4	0.4080	0.5418	0.6538

Table 9.7 Impact upon Sherwood number when $\phi_1 = \phi_2 = \phi_3 = 0.05$, $c = 0.1$ and $Pr = 7$

Le	K_c	Nb	Nt	$Sh_x Re_x^{-\frac{1}{2}}$ for Ag/H_2O	$Sh_x Re_x^{-\frac{1}{2}}$ for $Ag\text{-}Au/H_2O$	$Sh_x Re_x^{-\frac{1}{2}}$ for $Ag\text{-}Au\text{-}Cu/H_2O$
0.2	0.2	0.1	0.1	1.7856	1.7835	1.7781
0.4				1.9395	1.9434	1.9384
0.6				2.0652	2.0736	2.0682
	0.4			2.2085	2.2123	2.2060
	0.6			2.3296	2.3313	2.3237
	0.8			2.4353	2.4355	2.4271
		0.2		2.2079	2.2100	2.2044
		0.3		2.1261	2.1297	2.1258
		0.4		2.0810	2.0861	2.0834
			0.2	2.3104	2.3115	2.3035
			0.3	2.5538	2.5528	2.5401
			0.4	2.8058	2.8049	2.7888

9.5 CONCLUSIONS

In this numerical study, we have considered ternary hybrid nanofluid flow over a moving needle geometry and the influence of hall current and several other parameters over the flow. H_2O-based nanofluid consisting of silver (Ag), gold (Au) and copper (Cu) has been utilized. Keller box method is used to carry out the numerical analysis. The significant conclusions are outlined as follows:

- The velocity profile increases in the order of Ag-H_2O nanofluid, Ag-Au/H_2O hybrid nanofluid and Ag-Au-Cu/H_2O ternary hybrid nanofluid flow. This shows that ternary hybrid nanofluid is faster when compared with the other two flows.
- The temperature distribution decreases sequentially from Ag-H_2O nanofluid to Ag-Au/H_2O hybrid nanofluid and further to the Ag-Au-Cu/H_2O ternary hybrid nanofluid flow.
- No significant difference is observed in the concentration profile of the three different flows.
- The rate of heat transfer in Ag-Au-Cu/H_2O ternary hybrid nanofluid flow is greater as evidenced by the higher Nusselt number values when compared to the other two flows.
- The enhancement of Hall parameter (m) increases the velocity profile.
- Increasing the porosity parameter (λ) and Forchheimer number (Fr) results in a decrease in the velocity profile.
- Larger size of the needle leads to a reduction in the velocity profile while simultaneously raising the temperature profile.
- The concentration profile shows a decline when chemical reaction parameter (K_c) is increased.

NOMENCLATURE

ρ	density
μ	dynamic viscosity
σ	electrical conductivity
υ	kinematic viscosity
κ	coefficient of thermal conductivity
σ^*	Stefan- Boltzmann coefficient
B_0	magnetic field strength
C_b	drag coefficient
C_p	specific heat at constant pressure
D_B	Brownian diffusion coefficient
D_T	thermophoretic diffusion coefficient
k	permeability of porous medium
k^*	mean absorption coefficient
K_1	reaction coefficient
m	Hall parameter

REFERENCES

Ahmad, R., Mustafa, M., & Hina, S. (2017, August). Buongiorno's model for fluid flow around a moving thin needle in a flowing nanofluid: A numerical study. Chinese Journal of Physics, 55(4), 1264–1274.

Aladdin, N. A. L., Bachok, N., & Pop, I. (2021). Boundary layer flow and heat transfer of cu-Al2O3/water over a moving horizontal slender needle in presence of hydromagnetic and slip effects. International Communications in Heat and Mass Transfer, 123, 105213.

Alharbi, K. A. M., Ahmed, A. E. S., Ould Sidi, M., Ahammad, N. A., Mohamed, A., El-Shorbagy, M. A., Bilal, M., & Marzouki, R. (2022, April 9). Computational valuation of Darcy ternary-hybrid nanofluid flow across an extending cylinder with induction effects. Micromachines, 13(4), 588.

Asghar, A., Ying, T. Y., & Zaimi, K. (2022, April 1). Two-dimensional mixed convection and radiative Al_2O_3-Cu/H_2O hybrid nanofluid flow over a vertical exponentially shrinking sheet with partial slip conditions. *CFD Letters*, *14*(3), 22–38.

Brewster, Q. M. (1992). Thermal Radiative Transfer Properties Solutions Manual. Wiley-Interscience.

Buongiorno, J. (2006). Convective transport in nanofluids. Journal of Heat Transfer, 128(3), 240–250.

Cebeci, T., & Bradshaw, P. (1988). Physical and Computational Aspects of Convective Heat Transfer. Springer-Verlag.

Choi, S. U. S., & Eastman, J. A. (1995). Enhancing thermal conductivity of fluids with nanoparticles. ASME International Mechanical Engineering Congress and Exposition, 66, 99–105.

Devi, S. A., & Devi, S. S. U. (2016). Numerical investigation of hydromagnetic hybrid Cu–Al_2O_3/water nanofluid flow over a permeable stretching sheet with suction. International Journal of Nonlinear Sciences and Numerical Simulation, 17(5), 249–257.

Grosan, T., & Pop, I. (2011, February 2). Forced convection boundary layer flow past nonisothermal thin needles in nanofluids. Journal of Heat Transfer, 133(5), 054503.

Hamilton, R. L., & Crosser, O. K. (1962). Thermal conductivity of heterogeneous two component systems. Industrial & Engineering Chemistry Fundamentals, 1(3), 187–191.

Huda, N. U., Hamid, A., & Khan, M. (2020). Impact of Cattaneo-Christov model on Darcy–Forchheimer flow of ethylene glycol base fluid over a moving needle. Journal of Materials Research and Technology, 9(3), 4139–4146.

Khan, A., Kumam, W., Khan, I., Saeed, A., Gul, T., Kumam, P., & Ali, I. (2021, April 15). Chemically reactive nanofluid flow past a thin moving needle with viscous dissipation, magnetic effects and hall current. PLOS ONE, 16(4), e0249264.

Lee, L. L. (1967, April 1). Boundary layer over a thin needle. The Physics of Fluids, 10(4), 820–822.

Lee, S., Choi, S. S., Li, S. A., & Eastman, J. A. (1999). Measuring thermal conductivity of fluids containing oxide nanoparticles. Journal of Heat Transfer, 121(2), 280–289.

Masuda, H., Ebata, A., Teramae, K., & Hishinuma, N. (1993). Alteration of thermal conductivity and viscosity of liquid by dispersing ultra-fine particles (dispersions of c-Al_2O_3, SiO_2, and TiO_2 ultrafine particles). Netsu Bussei, 4, 227–233.

Nasir, S., Shah, Z., Islam, S., Bonyah, E., & Gul, T. (2019, January 1). Darcy Forchheimer nanofluid thin film flow of SWCNTs and heat transfer analysis over an unsteady stretching sheet. AIP Advances, American Institute of Physics.

Nazar, T., Bhatti, M. M., & Michaelides, E. E. (2021, November 29). Hybrid (Au-TiO2) nanofluid flow over a thin needle with magnetic field and thermal radiation: Dual solutions and stability analysis. Microfluidics and Nanofluidics, 26(1), 2.

Ramya, D., Raju, R. S., Rao, J. A., & Chamkha, A. (2018, June). Effects of velocity and thermal wall slip on magnetohydrodynamics (MHD) boundary layer viscous flow and heat transfer of a nanofluid over a non-linearly-stretching sheet: A numerical study. Propulsion and Power Research, 7(2), 182–195.

Ramzan, M., Shaheen, N., Kadry, S., Ratha, Y., & Nam, Y. (2020, January 7). Thermally stratified Darcy Forchheimer flow on a moving thin needle with homogeneous heterogeneous reactions and non-uniform heat source/sink. Applied Sciences, 10(2), 432.

Rao, S., & Deka, P. (2022, August 22). A numerical solution using EFDM for unsteady MHD radiative Casson nanofluid flow over a porous stretching sheet with stability analysis. Heat Transfer, 51(8), 8020–8042.

Rao, S., & Deka, P. N. (2023). A study on MHD flow of SWCNT-Al_2O_3/water hybrid nanofluid past a vertical permeable cone under the influence of thermal radiation and chemical reactions. Numerical Heat Transfer, Part A: Applications, 85(10), 1591–1611.

Raptis, A. (1998). Radiation and free convection flow through a porous medium. International Communications in Heat and Mass Transfer, 25(2), 289–295.

Rashid, U., Iqbal, A., & Alsharif, A. (2021, December 10). Numerical study of (Au-Cu)/Water and (Au-Cu)/Ethylene glycol hybrid nanofluids flow and heat transfer over a stretching porous plate. Energies, 14(24), 8341.

Salleh, S. N. A., Bachok, N., Arifin, N. M., Ali, F. M., & Pop, I. (2018). Magnetohydrodynamics flow past a moving vertical thin needle in a nanofluid with stability analysis. Energies, 11(12), 3297.

Sarkar, J., Ghosh, P., & Adil, A. (2015). A review on hybrid nanofluids: Recent research, development and applications. Renewable and Sustainable Energy Reviews, 43, 164–177.

Souayeh, B., & Ramesh, K. (2023, February 14). Numerical scrutinization of ternary nanofluid flow over an exponentially stretching sheet with gyrotactic microorganisms. Mathematics, 11(4), 981.

Souayeh, B., Reddy, M. G., Sreenivasulu, P., Poornima, T., Rahimi-Gorji, M., & Alarifi, I. M. (2019, June). Comparative analysis on non-linear radiative heat transfer on MHD Casson nanofluid past a thin needle. Journal of Molecular Liquids, 284, 163–174.

Sparrow, E. M., & Cess, R. D. (1978). Radiation Heat Transfer (Augmented Edition (1st) ed.). Routledge.

Suresh, S., Venkitaraj, K. P., Selvakumar, P., & Chandrasekar, M. (2011). Synthesis of Al_2O_3–Cu/water hybrid nanofluids using two step method and its thermophysical properties. Colloids and Surfaces A: Physicochemical and Engineering Aspects, 388(1–3), 41–48.

Tiwari, R. K., & Das, M. K. (2007). Heat transfer augmentation in a two-sided lid-driven differentially heated square cavity utilizing nanofluids. International Journal of Heat and Mass Transfer, 50(9–10), 2002–2018.

Chapter 10

Natural convection in a triangular cavity filled with MoS_2 - *water* nanofluid and microorganisms

Finite element method

Bikash Das, Sahin Ahmed, and Anil Nangkar

10.1 INTRODUCTION

An extensive variety of engineering implications, including storing energy and solar collectors, processing food, nuclear reactors, and maintaining the temperature of electronic devices for efficient output, has gained the interest of many scholars to study the convective flow in or through porous regimes. Prominent academicians have explored the phenomenon of transmission in porous media (Ingham and Pop, 1998; Nield and Bejan, 2006; Vafai 2015). Magnetohydrodynamic (MHD) flow in porous media has attracted many authors (Nield and Bejan, 2006; Pelekasis and Benos, 2017) due to its numerous uses in scientific and engineering domains. The concept of MHD flow in porous materials is used in nuclear reactors (Pelekasis and Benos, 2017) and in the extraction of geothermal energy (Nield and Bejan, 2006). Ahmed et al. (2023) deliberated the consequences of endothermic and exothermic reactions on the flow of biofluid in permeable arteries under an inclining magnetic field. The solution was obtained analytically, and it was found that the porosity of the permeable artery widens the momentum boundary layer, whereas the endothermic reaction decreases its thickness. This research helps us understand how blood flow in branching arteries is affected by a magnetic field, surrounding tissue properties, and chemical reactions within the blood. This knowledge can be valuable for studying different blood flow scenarios, including simple flow, rhythmic flow (peristaltic flow), pulsating flow, and even drug delivery applications.

Due to their superior thermal conductivity over base fluids, nanofluids have been recognised over the past two decades as one of the most effective heat transfer fluids. As a result, they have been utilised in many applications, including cooling electronics, delivering drugs, manufacturing currency, and powering automobile engines. Nanoparticles of different types like carbon nanotubes (CNTs), metal, oxides of metals, non-metals, and graphene are typically added with base fluids such as water, ethylene glycol, lubricating oil, or blood to form nanofluids. The groundbreaking study on nanofluids (Choi and Eastman, 1995) was presented. Several researchers

 DOI: 10.1201/9781003527251-10

were motivated to explore nanofluids in greater detail by the outcome of their experiment. The impact of Lorentz force on the flow of a viscous fluid over vertical surface of semi-infinite length by considering the radiation effect was discussed by Azzam (2002). The fluid density decreases exponentially with thermal energy, and the Boussinesq approximation is neglected due to significant temperature differences. The Rosseland approximation describes the radiative heat flux. Using similarity transformations, the boundary layer equations are transmuted into ordinary differential equations (ODEs), which are mathematically solved by means of the Runge-Kutta-shooting technique. The study investigates the influence of thermal radiation, Lorentz force, and density/temperature parameter on momentum and thermal distributions, as well as shear stress and heat transmission rate. The impact of the presence of nanoparticles and the Reynolds number on the heat transmission in a tube was experimentally studied by Xuan and Li (2003), and this experimental setup was constructed to study the flow characteristics of nanofluids in a tube. The rate of heat transmission from the surface and shear stress were measured for turbulent flow conditions. The study examined the effect of nanoparticle volume fraction and Reynolds number on heat transfer and flow behaviour. A novel correlation for convective heat transfer was developed to relate experimental data for nanofluids. Kalita et al. (2020) examined the use of CNT in heat transfer and blood mobility in a vertical channel under the action of thermal radiation and a magnetic field. They found that an increment in the solid volume fraction of nanoparticles resulted in reduced velocity distribution while raising the temperature profile. Further, Hazarika and Ahmed (2021) analysed the flow behaviour of non-Newtonian fluid-based nanofluid in a vertical channel embedded in a porous material using Darcy's model. They explore heat and mass transfer in a vertical channel containing a saturated porous medium and Casson fluid with nano inclusion of CNTs, including single-walled and multiple-walled CNT types, with engine oil as the base fluid. The study investigates the effect of CNTs on momentum, thermal energy, shear stress, and heat transfer rate, considering various key physical parameters. The model's validity is confirmed through comparison with existing results, which show satisfactory agreement. The results show that heat generation and thermal radiation have a major impact on CNT behaviour and that fluid velocity increases with increasing Casson fluid parameter and nanoparticle volume fraction.

Bioconvection phenomena encompass the self-organised motion of microorganisms in fluid environments, giving rise to intriguing patterns and dynamics with implications for various fields, including ecology, biotechnology, and biophysics. Understanding the underlying mechanisms of bioconvection is essential for elucidating biological processes and engineering applications. Hazarika and Ahmed (2023) investigated the impression of bioconvection on the flow of Prandtl nanofluid under the influence of

Lorentz force. They observed that the density ratio of motile microorganisms plays a vital role in assessing their impact on movement of the microorganisms, affecting velocity and fluid flow patterns. Higher concentrations lead to increased microorganism density, intensifying bio-convection dynamics and significantly contributing to convective heat transfer, given their excessive density compared to the surrounding fluid. Balla et al. (2020) illustrated the impression of heat radiation on the bioconvective flow in a square cavity in which motion of the fluid is induced due to the temperature difference. The investigation analyses the effects of various key variables on streamlines, isothermal contours, oxygen, contours of microorganism concentrations, and the average heat transfer rate. Results indicate that bioconvection intensity increases with higher Rayleigh numbers but decreases with thermal radiation. The isothermal contours is significantly influenced by the Rayleigh number, while thermal radiation accelerates heat transfer in the cavity. Oxygen density and motile microorganism concentrations are enhanced with increasing Rayleigh numbers, and the stability of microorganisms improves with higher radiation parameters. Alluguvelli et al. (2022) demonstrated the significance of viscous dissipation in heat transmission in a porous enclosure. The bioconvection in a complex-shaped geometry was reported by Hussain et al. (2022a). The investigation analyses how parameters such as Rayleigh number and radiation parameter affect streamlines, isotherms, oxygen and microorganism concentrations, and the average heat transfer rate. Oxygen density and microorganism concentrations increase with higher Rayleigh numbers, and microorganism stability improves with higher radiation parameters. Izadpanah et al. (2022) considered two heaters in a square cavity to study the variation in density of motile microorganisms and flow behaviour. The impacts of Lewis, Peclet, and thermal and bio-convection Rayleigh numbers on heat transmission, entropy generation, and microbe concentration are the main topics of discussion. The diffusion equation for microorganism mobility and the energy equation for temperature impacts are the two discretized governing equations that are solved by help of a finite element approach. The key finding is that higher Rayleigh numbers lead to better thermo-bio-convection performance. For a high thermal Rayleigh number $\left(Ra = 10^5\right)$, increasing the bio-convection Rayleigh number from 10 to 100 results in a 4.5% rise in the average heat flux but a corresponding decrease in the average mass flux. These insights have applications in a variety of sectors, including fuel cells, oil recovery, and ocean ecosystems. Hussain et al. (2022b) considered a wavy shaped geometry filled with porous material to study the natural convection under the impact of thermal radiation. The findings revealed how thermal radiation helps to strengthen streamlines. The bioconvection flow, streamline shapes, and heat capacity inside a wavy cavity are all impacted by the number of waves of a wavy wall. The streamlines are strengthened by the Rayleigh number. The isotherms, oxygen iso-concentration, and microbe iso-concentration are all improved by raising the Darcy parameter. As a result, the phase change material moves with the cavity towards its right side, and the Darcy parameter increases.

Elshehabey et al. (2014) address the fluid flow due to temperature difference in a L-shaped subjected to differentially heated walls filled with Cu/water nanofluid under the impact of an inclined magnetic field. The fully implicit finite difference method was employed to acquire the solution of the non-linear governing equations, and the solutions were plotted for the variation of various variables, including the Hartmann number, nanoparticle volume fraction, cavity inclination angle, magnetic field inclination angle, cavity aspect ratio, and Rayleigh number. According to the findings, fluid flow and heat transfer properties are greatly reduced by the magnetic field. Furthermore, adding copper nanoparticles enhances heat transfer rates in the nanofluid. The effect of one wall's waviness on the flow and heat transfer properties of an incompressible viscous fluid driven by temperature differences between vertical plates was investigated by Vajravelu and Sastri (1978). The authors solve governing equations subject to relevant constraints in the boundary, dividing the solution into mean and perturbed parts. The perturbed part, accounting for wall waviness, is derived using the long-wave approximation. The mean solution closely aligns with previous findings after adjustments. The properties of fluid pressure variations, heat transfer, and flow characteristics on both flat and wavy walls at various parameter values are revealed through numerical evaluation. Ashorynejad and Shahriari's (2018) study uses the lattice Boltzmann method to examine the natural convection heat transmission of an $Al_2O_3 - Cu\,/\,water$ hybrid nanofluid in an open wavy geometry under the action of uniform magnetic field. With the right wall exposed to the outside environment, the left wall experiences sinusoidal heating. They have smooth, insulated top and bottom walls. The Rayleigh number, Hartmann number, phase deviation, and nanoparticle volume fraction are among the parameters considered. The findings indicate that while Ra and the nanoparticle volume fraction raise the heat flux, it decreases as the Hartmann number increases. Based on *Ra*, the nanoparticle effect is either amplified or attenuated by the Lorentz force. Phase deviation impacts the heat transfer inconsistently at low *Ra* but favourably at high Ra, where it enhances the efficiency of nanoparticles. Ghasemi et al. (2011) probed natural convection in a water-Al_2O_3 nanofluid geometry subjected to a magnetic field. The enclosure consists of two horizontal adiabatic walls and two vertical isothermal walls at temperatures. Using a fixed Prandtl number, numerical predictions investigate the effects of the Rayleigh number, solid volume fraction, and Hartmann number on heat transfer performance, flow, and thermal energy. The rate of heat transmission rises with *Ra* but falls with *Ha*, according to the results. Depending on the values of *Ha* and *Ra*, the effect of the solid volume fraction on heat transfer performance varies.

The prime objective of this investigation is to analyse the flow behaviour of $MoS_2 - water$ nanofluid with the inclusion of motile microorganisms under the influence of Lorentz force. The novelty of the analysis is to study the impact of Darcy number, buoyancy ratio, and Hartmann number on the isothermal contours, vortices, concentration of nanoparticles, and density of motile microorganisms within the triangular cavity. The solution of current mathematical model has been obtained by using COMSOL Multiphysics via the FEM. The analysis has been well validated with existing literature and found to be in good agreement.

10.2 PROBLEM STATEMENT AND MATHEMATICAL METHOD

A 2-D steady flow within a triangular cavity occupied with porous cavity and $MoS_2 - water$ nanofluid has been considered. The vertical wall of the triangle is maintained at T_H temperature as shown in Figure 10.1, the base wall is insulated, and the inclined wall of the triangular enclosure is maintained at T_c temperature. The fluid flow inside the porous medium has been simulated using the Darcy-Brinkman-Forchheimer model, and the Boussinesq approximation is implicated in fluid properties, which include density variations. Furthermore, species diffusion with the Soret effect has been considered, and the contributions of ohmic heating, viscous dissipation, Brownian motions, and thermophoresis motions are included in the energy equation. Also, uniform magnetic field $\boldsymbol{B} = \boldsymbol{B}_x e_x + \boldsymbol{B}_y e_y$ of constant magnitude $\boldsymbol{B}_0 = \sqrt{\boldsymbol{B}_x^2 + \boldsymbol{B}_y^2}$ has been applied at different angles, where e_x and e_y are the unit vectors along the coordinate axis.

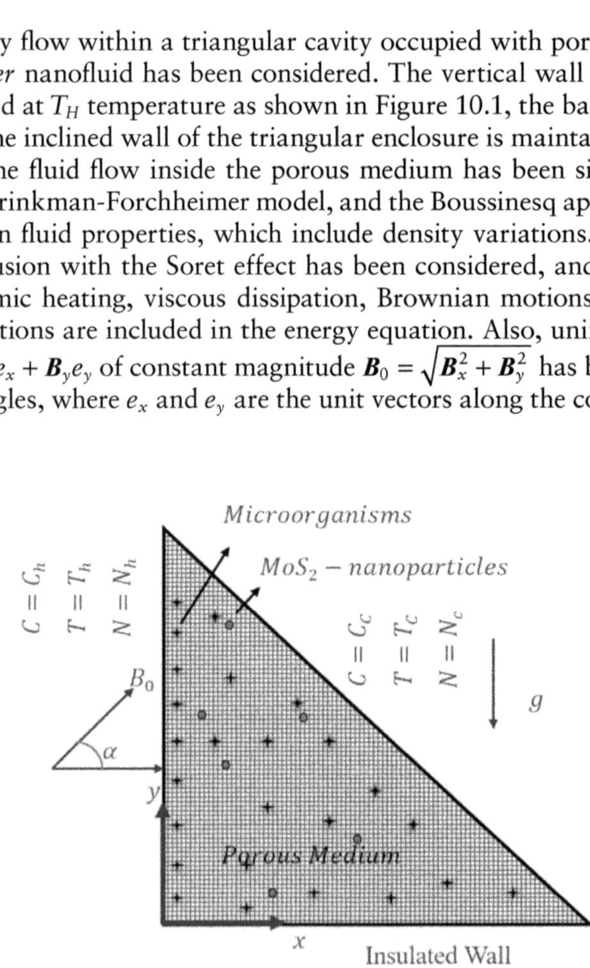

Figure 10.1 The geometrical description of the model.

The electric current density and electromagnetic force are determined by the relations given by Ohm's law, the law of conservation of charge, and Lorentz force and is given by Elshehabey et al. (2014):

$$\boldsymbol{J} = \sigma_l \left(-\nabla \varphi + \boldsymbol{V} \times \boldsymbol{B} \right), \tag{10.1}$$

$$\nabla . \boldsymbol{J} = 0, \tag{10.2}$$

$$\boldsymbol{F} = \boldsymbol{J} \times \boldsymbol{B}. \tag{10.3}$$

Thus Equations (10.1)–(10.3) reduce to:

$$\boldsymbol{J} = \sigma_l \left(\boldsymbol{V} \times \boldsymbol{B} \right), \tag{10.4}$$

$$\boldsymbol{F} = \sigma_l \left(\boldsymbol{V} \times \boldsymbol{B} \right) \times \boldsymbol{B}. \tag{10.5}$$

Following are the basic assumptions used in the governing equations for natural convection in a porous triangular cavity throughout the analysis:

- The cavity is considered to be impermeable;
- The flow inside the cavity is Newtonian, laminar, incompressible and time independent;
- Apart from the density, which is adjusted using the Boussinesq approximation, all the fluid variables remain constant;
- Size and shape of the nanoparticles are considered to be uniform;
- The nanofluid is electrically conducting;
- The porous medium is uniform permeability and thermally isotropic; the local thermal equilibrium conditions between porous material and nanofluid exists;
- Darcy-Forchheimer model is employed for momentum analysis in the porous medium;
- The induced magnetic field can be disregarded by assuming a small magnetic Reynolds number;
- There is no slip between the fluid and the walls;
- Heating due to the presence of porous medium has been neglected.

Under the above assumptions, the 2-D and time-independent equations of continuity, modified Navier-stokes equation for the natural convection of nanofluid flow in x and y directions, heat transmission and nanoparticles, and microorganisms density equations are as follows (Vajravelu and Sastri, 1978):

$$\frac{\partial u}{\partial x} + \frac{\partial v}{\partial y} = 0, \tag{10.6}$$

$$u\frac{\partial u}{\partial x}+v\frac{\partial u}{\partial y}=-\frac{\partial p}{\partial x}+\nu_{nf}\left[\frac{\partial^2 u}{\partial x^2}+\frac{\partial^2 u}{\partial^2 y}\right]-\frac{\sigma_{nf}B_0^2}{\rho_{nf}}\left(usin^2\alpha-vsin\alpha cos\alpha\right)$$
$$-\frac{\nu_{nf}}{k}u-\frac{C_b}{\rho_{nf}\sqrt{k}}u\sqrt{u^2+v^2}, \tag{10.7}$$

$$u\frac{\partial v}{\partial x}+v\frac{\partial v}{\partial y}=-\frac{\partial p}{\partial y}+\nu_{nf}\left[\frac{\partial^2 v}{\partial x^2}+\frac{\partial^2 v}{\partial^2 y}\right]-\frac{\sigma_{nf}B_0^2}{\rho_{nf}}\left(vcos^2\alpha-usin\alpha cos\alpha\right)$$
$$-\frac{\nu_{nf}}{k}v-\frac{C_b}{\rho_{nf}\sqrt{k}}v\sqrt{u^2+v^2}+g\beta_{nf}\left[(T-T_c)+(C-C_c)+(N-N_c)\right], \tag{10.8}$$

$$(\rho C_p)_{nf}\left(u\frac{\partial T}{\partial x}+v\frac{\partial T}{\partial y}\right)=\kappa_{nf}\left[\frac{\partial^2 T}{\partial x^2}+\frac{\partial^2 T}{\partial^2 y}\right]$$
$$+\mu_{nf}\left[2\left(\frac{\partial u}{\partial x}\right)^2+2\left(\frac{\partial u}{\partial y}\right)^2+\left(\frac{\partial u}{\partial y}+\frac{\partial v}{\partial x}\right)^2\right],$$
$$+\frac{\sigma_{nf}B_0^2}{(\rho C_p)_{nf}}(usin\alpha-vcos\alpha)^2+(\rho C_p)_{np}\left(\begin{array}{c}D_B\left(\frac{\partial C}{\partial x}\frac{\partial T}{\partial x}+\frac{\partial C}{\partial y}\frac{\partial T}{\partial y}\right)+\\ \frac{D_T}{T_c}\left(\left(\frac{\partial T}{\partial x}\right)^2+\left(\frac{\partial T}{\partial Y}\right)^2\right)\end{array}\right), \tag{10.9}$$

$$\left(u\frac{\partial C}{\partial x}+v\frac{\partial C}{\partial y}\right)=D_B\left(\frac{\partial^2 C}{\partial x^2}+\frac{\partial^2 C}{\partial y^2}\right)+\frac{D_T}{T_c}\left[\frac{\partial^2 T}{\partial x^2}+\frac{\partial^2 T}{\partial y^2}\right], \tag{10.10}$$

$$\frac{\partial}{\partial x}\left(un+\tilde{u}n-D_n\frac{\partial n}{\partial x}\right)+\frac{\partial}{\partial y}\left(vn+\tilde{v}n-D_n\frac{\partial n}{\partial y}\right)=0, \tag{10.11}$$

where $\tilde{u}=\frac{bWc}{\Delta C}\frac{\partial C}{\partial x}$ and $\tilde{v}=\frac{bWc}{\Delta C}\frac{\partial C}{\partial y}$.

Subjected to the boundary conditions:

$$\text{Hot wall}: u=v=0,\ T=T_h=T_H,\ C=C_h,\ N-N_h, \tag{10.12}$$

$$\text{Cold Wall}: u=v=0,\ T=T_c,\ C=C_c,\ N=N_c, \tag{10.13}$$

$$\text{Insulated wall}: u=v=0,\ \frac{\partial T}{\partial y}=0,\ \frac{\partial C}{\partial y}=0,\ \frac{\partial N}{\partial y}=0. \tag{10.14}$$

Dimensionless parameters are as follows:

$$\left\{\begin{array}{l} X=\dfrac{x}{L},\ Y=\dfrac{y}{L},\ U=\dfrac{uL}{\nu_f},\ V=\dfrac{vL}{\nu_f},\ \theta=\dfrac{T-T_c}{\Delta T},\ \Delta T=T_h-T_c,\ \phi=\dfrac{C-C_c}{C_h-C_c}, \\ \chi=\dfrac{N-N_c}{N_h-N_c},\ P=\dfrac{pL^2}{\rho_f\nu_f^2},\ Pr=\dfrac{\mu_f C_p}{\kappa_f},\ Ra=\dfrac{g\beta_f\Delta TL^3\rho_f\left(C_p\right)_f}{\kappa_f\nu_f},\ Ec=\dfrac{g\beta_f L}{\left(C_p\right)_f}, \\ Fr=\dfrac{C_b L}{\sqrt{k}\rho_f},Da=\dfrac{L^2}{k},\ Ha^2=\dfrac{\sigma_f B_0^2L^2}{\mu_f},\ Sc=\dfrac{\nu_f}{D_B},\ Nt=\dfrac{\left(\rho C_p\right)_{np}\left(T_h-T_c\right)D_T}{\alpha\left(\rho C_p\right)_f T_c}, \\ Nb=\dfrac{\left(\rho C_p\right)_{np}\left(C_h-C_c\right)D_B}{\alpha\left(\rho C_p\right)_f},\ Nr=\dfrac{\Delta C}{\Delta T\beta_f\rho_f},\ \delta_T=T_h-T_c,\ \delta_C=C_h-C_c \end{array}\right\}, \tag{10.15}$$

where L is the characteristic length of the geometry and is given by $L=\frac{\sqrt{2}}{2}l$.

Employing the above dimensionless parameters, we get

$$\frac{\partial U}{\partial X}+\frac{\partial V}{\partial Y}=0, \tag{10.16}$$

$$\frac{\rho_{nf}}{\rho_f}\left(U\frac{\partial U}{\partial X}+V\frac{\partial U}{\partial Y}\right)=-\frac{\partial P}{\partial X}+\frac{\mu_{nf}}{\mu_f}\left[\frac{\partial^2 U}{\partial X^2}+\frac{\partial^2 U}{\partial^2 Y}\right]-\frac{\sigma_{nf}}{\sigma_f}M^2\left(Usin^2\alpha-Vsin\alpha cos\alpha\right)$$
$$-\frac{\mu_{nf}}{\mu_f}DaU-Fr.U\sqrt{U^2+V^2}, \tag{10.17}$$

$$\frac{\rho_{nf}}{\rho_f}\left(U\frac{\partial V}{\partial X}+V\frac{\partial V}{\partial Y}\right)=-\frac{\partial P}{\partial Y}+\frac{\mu_{nf}}{\mu_f}\left[\frac{\partial^2 V}{\partial X^2}+\frac{\partial^2 V}{\partial^2 Y}\right]-\frac{\mu_{nf}}{\mu_f}DaV-Fr.V\sqrt{U^2+V^2}$$
$$+\frac{Ra}{Pr}\frac{\left(\rho\beta\right)_{nf}}{\left(\rho\beta\right)_f}\left(\theta+Nr\phi+Rb\chi\right)-Ha^2\left(Vcos^2\alpha-Usin\alpha cos\alpha\right), \tag{10.18}$$

$$U\frac{\partial\theta}{\partial X}+V\frac{\partial\theta}{\partial Y}=-\frac{\kappa_{nf}}{\kappa_f}\frac{\left(\rho C_p\right)_f}{\left(\rho C_p\right)_{nf}}\frac{1}{Pr}\left[\frac{\partial^2\theta}{\partial X^2}+\frac{\partial^2\theta}{\partial^2 Y}\right]+\frac{\left(\rho C_p\right)_f}{\left(\rho C_p\right)_{nf}}\frac{\mu_{nf}}{\mu_f}\frac{Ec.Pr}{Ra}\left[\begin{array}{l}2\left(\dfrac{\partial U}{\partial X}\right)^2+\left(\dfrac{\partial U}{\partial Y}\right)^2\\+\left(\dfrac{\partial U}{\partial Y}+\dfrac{\partial V}{\partial X}\right)^2\end{array}\right]$$
$$+\frac{Nb}{Pr}\left(\frac{\partial\phi}{\partial X}\frac{\partial\theta}{\partial X}+\frac{\partial\phi}{\partial Y}\frac{\partial\theta}{\partial Y}\right)+\frac{Nt}{Pr}\left(\left(\frac{\partial\theta}{\partial x}\right)^2+\left(\frac{\partial\theta}{\partial y}\right)^2\right)+\frac{\sigma_{nf}}{\sigma_f}\frac{EcHa^2}{RaPr}\left(Usin\alpha-Vcos\alpha\right)^2, \tag{10.19}$$

$$\left(U\frac{\partial \phi}{\partial X}+V\frac{\partial \phi}{\partial Y}\right)=\frac{1}{Sc}\left(\frac{\partial^2 \phi}{\partial X^2}+\frac{\partial^2 \phi}{\partial Y^2}\right)+\frac{Nt}{Nb}\frac{1}{Sc}\left[\frac{\partial^2 \theta}{\partial X^2}+\frac{\partial^2 \theta}{\partial Y^2}\right], \tag{10.20}$$

$$\left(U\frac{\partial \chi}{\partial X}+V\frac{\partial \chi}{\partial Y}\right)=\frac{1}{Lb}\left(\frac{\partial^2 \chi}{\partial \chi^2}+\frac{\partial^2 \chi}{\partial^2 Y}\right)-\frac{Pe}{Lb}\left(\chi\frac{\partial^2 \chi}{\partial X^2}+\chi\frac{\partial^2 \phi}{\partial Y^2}+\frac{\partial \chi}{\partial X}\frac{\partial \phi}{\partial X}+\frac{\partial \chi}{\partial Y}\frac{\partial \phi}{\partial Y}\right). \tag{10.21}$$

Subjected to the boundary conditions:

$$\text{Hot wall}: U=V=0,\ \theta=1,\ \phi=1,\ \chi=1, \tag{10.22}$$

$$\text{Cold Wall}: U=V=0,\ \theta=0,\ \phi=0,\ \chi=0, \tag{10.23}$$

$$\text{Insulated wall}: U=V=0,\ \frac{\partial \theta}{\partial Y}=0,\ \frac{\partial \phi}{\partial Y}=0,\ \frac{\partial \chi}{\partial Y}=0. \tag{10.24}$$

The material properties of nanofluid (Brinkman and Luis, 1952; Maxwell and Thompson, 1904; Xuan and Li, 2003) may be highlighted as –

$$\left\{\begin{array}{c}
\mu_{nf}=\mu_f(1-\varphi)^{-2.5},\ \rho_{nf}=(1-\varphi)\rho_{bf}+\varphi\rho_{np},\ \varphi=\dfrac{\dfrac{W_{np}}{\rho_{np}}}{\dfrac{W_{np}}{\rho_{np}}+\dfrac{W_{bf}}{\rho_{bf}}}\\[3ex]
(\rho\beta)_{nf}=(\rho\beta)_{bf}(1-\varphi)+\varphi\dfrac{(\rho\beta)_{np}}{(\rho\beta)_{bf}},\ \sigma_{nf}=\sigma_{bf}\left[1+\dfrac{3\left(\dfrac{\sigma_{np}}{\sigma_{bf}}-1\right)\varphi}{\left(\dfrac{\sigma_{np}}{\sigma_{bf}}+2\right)-\left(\dfrac{\sigma_{np}}{\sigma_{bf}}-1\right)\varphi}\right],\\[3ex]
\kappa_{nf}=\kappa_{bf}\left(\dfrac{\kappa_{np}+\kappa_{bf}-\varphi(\kappa_{bf}-\kappa_{np})}{\kappa_{np}+\kappa_{bf}+\varphi(\kappa_{bf}-\kappa_{np})}\right),\ (\rho C_p)_{nf}=(\rho C_p)_{bf}(1-\varphi)+\varphi(\rho C_p)_{np}
\end{array}\right\}.$$

Table 10.1 can be used to calculate the thermophysical characteristics of base fluids and nanoparticles.

Table 10.1 The thermophysical properties of MoS_2 and water

Physical Properties	$\rho\left(\frac{kg}{m^3}\right)$	$C_p\left(\frac{J}{kgK}\right)$	$\beta\times10^{-5}$	$\kappa\left(\frac{W}{mK}\right)$	σ
Water	997.1	4179	21	0.613	0.05
MoS_2	5060	397.21	2.8424	35.5	2.09×10^4

10.3 NUMERICAL SOLUTION, VALIDATION, AND MESH INDEPENDENCY

To obtain the solution to the current flow problem, the FEM has been implemented via COMSOL Multiphysics, and the solutions are deliberated with the help of contours.

10.4 RESULTS AND DISCUSSIONS

The above analysis of the natural convection in the triangular cavity filled with porous medium and $MoS_2 - water$ nanofluid has been analysed. The impact of buoyancy force and Lorentz force and the permeability of the porous material on physical phenomenon have been considered. The solution of the current analysis has been obtained and illustrated with the help of COMSOL Multiphysics. Current analysis has been compared with already existing literatures and found to be in well agreement. Table 10.2 demonstrates the validity of the current investigation.

The impression of Nr on the flow pattern has been illustrated in Figure 10.3 (a–d). The relative significance of species buoyancy force and thermal buoyancy force is determined by the buoyancy parameter. The zero-velocity region is represented by the dark zones in the cavity's core; larger dark zones denote immobile fluid molecules and clockwise motion of the nanofluid along the cavity's walls. There are three major high-velocity regions, and among these regions, fluid attains its peak velocity near the slant wall. The implication of Nr on isothermal contours has been presented in Figure 10.2(b). The temperature of the nanofluid has been elevated in the central zone due to the motion of the nanofluid induced due to the temperature difference. The nanoparticles transport from vertical wall towards the slant wall, and the iso-concentration lines take parabolic shape within the enclosure which is reported in Figure 10.2(c). The density of microorganisms increases at the central part of the enclosure, and the contours of the density of microorganisms takes the parabolic shape. Figure 10.2(d) demonstrates the migration of microorganisms from the vertical wall within the cavity.

The impression of Da over the velocity magnitude, temperature, concentration, and density of microorganism has been portrayed in Figure 10.3. The Darcy number (Da) has an inverse relationship with the permeability of the porous medium. Higher values of the Darcy number result in a reduction in the permeability of

Table 10.2 Validity of the analysis

Ha	*Ashorynejad and Shahriari (2018)*	*Ghasemi et al. (2011)*	*Current analysis*
0	4.6865	4.738	4.0180
15	4.1006	4.143	3.9180
30	3.1203	3.150	3.8747
45	2.3461	2.369	3.8602

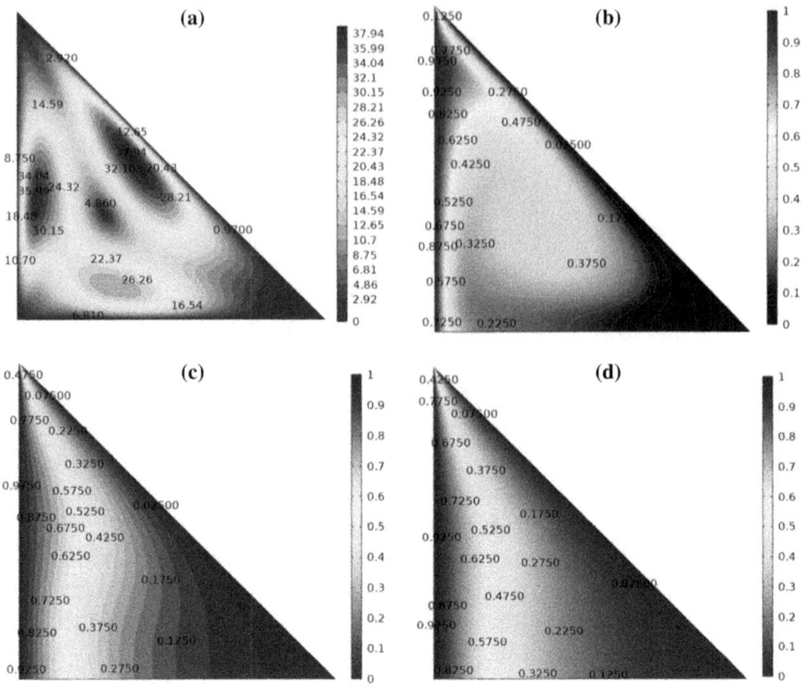

Figure 10.2 Influence of $Nr = 0.5$ on (a) velocity magnitude, (b) temperature, (c) concentration, and (d) density of microorganism.

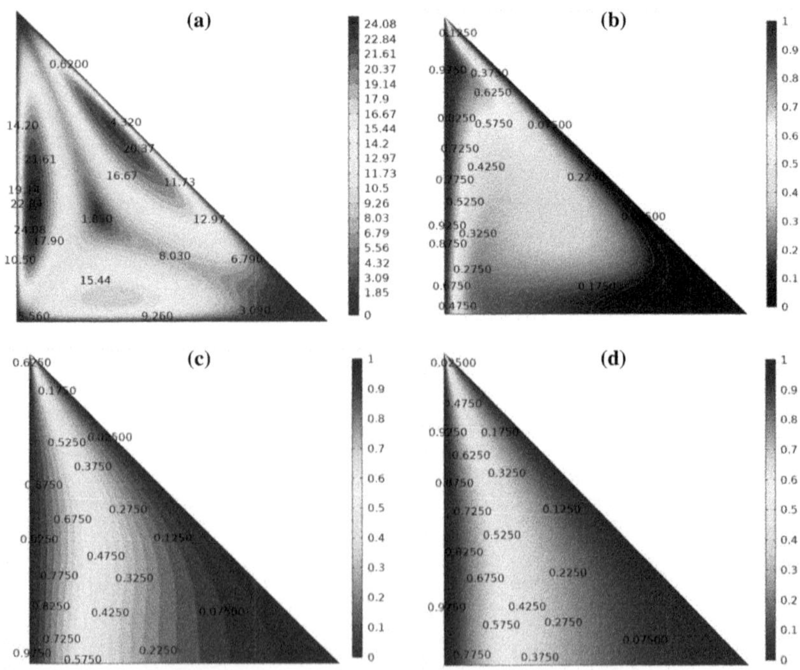

Figure 10.3 Influence of $Da = 5$ on (a) velocity magnitude, (b) temperature, (c) concentration, and (d) density of microorganism.

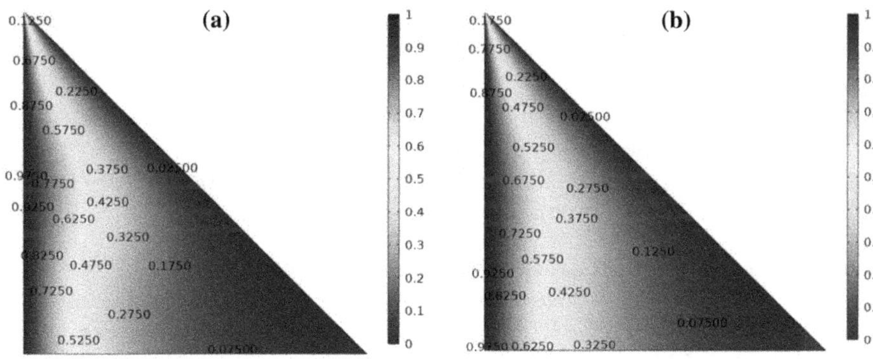

Figure 10.4 Influence of *Ha* on density of microorganisms. (a) $Ha = 0$ and (b) $Ha = 20$.

the porous medium. With an increase in *Da*, the flow becomes more restricted and constrained along the walls of the cavity, resulting in a decrease in overall fluid motion. This illustrates the substantial influence of the Darcy number on the flow behaviour and distribution of the nanofluid within the enclosure, emphasising the role of permeability in determining the flow pattern.

Figure 10.4 demonstrates the impact of *Ha* on the density of microorganism within the cavity. The augmented values of *Ha* result in the higher Lorentz force which resists the motion of the nanofluid due to which the microorganisms slow, and hence the density of microorganisms is higher at the central region of the cavity for $Ha = 0$. Since the Lorentz force resists the motion of the fluid, the resistive force tends to slow down the diffusion of microorganisms, which results in the uniform distribution of the contours of density of microorganisms.

10.5 CONCLUSION

The combination of nanofluids with enclosure problems opens a promising prospect for improving thermal efficiency and addressing heat-related concerns in various engineering fields. Natural convection of $MoS_2 - water$ nanofluid in a triangular enclosure filled with non-Darcian material has been investigated in this analysis. The combined effect of Brownian motion and thermophoretic force has been considered in mass and heat diffusion phenomena. The flow of nanofluid has been induced due to the temperature difference within the enclosure. A finite element simulation employing COMSOL Multiphysics has yielded the correct solution to the flow problem. The crucial findings in this investigation are summarised below:

1. The buoyancy ratio boosts the nanofluid while reducing the temperature of the nanofluid. The concentration of nanoparticles and density of microorganism elevates for higher *Nr* values.

2. The *Da* reduces the motion of the nanofluid, while increasing temperature of the nanofluid. The isothermal lines become more uniform for higher *Da* values.
3. The parabolic nature of the iso-concentration and density of microorganism diminishes for higher *Da*.
4. Higher Lorentz force resulted in the uniform distribution of iso-concentration and density of microorganisms.

In many engineering appliances, friction is the main cause of energy loss, and lubrication is the way to reduce the friction; hence, using lubricants in place of water as a base fluid can give fruitful results. In this investigation, water is used as a base fluid and MoS_2 nanoparticles for the preparation of nanofluid. Due to some limitations and the better scope of non-Newtonian fluid, this study can be conducted in the future in the future using non-Newtonian fluid as the base fluid. This study can be beneficial in developing mathematical models to study the effects of other physical phenomena by incorporating minor changes in the governing equations. In this investigation, only one type of nanoparticle is added to the base fluid; however, for an advanced rate of heat transmission and an effective cooling process, one can explore the hybrid nanofluid.

NOMENCLATURE

B_0	$A.m^{-1}$	Magnetic field strength	T	K	Temperature
C	$mole.m^{-3}$		T_c	K	Temperature of the cold surface
C_b		Drag coefficient of porous medium	T_h	K	Temperature of the hot surface
C_p	$J.kg^{-1}.K^{-1}$	Specific heat	(u, v)	$m.s^{-1}$	Dimensional velocity field
Da		Darcy number	(U,V)		Non-dimensional velocity field
Ec		Eckert number	U_0	$m.s^{-1}$	Characteristic velocity
Fr		Forchheimer parameter	(x, y)		Dimensional Cartesian coordinates
g	$m.s^{-2}$	Acceleration due to gravity	(X,Y)		Non-dimensional Cartesian coordinates
Gr		Grashof number	β	K^{-1}	Coefficient of thermal expansion
k		Permeability of porous medium	θ		Non-dimensional temperature
L	m	Characteristic length	κ	$W.m^{-1}.K^{-1}$	Thermal conductivity

Lb		Bioconvection Lewis number	ν	$m^2.sec^{-1}$	Kinematic viscosity
M		Magnetic field parameter	ρ	$kg.m^{-3}$	Density
p	*Pa*	Dimensional pressure	σ	$sec.m^{-1}$	Electrical conductivity
P		Non-dimensional pressure	φ		Volume fraction
Pe		Peclet number	μ	$kg^{-1}.sec^{-1}$	Dynamic Viscosity
Pr		Prandtl number	*f*		base fluid
			nf		Nanofluid
Ra		Rayleigh number			
Rb		Bioconvection Rayleigh number	FEM		Finite element method
Re		Reynolds number			

REFERENCES

Ahmed, K. K., Ahmed, S., & Chamkha, A. J. (2023). Impact of inclined magnetic force on bio-fluid in permeable bifurcated arteries: Analytical approach. *Journal of Nanofluids*, *12*(2), 332–340.

Alluguvelli, R., Balla, C. S., Naikoti, K., & Makinde, O. D. (2022). Nanofluid bioconvection in porous enclosure with viscous dissipation. *Indian Journal of Pure & Applied Physics (IJPAP)*, *60*(1), 78–89.

Ashorynejad, H. R., Shahriari, A. (2018). MHD natural convection of hybrid nanofluid in an open wavy cavity. *Results in Physics*, *9*, 440–455.

Azzam, G. E. D. A. (2002). Radiation effects on the MHD mixed free-forced convective flow past a semi-infinite moving vertical plate for high temperature differences. *Physica Scripta*, *66*(1), 71.

Balla, C. S., Ramesh, A., Kishan, N., Rashad, A. M., & Abdelrahman, Z. M. A. (2020). Bioconvection in oxytactic microorganism-saturated porous square enclosure with thermal radiation impact. *Journal of Thermal Analysis and Calorimetry*, *140*(5), 2387–2395.

Brinkman, H. C., & Luis, M. (1952). The viscosity of concentrated suspensions and solutions. *The Journal of Chemical Physics*. 20, 571. Doi: https://doi.org/10.1063/1.1700493.

Choi, S. U., & Eastman, J. A. (1995). Enhancing thermal conductivity of fluids with nanoparticles (No. ANL/MSD/CP-84938; CONF-951135-29). Argonne National Lab (ANL), Argonne, IL (United States).

Elshehabey, H. M., Hady, F. M., Ahmed, S. E., & Mohamed, R. A. (2014). Numerical investigation for natural convection of a nanofluid in an inclined L-shaped cavity in the presence of an inclined magnetic field. *International Communications in Heat and Mass Transfer*, *57*, 228–238.

Ghasemi, B., Aminossadati, S. M., Raisi, A. (2011). Magnetic field effect on natural convection in a nanofluid-filled square enclosure. *International Journal of Thermal Sciences*, *50*(9), 1748–1756.

Hazarika, S., & Ahmed, S. (2021). Study of carbon nanotubes with a Casson fluid in a vertical channel of porous media under MHD and Dufour effect. *Journal of Scientific Research, 13*(1), 31–45.

Hazarika, S., & Ahmed, S. (2023). Physical insights on bio-convection in Prandtl nanofluid over an inclined stretching sheet in non-Darcy medium: Numerical simulation. *Scientia Iranica*. DOI: https://doi.org/10.24200/sci.2023.62949.8125

Hussain, S., Aly, A. M., &Alsedias, N. (2022a). Bioconvection of oxytactic microorganisms with nano-encapsulated phase change materials in an omega-shaped porous enclosure. *Journal of Energy Storage, 56*, 105872.

Hussain, S., Raizah, Z., & Aly, A. M. (2022b). Thermal radiation impact on bioconvection flow of nano-enhanced phase change materials and oxytactic microorganisms inside a vertical wavy porous cavity. *International Communications in Heat and Mass Transfer, 139*, 106454.

Ingham, D. B., & Pop, I. (1998). *Transport phenomena in porous media*. Oxford: Elsevier.

Izadpanah, F., Sadegh Sadeghi, M., Ghodrat, M., & Behnia, M. (2022). Natural thermo-bio convection of gyrotactic micro-organisms in a square cavity with two heaters inside: Application to ocean ecosystems. *International Journal of Environmental Studies, 81*(3), 1390–1412.

Kalita, D., Hazarika, S., & Ahmed, S. (2020). Applications of CNTs in a vertical channel of porous medium for human blood flow: A rheological model. *JP Journal of Heat and Mass Transfer, 20*(2), 105–120.

Maxwell, J. C., & Thompson, J. J. (1904). *A treatise on electricity and magnetism*. Oxford, UK: Oxford University Press.

Nield, D. A., & Bejan, A. (2006). *Convection in porous media* (Vol. *3*, pp. 629–982). New York: springer.

Pelekasis, N., & Benos, L. (2017). Static arrangement of a capillary porous system (CPS): Modelling. *Fusion Engineering and Design, 117*, 180–187.

Vafai, K. (Ed.). (2015). *Handbook of porous media*. New York: CRC Press.

Vajravelu, K., & Sastri, K. S. (1978). Free convective heat transfer in a viscous incompressible fluid confined between a long vertical wavy wall and a parallel flat wall. *Journal of Fluid Mechanics, 86*(2), 365–383.

Xuan, Y., & Li, Q. (2003). Investigation on convective heat transfer and flow features of nanofluids. *Journal of Heat Transfer, 125*(1), 151–155.

Chapter 11

Activation energy and inclined magnetization on convective transportation of Casson nanofluid with thermal radiation

Vivek Kumar, Paras Ram, Devender, and Kushal Sharma

11.1 INTRODUCTION

During the 1995 annual conference of the American Society of Mechanical Engineers (Choi and Eastman, 1995) coined the term "nanofluid" to define a particular category of conventional transport fluids that utilize nanoscales and exhibit superior heat transfer phenomena compared to their traditional base fluids. A nanofluid is a blend comprising nanoparticles, which are characterized by their exceedingly minute dimensions, usually less than one nanometer. Fluids are transported via a medium that contains suspended colloidal particles. The nanoparticles included in nanofluidic systems come from several sources, including carbon nanotubes, carbides, oxides, and metals. Researchers have found that employing nanoparticles can improve thermal conductivity and heat transfer characteristics. Nanofluids research primarily centers on the application of colloidal theory and surface science in materials.

Andreas *et al.* (1960) aimed to investigate the transfer of momentum and heat in non-Newtonian fluids as they flow in a laminar fashion across exterior surfaces. The concept of boundary layers is widely acknowledged as a highly efficient simplification in the field of mechanics for Newtonian fluids, as evidenced by Herrmann and Gersten (2017). Contemporary perspectives generally agree that non-Newtonian fluids are more suitable for practical industrial applications compared to Newtonian fluids. These fluids, known for their departure from the dynamic assumptions of traditional fluids, are widely used in numerous industrial areas. The practical uses of these technologies include improving the efficiency of reducing resistance, thrust bearings, and rotating diffusion devices; adopting cooling ways employing perspiration techniques; and strengthening thermal oil recovery activities. Recently, many authors (Khan, 2006; Rehman *et al.*, 2017; Vijay and Sharma, 2023; Sharma *et al.*, 2022) are working on these types of problems.

The process of heat transmission in engineering and scientific procedures heavily depends on the structure from which thermal energy moves to the working fluid. The phenomena of heat transmission arise from variations in temperature. A heat exchanger is a primary thermal device utilized for heat transmission and is extensively employed in various thermal applications. Using

DOI: 10.1201/9781003527251-11

a variety of nanoparticles with varying shapes, Hussanan *et al.* (2017, 2018) and Hussanan and Trung (2019) investigated convective heat transfer phenomena. The most current research on heat transport and nanofluid can be found in studies (Ashraf *et al.*, 2017; Rafque *et al.*, 2019; Naik *et al.*, 2020; Shaiq *et al.*, 2023; Sharanayya and Biradar, 2023; Sharma *et al.*, 2023; Kanchi *et al.*, 2024).

There are important applications of an inclined magnetization in many fields of science and technology. This has important implications for fluid dynamics research as well as for the advancement of energy production, propulsion, and plasma control technologies. When heat generation, chemical reactions, and radiation all happen at the same time. The references (Devi *et al.*, 2020; Aamir Abbas *et al.*, 2021; Sahoo and Nandkeolyar, 2022; Mahato *et al.*, 2022; Sarwar *et al.*, 2022; Kumar and Sharma, 2023; Kosar *et al.*, 2024; Leelavathi *et al.*, 2024) include some recent work with an angled magnetic field.

In the present analysis, we have investigated the impact of inclined magnetization and activation energy on the convective transportation of Casson nanofluid across a stretched surface with thermal radiation and heat absorption/generation. We used BVP4C scheme to find out the solution of similarity converted system of coupled nonlinear ordinary differential equations (ODEs). Here, using the graphical representation, we have examined the impact of various prominent flow parameters on the velocity, thermal, and solute distribution in the boundary regime.

11.2 PROBLEM FORMULATION

In this investigation, we have considered viscous incompressible convective transportation of Casson nanofluid through a stretched surface with activation energy and thermal radiation. The velocity change along the x-axis is $U_w = ax$ and in y-axis is $u_2 = u_{2w}$. Here, the convective heating is considered at the place of constant temperature at the wall. The temperature and concentration value in the free stream are represented by T_∞ *and* C_∞, respectively. A magnetic field of strength $B(x) = B_0^2(x)\sin^2\beta$ is applied at an angle β perpendicular to the fluid flow. It is crucial to consider that the magnetization frequently influences the electrical conductivity of the liquid. Neglecting the induced magnetization strength, the variation in Reynolds number has an insignificant effect. In addition, we considered the utilization of heat sources and chemical reaction, both significantly increased the flow.

The basic equation of rheological that governs the flow of an incompressible and non-isotropic Casson fluid is defined as a function of the Cauchy stress tensor as follows:

$$\tau_{ij} = \begin{cases} 2\left(\mu_B + \dfrac{p_y}{\sqrt{2\pi}}\right)e_{ij},\ \pi > \pi_c \\[2ex] 2\left(\mu_B + \dfrac{p_y}{\sqrt{2\pi_c}}\right)e_{ij},\ \pi_c > \pi. \end{cases} \tag{11.1}$$

Here, P_y represents the yield stress, the rate of deformation is denoted by π, and at critical level is π_c. μ_B stands for dynamic value of plastic viscosity.

Using the above assumption, the governing equations are as follows:

$$\frac{\partial u_1}{\partial x}+\frac{\partial u_2}{\partial y}=0, \tag{11.2}$$

$$u_1\frac{\partial u_1}{\partial x}+u_2\frac{\partial u_1}{\partial y}=v\left(1+\frac{1}{\gamma}\right)\frac{\partial^2 u_1}{\partial y^2}-\frac{\sigma B_0^2(x)\sin^2\beta}{\rho_f}u_1, \tag{11.3}$$

$$u_1\frac{\partial T}{\partial x}+u_2\frac{\partial T}{\partial y}=\alpha^*\frac{\partial^2 T}{\partial y^2}+\frac{(\rho c)_p}{(\rho c)_f}\left(\frac{D_T}{T_\infty}\left(\frac{\partial T}{\partial y}\right)^2+D_B\frac{\partial C}{\partial y}\frac{\partial T}{\partial y}\right)+\frac{v}{c_p}\left(1+\frac{1}{\gamma}\right)\left(\frac{\partial u_1}{\partial y}\right)^2$$
$$+\frac{\sigma B^2(x)\sin^2\beta}{\rho_f}u_1^2+\frac{16\sigma^* T^3}{3k^*\rho_f C_p}\frac{\partial^2 T}{\partial y^2}, \tag{11.4}$$

$$u_1\frac{\partial C}{\partial x}+u_2\frac{\partial C}{\partial y}=D_B\frac{\partial^2 C}{\partial y^2}+\frac{D_T}{T_\infty}\frac{\partial^2 T}{\partial y^2}--k_r^2(C-C_\infty)\left(\frac{T}{T_\infty}\right)^m\exp\left(-\frac{E_a}{k^*T}\right), \tag{11.5}$$

and corresponding boundary constraints are as follows:

$$u_1=U_w,\ -k_1\frac{\partial T}{\partial y}=h_1(T_w-T),\ u_2=u_{2w},\ -k_2\frac{\partial T}{\partial y}=h_2(C_w-C)\ \text{at}\ y=0, \tag{11.6}$$

$$u_2=0,\ T\to T_\infty,\ u_1\to 0,\ C\to C_\infty\ \text{as}\ y\to\infty. \tag{11.7}$$

Here, $B(x)=B_0^2(x)\sin^2\beta$ is assumed as variable inclined magnetic field.

Now to simplify the partial differential equations (PDEs), we use appropriate scaling transformation defend as follows:

$$\begin{gathered}\eta=y\sqrt{\frac{a}{v}},\ \psi=\sqrt{a\,v}\ \ x\,f(\eta),\\ u_1=\frac{\partial\psi}{\partial y},\ \ u_2=-\frac{\partial\psi}{\partial x}\\ \Phi(\eta)=\frac{C-C_\infty}{C_w-C_\infty},\ \Theta(\eta)=\frac{T-T_\infty}{T_w-T_\infty}.\end{gathered} \tag{11.8}$$

Using Eq. (11.8), the converted system of coupled nonlinear ODEs is as follows:

$$\left(1+\frac{1}{\gamma}\right)f'''+ff''-f'^2-M\cdot\sin^2\beta f'=0, \tag{11.9}$$

$$(1+\mathrm{R})\Theta'' + \Pr\left(f\ \Theta' + Nb\ \Theta'\Phi' + Nt\Theta'^2 + \left(1+\frac{1}{\gamma}\right)Ecf''^2 + +M\sin^2\beta f'^2\right) = 0, \tag{11.10}$$

$$\Phi'' + Sc\ f\ \Phi' + \frac{Nt}{Nb}\Theta'' - \sigma^*(1+\delta\theta)^m \exp\left(\frac{-E}{(1+\delta\theta)}\right)\Phi = 0. \tag{11.11}$$

And modified boundary constraints are

$$\left.\begin{array}{c} f = s,\ f' = 1,\ \Phi'(0) = -Bi_2(1-\Phi(0)),\ \Theta'(0) = -Bi_1(1-\Theta(0))\text{at } \eta = 0 \\ f' = 0,\ \Phi = 0,\ \Theta = 0 \text{ as } \eta \to \infty \end{array}\right\}. \tag{11.12}$$

11.2.1 The quantity of physical interest

The quantity of physical interest "the skin friction C_{fx}", "the local Nusselt number Nu_x", and "the local Sherwood number Sh_x" are well-defined as follows:

$$C_{fx} = \frac{\tau_w}{\rho U_w^2},\ Nu_x = \frac{xq_w}{k(T_w - T_\infty)},\ Sh_x = \frac{xq_m}{D_B(C_w - C_\infty)}. \tag{11.13}$$

After simplification, the skin friction, the reduced Nusselt, and Sherwood numbers are written as follows:

$$Re_x^{1/2}Cf_x = \left(1+\frac{1}{\gamma}\right)f''(0), \tag{11.14}$$

$$Re_x^{-1/2}Nu_x = -(1+\mathrm{R})\Theta'(0), \tag{11.15}$$

$$Re_x^{-1/2}Sh_x = -\Phi'(0). \tag{11.16}$$

where $Re_x = \dfrac{xU_w}{\nu}$ is the local Reynold number.

11.3 RESULT AND DISCUSSIONS

The system of nonlinear ODEs (11.9)–(11.11) along with the Eq. (11.12) is computed using the finite difference method BVP4C method.

In this section, we use the graphic depiction to examine the effects of several variables on the velocity, temperature, and concentration variations. The predetermined values of each parameter in the equations that control the flow in the boundary region are as follows:

$$Pr = 0.71,\ Sc = 0.6,\ \gamma = 1.5,\ M = 0.2,\ Nb = Nt = Ec = 0.1,$$

$$\beta = \frac{\pi}{3},\ \sigma = 0.1,\ \ t_w = 0.2,\ m = 0.1 = Bi_1 = Bi_2.$$

11.3.1 Velocity distribution

In Figure 11.1, we have discussed the physical impact of Hartmann number M and angle of field inclination β on the velocity $f'(\eta)$ in the fluid zone.

Figure 11.1 (a) illustrates the influence of the Hartmann number M on the velocity of the working fluid $f'(\eta)$. The figure illustrates that the velocity field diminishes as the Hartmann number Mincreases, in direct proportion to the parameter η. Higher magnetic intensity results in a stronger Lorentz force in the fluid flow region, leading to a reduction in velocity. The correlation between the inclination angle β and velocity curve can be shown in Figure 11.1(b). The greater β value results in the amplification of the velocity distribution in the boundary regime.

11.3.2 Temperature distribution

Figure 11.2 examines the influence of distinct fluid parameters such as thermal radiation R, Prandtl number Pr, and the convection Biot number Bi_1.

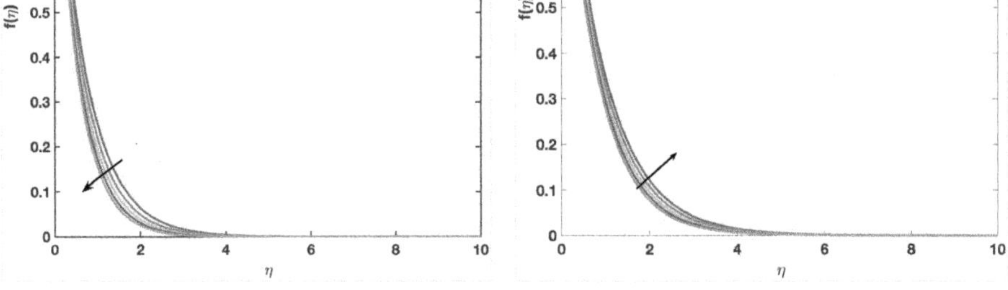

Figure 11.1 Percussion of distinct flow variable on velocity distribution (a) with M and (b) with β.

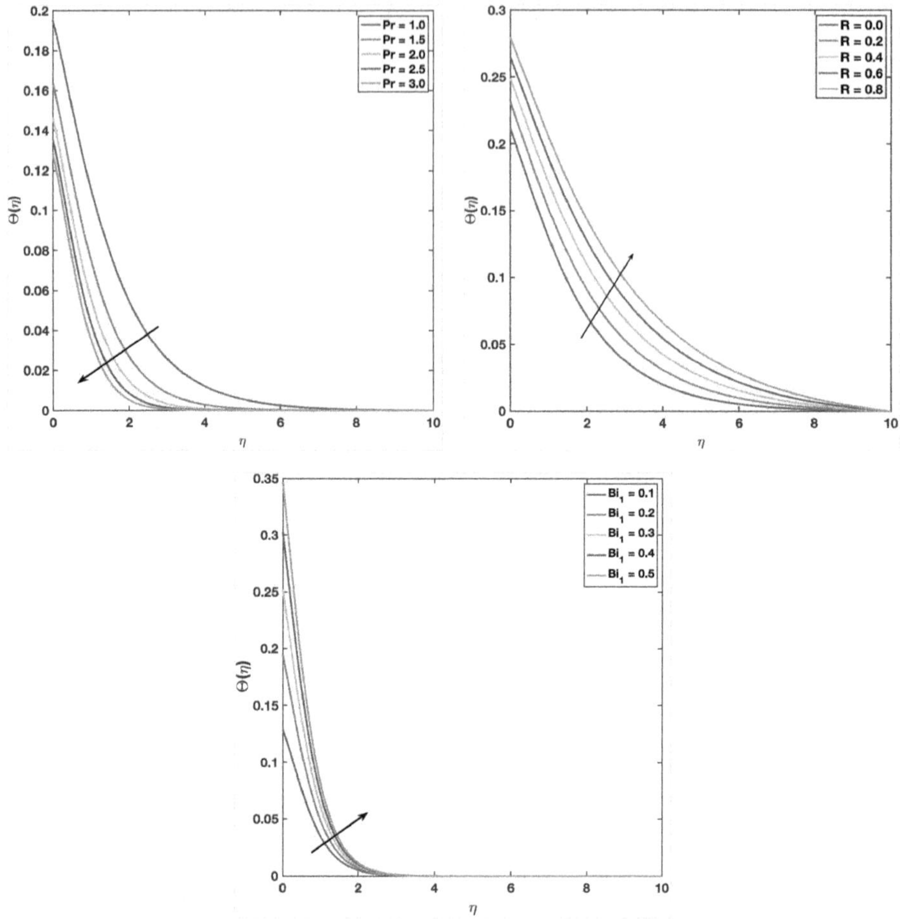

Figure 11.2 Percussion of distinct flow variable on temperature distribution (a) with *Pr*, (b) with *R*, and c) with Bi_1.

Figure 11.2(a) displays how the Prandtl number *Pr* affects the thermal diffusion. Raising the Prandtl value resulted in a diminution of the thermal diffusion field. In physical manner, *Pr* characterizes relationship between thermal and momentum diffusion. In this case, rising *Pr* reduces the thermal conductivity in the boundary region, which leads to decline in thermal diffusion. Additionally, there exists an inverse relationship between thermal diffusivity and kinematic viscosity, which has a major impact on the thermal area. Figure 11.2(b) demonstrates the effect of the thermal radiation *R* on the temperature profile. The heat transfer fields are expedited by increasing the values of R. By increasing the thermal radiation, additional heat is introduced into the flow system. The impact of thermal Biot number Bi_1 on thermal field

$\Theta(\eta)$ is examined in Figure 11.2(c). We found that increasing Bi_1 accelerates the temperature $\Theta(\eta)$ in boundary regime due to a larger value of heat transfer coefficient.

11.3.3 Concentration distribution

Figure 11.3 shows the impact of distinct fluid variables such as the activation energy E, the concentration Biot number Bi_2, and the Brownian diffusion parameter Nb.

Figure 11.3(a) illustrates the impact of E on mass transfer $\Phi(\eta)$. An elevation in E enhances the concentration gradient. The Arrhenius equation is a mathematical expression that defines how activation energy affects a system.

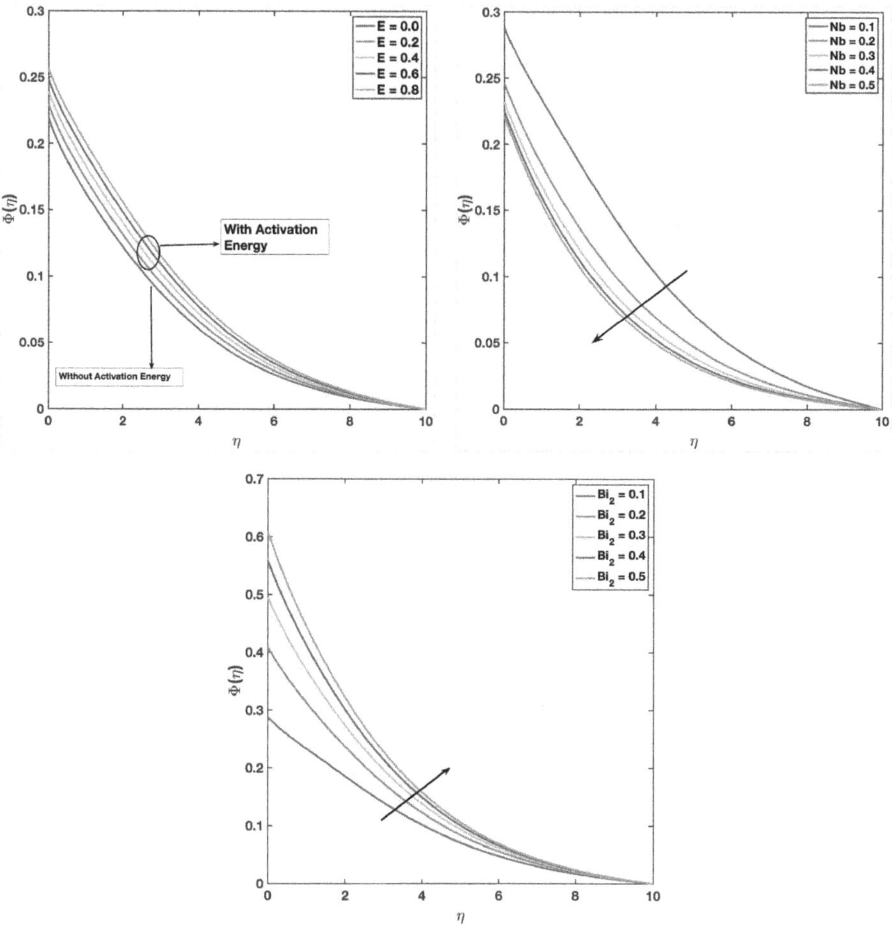

Figure 11.3 Percussion of distinct flow variable on concentration distribution (a) with E, b) with Nb, and c) with Bi_2.

It indicates that diminish in heat and acceleration in E lead to a lower response rate constant. This leads to a deceleration of the chemical process, which raises the particle concentration. With an increasing activation energy E, the modified Arrhenius process becomes less robust. This ultimately triggers the generative chemical process, leading to a rise in the concentration of nanoparticles. Figure 11.3(b) depicts the fluctuations in mass diffusion when the Nb is altered. The thickness of the concentrate boundary layer decimates with the increased values of Nb. An increase in Nb values has led to a corresponding escalation in mass diffusion within the boundary regime. Figure 11.3(c) reveals the impact of Bi_2 on $\Phi(\eta)$. It is noticed that concentration distribution and its thickness of the boundary layer accelerate with higher Bi_2. As the rate of mass transfer increases with an improvement in Bi_2, this leads to an enhancement in mass distribution in the boundary regimes.

11.4 CONCLUSIONS

In the present analysis, we effectively utilized the BVP4C scheme to conduct a comparative study of the impact of inclined magnetization and activation energy on convective transportation of Casson nanofluid with thermal radiation across a stretched plate using the convective boundary condition. The analysis focuses on examining the representation of velocity, temperature, and concentration under various physical conditions. In summary, the findings of this investigation can be described as follows:

- Velocity in boundary regime reduced with M and β.
- Temperature distribution accelerates with R and Bi_1, whereas decline with Pr.
- Mass diffusion in fluid flow zone enhanced with E *and* Bi_2, whereas a significant decline with Nb.

NOMENCLATURE

∞	Condition at the free stream
ψ	Stream function
η	Non-dimensional variable
σ	Electrical conductivity
γ	Casson fluid parameter
τ_w	Wall shear stress
β	Angle of field inclination
α^*	Thermal diffusivity
σ^*	Chemical reaction rate parameter
ρc_p	Nanoparticle heat capacity

Bi_1	Thermal Biot number
Bi_2	Concentration Biot number
D_B	Brownian diffusion coefficient
E	Dimensionless activation energy
Ec	Eckert number
k_r	Chemical reaction coefficient
m	Fitted rate constant
M	Hartmann number
Nb	Brownian diffusion coefficient
Nt	Thermophoresis parameter
Pr	Prandtl number
q_m	Mass flux
q_w	Thermal flux
Sc	Schmidt number
(u_1, u_2)	Velocity Component
w	Condition at the surface

REFERENCES

Aamir Abbas, K., Awais, A., Sameh, A., Ashrafa, M., Hijaz, A., & Muhammad Naveed, K. (2021). Influence of the induced magnetic field on second-grade nanofluid flow with multiple slip boundary conditions. Waves Random Complex Media, 16(1), 1745–5030.

Andreas, A., Shah, M. J., & Petersen, E. E. (1960). Momentum and heat transfer in laminar boundary-layer flows of non-Newtonian fluids past external surfaces. AIChE Journal, 6(2), 312–317.

Ashraf, B. M., Hayat, T., & Alsaedi, A., (2017). Mixed convection flow of Casson fluid over a stretching sheet with convective boundary conditions and Hall effect. Boundary Value Problems. 137, https://doi.org/10.1186/s13661-017-0869-7.

Choi, S., & Eastman, J. (1995). Enhancing thermal conductivity of fluids with nanoparticles. ASME-Publications-Fed, 231, 99–106.

Devi, R., Poply, V., & Manimala (2020). Impact of aligned MHD flow with inclined outer velocity for a Casson nanofluid over a stretching sheet. Heat Transfer, 49(6), 3702–3725.

Herrmann, S., & Gersten, K. (2017). Boundary-layer control (suction/blowing). Boundary-Layer Theory, 291–320. https://doi.org/10.1007/978-3-662-52919-5_11.

Hussanan, A., Salleh, M. Z., Taha, T. A., & Khan, I. (2018). MHD flow and heat transfer in a Casson fluid over a nonlinearly stretching sheet with Newtonian heating. Heat Transfer Research, 49(12), 1185–1198.

Hussanan, A., & Trung, N. T. (2019). Heat transfer analysis of sodium carboxymethyl cellulose based nanofluid with titania nanoparticles. Journal of Advanced Research in Fluid Mechanics and Thermal Sciences, 56(2), 248–256.

Hussanan, A., Zuki, M. S., Khan, I., & Shafe, S. (2017). Convection heat transfer in micropolar nanofluids with oxide nanoparticles in water, kerosene and engine oil. Journal of Molecular Liquids, 229, 482–488.

Kanchi, S., Gaddala, P. R., & Gurram, S. (2024). Impact of activation energy, diffusion thermo, thermal diffusion and Hall current on MHD Casson fluid flow with inclined plates. CFD Letters, 16(6), 90–108.

Khan, S. K. (2006). Boundary layer viscoelastic fluid flow over an exponentially stretching sheet. International Journal of Applied Mechanics and Engineering, 11(2), 321–335.

Kosar, S., Sagheer, M., & Hussain, S. (2024). Effect of inclined magnetic field and chemical reaction on Casson fluid flow over a stretching surface with Cattaneo–Christov double diffusion. International Journal of Modern Physics B, 2550002. https://doi.org/10.1142/S021797922550002X.

Kumar, S. and Sharma, K. (2023). Entropy optimization analysis of Marangoni convective flow over a rotating disk moving vertically with an inclined magnetic field and nonuniform heat source. Heat Transfer, 52(2), 1778–1805.

Leelavathi, R., Vyakaranam, S., Rao, T. S., Gurrampati, V. R. R., & Oke, A. S. (2024). MHD Casson fluid flow in stagnation-point over an inclined porous surface. CFD Letters, 16(4), 69–84.

Mahato, R., Das, M., Sen, S. S. S., & Shaw, S. (2022). Entropy generation on unsteady stagnation-point Casson nanofluid flow past a stretching sheet in a porous medium under the influence of an inclined magnetic field with homogeneous and heterogeneous reactions. Heat Transfer, 51(6), 5723–5747.

Naik, H. S., Goud, B. S., Suresh, P., & Ramana Murthy, M. V. (2020). Radiation and Hall effect on MHD mixed convection of Casson fluid over a stretching sheet. International Journal of Advanced Science and Technology, 29, 1121–1131.

Rafque, K. et al. (2019). Numerical solution of Casson nanofluid flow over a non-linear inclined surface with Soret and Dufour effects by Keller-Box method. Frontiers in Physics, 7, 139, https://doi.org/10.3389/fphy.2019.00139.

Rehman, K. U., Malik, A. A., Malik, M. Y., Sandeep, N., & Ul Saba, N. (2017). Numerical study of double stratification in Casson fluid flow in the presence of mixed convection and chemical reaction. Results in Physics, 7, 2997–3006.

Sahoo, A., & Nandkeolyar, R. (2022). Entropy generation analysis on convective radiative stagnation point flow of a Casson nanofluid in a non-Darcy porous medium with induced magnetic field and activation energy: The multiple quadratic regression model. Waves Random Complex Media, 402, 1–45.

Sarwar, N., Asjad, M. I., Hussain, S., Alam, M. N., & Inc, M. (2022). Inclined magnetic field and variable viscosity effects on bio-convection of Casson nanofluid slip flow over non linearly stretching sheet. Propulsion and Power Research, 11(4), 565–574 https://doi.org/10.1016/j.jppr.2022.09.002.

Shaiq, S., Maraj, E. N., & Shahzad, A. (2023). An unsteady instigated induced magnetic field's influence on the axisymmetric stagnation point flow of various shaped copper and silver nanomaterials submerged in ethylene glycol over an unsteady radial stretching sheet. Numerical Heat Transfer, Part A: Applications, 85(6), 822–844.

Sharanayya, & Biradar, S. (2023). Magnetized dissipative Casson nanofluid flow over a stretching sheet with heat source/sink and Soret effect under porous medium. Bio-NanoScience, 13(4), 2103–2121.

Sharma, K., Kumar, L., Singh, A., & Joshi, V. K. (2023). Dynamics of MHD Casson fluid in non-Darcy porous medium: The impact of thermal radiation, Dufour–Soret, and chemical reaction. Modern Physics Letters B, 38(16), 2341007.

Sharma, K., Vijay, N., & Kumar, S. (2022). Significance of geothermal viscosity and heat generation/absorption on magnetic nanofluid flow and heat transfer. Numerical Heat Transfer, Part A: Applications, 82(3), 70–81.

Vijay, N., & Sharma, K. (2023). Dynamics of stagnation point flow of Maxwell nanofluid with combined heat and mass transfer effects: A numerical investigation. International Communications in Heat and Mass Transfer, 141, 106545.

Chapter 12

Numerical computation of bioconvection flow in nanofluid over stretching sheet in non-Darcy medium with Forchheimer correction

Bikash Das, Mayzul Alom Hussain, and Sahin Ahmed

12.1 INTRODUCTION

Modern technology and machinery are essential to our survival. We are improving them in a number of ways for efficient utilization. So, an innovative method was discovered a few decades ago by Choi and Eastman (1995) to add nanosized (diameter < 100 nm) particles into base liquids to improve the capacity of machines to cool. The dispersed particles, which are typically metallic or metallic oxides, offer greater heat transfer qualities when compared to ordinary fluids, and it has been shown that adding nanoparticles to a base fluid increases the ability to transfer heat of the resulting nanofluid. Nanofluids improved heat transfer properties which have made them effective in a number of heat exchange processes with several industrial uses, including the pasteurization of food, fuel cells, biomedicine/microelectronics, and cancer treatment. In comparison to base fluids like water, lubricating oil, and ethylene glycol, they found that nanofluids provide better heat conductivity. Bejan (1979) and Bejan and Kestin (1983) conducted an investigation on entropy formation in convective heat transport. Later, Choi *et al.* (2007), Hai *et al.* (2008), Lee *et al.* (1999), and Yoo *et al.* (2007) used this idea while taking into account the various sizes and shapes of nanoparticles. An empirical examination of the suspension of copper nanoparticles in ethylene glycol was conducted by Eastman *et al.* (2001) and they found that suspended nanomaterials enhance the ability to transfer heat of ethylene glycol by up to 40%, and thus increasing the ability of heat transfer in the flow. Nanofluids have been used in a variety of applications, including power plant fuel minimization, vehicle cooling, and nanodrug delivery. Employing nanofluidic coolants in electronics saves energy and reduces pollutants, resulting in lower production costs as claimed by Wong and De (2010). The flow of nanofluid through a stretched cylinder while considering both radiation and natural convection has been investigated by Pandey and Kumar (2017). Bhattacharyya *et al.* (2022) investigated the effect of hybrid nanostructures in nanofluids and employed a numerical method to assess the features of copper- and graphene-type nanoparticles. Gowda *et al.* (2021) investigated the impact of heat and mass transfer on the flow of non-Newtonian Marangoni-driven nanofluid and also focused on the activation energy and

DOI: 10.1201/9781003527251-12

second-order chemical processes. The impact of thermal radiation due to heat transport, mechanism of slip flow, and wall suction on non-steady nanofluid with Lorentz force towards the wall in a stretched surface was investigated by Sreedevi *et al.* (2020). Notably the authors (Cui *et al.*, 2022; Gowda *et al.*, 2021; Hayat *et al.* 2020a; Rawat *et al.*, 2021) have scrutinized the boundary layer flow in nanofluid across stretched sheets (linear/non-linear) under various physical circumstances. The application of the Coriolis force on the dynamics of a Carreau-Yasuda spinning nanofluid subjected to the Darcy-Forchheimer model was investigated by Ahmad *et al.* (2022). Rehman *et al.* (2022) developed a mathematical model for two-phase flow, focusing on the relationship of Casson nanofluid and dust particles on a stretched surface and taking into account magnetohydrodynamics (MHD) Darcy-Forchheimer porous medium.

Bioconvection is the large-scale microbial motion in fluids, involving gyrotactic, oxytactic, gravitactic, or chemotactic microorganisms swimming upstream against gravity. It is used in various industries, including pharmaceuticals, green technologies, sustainable fuel cells, oil recovery, biosensors, and biotechnology. Gautam *et al.* (2021) studied MHD flow of Williamson nanofluid in Darcy-Forchheimer porous medium across a stretched cylinder. The impact of Ohmic heating on Darcy-Forchheimer flow of carbon nanotube (CNT) nanofluid over a stretching surface was explored by Upreti *et al.* (2020). Eminent researchers (Hosseinzadeh *et al.*, 2019; Khan *et al.*, 2019; Khan *et al.*, 2020; Liu *et al.*, 2019) investigated on the flow of nanofluid with entropy generation. Moreover, many significant research have been done on nanofluid by Hayat *et al.* (2020b), Iijima (1991), Joshi *et al.* (2022), Khan *et al.* (2014), and Shaiq *et al.* (2019).

In the present investigation, a steady boundary layer natural bioconvection flow from a horizontal surface embedded in a porous medium filled by a $TiO_2 - water$ nanofluid containing both nanoparticles and microorganisms has been considered. The novelty of the investigation is to study the effect of bioconvection on the energy diffusion as well as on the diffusion of nanoparticles from the horizontal surface under the influence of Lorentz force. The current study has potential for the investigation of microbial fuel cells, utilizing nanofluid thermo-bioconvection in porous media for efficient fuel cells based on haemophilic bacteria like Bacillus licheniformis and Bacillus thermoglucosidasius. The non-linear partial differential equations (PDEs) have been transformed into system of ordinary differential equations (ODEs) by using appropriate similarity transformations and finally solved via MATLAB BVP4C solver. The stability and validity of the study have been carried out with grid stability analysis and by comparing them with the existing literature, respectively.

Throughout the investigation, the prime objective will be to study:

1. How the microorganisms effect the heat diffusion process?
2. What influence does a nanoparticle's volume fraction have on species diffusion?
3. Which nanoparticles are optimal for better heat diffusion?
4. The effect of non-dimensional parameters on the physical processes.

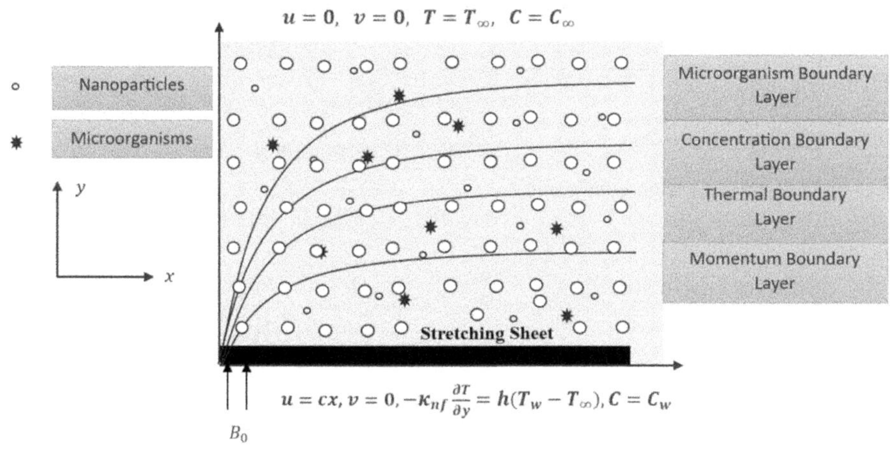

Figure 12.1 Schematic diagram of the flow model.

12.2 FORMULATION OF THE PROBLEM

A mathematical model has been developed to study the laminar, steady, and incompressible flow of MHD Forchheimer nanofluid $(TiO_2 - water)$ over a stretching surface with application of wall suction and wall injection. The effect of Soret and Dufour has been considered in the heat and species diffusion phenomenon. The momentum of the fluid is induced in the positive x-direction due to the motion of the stretching surface with velocity $U_w(x) = cx$, $(c > 0)$. The graphical representation of the present analysis has been demonstrated in Figure 12.1. Furthermore, it is considered that the surface is being heated by a nanofluid with a heat exchange coefficient h_f and a temperature T_w. T_∞ is supposed to represent the ambient temperature. The thermophysical properties of base fluid and nanoparticles are obtainable in **Table 12.1.** Moreover, the thermophysical properties of the nanofluids are mentioned in Table 12.2 as mentioned in (Ghasemi and Hatami, 2021). Under the above conventions, the resulting governing equations are given by Devi and Devi (2016) and (Ghasemi and Hatami, 2021).

Table 12.1 Thermophysical properties of base fluid and nanoparticles

Physical properties	$\rho\,(Kg/m^3)$	$C_p\,(J/Kg.K)$	$\kappa\,(W/mK)$	$\beta\,(K^{-1})$	$\sigma\,(s/m)$
Water	997.1	4179	0.613	21×10^{-5}	0.05
Cu	8933	385	401	1.67×10^{-5}	5.95×10^{6}
Ag	10500	235	429	1.89×10^{-5}	6.3×10^{7}
TiO_2	4250	686	8.95	0.90×10^{-5}	10^{-12}
Al_2O_3	3970	765	40	0.85×10^{-5}	10^{-10}

Table 12.2 Thermophysical properties of nanofluids

Properties	*Nanofluid*
Viscosity	$\mu_{nf} = \mu_f (1-\varphi)^{-2.5}$
Density	$\rho_{nf} = (1-\phi)\rho_{bf} + \phi\rho_{np}$
Thermal expansion coefficient	$(\rho\beta)_{nf} = (\rho\beta)_{bf}(1-\varphi) + \varphi\dfrac{(\rho\beta)_{np}}{(\rho\beta)_{bf}}$
Electrical conductivity	$\sigma_{nf} = \sigma_{bf}\left[1 + \dfrac{3\left(\dfrac{\sigma_{np}}{\sigma_{bf}} - 1\right)\varphi}{\left(\dfrac{\sigma_{np}}{\sigma_{bf}} + 2\right) - \left(\dfrac{\sigma_{np}}{\sigma_{bf}} - 1\right)\varphi}\right]$
Thermal conductivity	$\kappa_{nf} = \kappa_{bf}\left(\dfrac{\kappa_{np} + \kappa_{bf} - \varphi(\kappa_{bf} - \kappa_{np})}{\kappa_{np} + \kappa_{bf} + \varphi(\kappa_{bf} - \kappa_{np})}\right)$
Heat capacity	$(\rho C_p)_{nf} = (\rho C_p)_{bf}(1-\varphi) + \varphi(\rho C_p)_{np}$
Volume fraction	$\varphi = \dfrac{\dfrac{W_{np}}{\rho_{np}}}{\dfrac{W_{np}}{\rho_{np}} + \dfrac{W_{bf}}{\rho_{bf}}}$

Following are the basic assumptions used in the governing equations:

- The flow is Newtonian, laminar, incompressible, and steady;
- All the fluid properties are constant;
- Size and shape of the nanoparticles are considered to be uniform;
- The nanofluid is electrically conducting;
- The porous medium is uniform permeability and thermally isotropic; the local thermal equilibrium conditions between porous medium and nanofluid exist;
- Darcy-Forchheimer model is employed for momentum analysis in the porous medium;
- Magnetic Reynolds number is taken to be small so that the induced magnetic field can be neglected;
- There is no slip between the fluid and the walls;
- Heating due to the presence of porous medium has been neglected.

Under the above assumptions, the two-dimensional and steady equations of continuity, modified Navier-Stokes equation for the nanofluid flow in x and

y directions, heat diffusion and nanoparticles diffusion equations, and motile microorganism diffusion equation are as follows:

$$\frac{\partial u}{\partial x}+\frac{\partial v}{\partial y}=0, \tag{12.1}$$

$$u\frac{\partial u}{\partial x}+v\frac{\partial u}{\partial y}=\nu_{nf}\frac{\partial^2 u}{\partial y^2}-\frac{\nu_{nf}}{k_{fh}}u-\sigma_{nf}B_0^2u-Fu^2, \tag{12.2}$$

$$\left(\rho C_p\right)_{nf}\left(u\frac{\partial T}{\partial x}+v\frac{\partial T}{\partial y}\right)=\kappa_{nf}\frac{\partial^2 T}{\partial y^2}+\tau\left(D_B\frac{\partial C}{\partial y}+\frac{D_T}{T_\infty}\left(\frac{\partial T}{\partial y}\right)^2\right), \tag{12.3}$$

$$u\frac{\partial C}{\partial x}+v\frac{\partial C}{\partial y}=D_B\frac{\partial^2 C}{\partial y^2}-\frac{D_T}{T_\infty}\frac{\partial^2 T}{\partial y^2}, \tag{12.4}$$

$$u\frac{\partial N}{\partial x}+v\frac{\partial N}{\partial y}=D_m\frac{\partial^2 N}{\partial y^2}-\frac{bW_c}{C_w-C_\infty}\frac{\partial}{\partial y}\left(N\frac{\partial C}{\partial y}\right). \tag{12.5}$$

Boundary conditions associated with the above PDEs are as follows:

$$\begin{bmatrix} y=0:\left\{u=U_w=cx,\ v=v_w,\ -\kappa_{nf}\dfrac{\partial T}{\partial y}=h(T_w-T_\infty),\ C=C_w\right\} \\ y=\infty:\left\{u=0,\ v=0,\ T=T_\infty,\ C=C_\infty\right\} \end{bmatrix}, \tag{12.6}$$

where $v_w>0$ and $v_w<0$ represent injection and suction, respectively.
Now, using the following similarity transformations:

$$\left\{\begin{array}{c} \eta=\sqrt{\dfrac{c}{\nu_{bf}}}y,\ u=cxf'(\eta),\ v=-\sqrt{c\nu_{bf}}f(\eta), \\ T=T_\infty+\theta(\eta)(T_w-T_\infty),\ C=C_\infty+\varphi(\eta)(C_w-C_\infty) \end{array}\right\}. \tag{12.7}$$

The equation of continuity (12.1) is satisfied and equations (12.2)–(12.5) are transformed to the following non-dimensional form:

$$f'''-wf'-\lambda_1M^2f'-\lambda_2\left[(Fr+1)(f')^2-ff''\right]=0, \tag{12.8}$$

$$\lambda_6\frac{1}{\lambda_5 Pr}\theta''+Nb\theta'\phi'+Nt\theta'^2+f\theta'=0, \tag{12.9}$$

$$\phi''+Scf\phi'-\frac{ScNt}{Nb}\theta''=0, \tag{12.10}$$

$$\chi''-Pe(\Delta N\phi''-\chi\phi''-\chi'\phi')+Lbf\chi'=0. \tag{12.11}$$

The corresponding non-dimensional boundary conditions are as follows:

$$\left\{\begin{array}{c} f(0)=Fw,\ f'(0)=1,\ \theta'=\dfrac{Bi}{\lambda_6}(1-\theta),\ \phi(0)=1) \\ f'(\infty)=0,\ \theta(\infty)=0,\ \phi(\infty)=0 \end{array}\right\}, \tag{12.12}$$

where

$$\left\{\begin{array}{l} w=\dfrac{\nu_{bf}}{ck_{fh}},\ Fr=C_b k_{fh}^{-\frac{1}{2}}=F\cdot x,\ M^2=\dfrac{\sigma_{bf}B_0^2}{\rho_{bf}c},\ Sc=\dfrac{\nu_{bf}}{D_B},\ Pr=\dfrac{\mu_{bf}\left(C_p\right)_{bf}}{\kappa_{bf}} \\ Fw=-\dfrac{v_w}{\sqrt{c\nu_f}},\ Nt=\dfrac{\tau D_T\left(T_w-T_\infty\right)}{T_\infty \nu_{bf}},\ Nb=\dfrac{\tau D_B\left(C_w-C_\infty\right)}{\nu_{bf}}, \\ Ec=\dfrac{U_w^2}{\left(C_p\right)_{bf}\left(T_w-T_\infty\right)},\ Lb=\dfrac{\nu_{bf}}{D_m},\ Pe=\dfrac{bW_c}{D_m},\ \Delta N=\dfrac{N_\infty}{N_w-N_\infty}, \\ \lambda_1=\dfrac{\sigma_{nf}}{\sigma_{bf}}.\dfrac{\mu_{bf}}{\mu_{nf}},\ \lambda_2=\dfrac{\nu_{bf}}{\nu_{nf}},\ \lambda_5=\dfrac{\left(\rho C_p\right)_{nf}}{\left(\rho C_p\right)_{bf}},\ \lambda_6=\dfrac{\kappa_{nf}}{\kappa_{bf}},\ \lambda_7=\dfrac{\sigma_{nf}}{\sigma_{bf}} \end{array}\right\}. \tag{12.13}$$

12.2.1 Engineering interest quantities

The key quantities of physical importance related to this problem are as follows:

Nusselt Number: The Nusselt number is given by:

$$Nu_x=\frac{xq_w}{\kappa_{bf}\left(T_w-T_\infty\right)},$$

where $q_w=-\kappa_{nf}\dfrac{\partial T}{\partial y}$

Using (12.7), we get non-dimensional form of Nusselt number:

$$Nu_x Re_x^{-\frac{1}{2}}=-\lambda_6\theta'(0), \tag{12.14}$$

Local density of microorganism: Density of microorganisms gives the rate of microorganism diffusion from the surface $y = 0$ to the fluid and is given by:

$$Nn_x = \frac{xq_n}{D_m\left(N_w - N_\infty\right)},$$

where, $q_n = -D_m\left(\frac{\partial N}{\partial y}\right)_{y=0}$.

Using (12.7), we get non-dimensional form of Local density of micro-organism:

$$Nn_x Re_x^{-\frac{1}{2}} = -\chi'(0)\,. \tag{12.15}$$

12.3 RESEARCH METHODOLOGY

The non-linear system of ODEs given by (12.8)–(12.11) are solved numerically using BVP4C solver in MATLAB, which employs the finite difference code and the Lobatto IIIa formula. This method provides an explanation of fourth order which is precise and uniform in the range of integration. The problem is solved by taking $\Delta\eta = 0.001$, and the results of convergency are maintained within the tolerance limit of 10^{-6}. The obtained result satisfies the asymptotic boundary conditions that accentuate the accuracy of the numerical computations.

12.4 STABILITY ANALYSIS AND VALIDATION

By comparing the current numerical calculations by finite difference scheme of BVP4C methodology with the results obtained by Chakrabarti and Gupta (1979) for Nusselt number at the surface, **Table 12.3** cites the correctness of the suggested flow model. The behaviour of Nusselt number at the surface under the influence of Pr for the present work is comparable to earlier studies. This

Table 12.3 Comparison table for Nusselt number for $Pr = 0.73$, $Fw = 1$ and rest of the parameters are zero

M	*Chakrabarti and Gupta (1979)*	*Present*
5	0.9154	0.7176
10	0.8837	0.6731
15	0.8645	0.6554
20	0.8522	0.6461

Table 12.4 Stability test

	$h = 0.001$				$h = 0.0001$			
η	f'	θ	ϕ	χ	f'	θ	ϕ	χ
0	1	0.5139	1.0000	1	1	0.5139	1.0000	1
1	0.0736	0.0694	0.6257	0.5506	0.0736	0.0695	0.6257	0.5507
2	0.0055	0.0044	0.3424	0.2093	0.0055	0.0044	0.3425	0.2094
3	0.0004	0.0002	0.1665	0.0383	0.0004	0.0002	0.1666	0.0383
4	0.0000	0.0000	0.0619	−0.0119	0.0000	0.0000	0.0620	−0.0119
5	0.0000	0.0000	0.0000	0	0.0000	0.0000	0.0000	0

comparison validates the present flow model. **Table 12.4** provides evidence for the convergence and stability of the velocity, temperature, and concentration solutions for $h = 0.001$ and $h = 0.0001$. Thus, it can be concluded that the result is stable and convergent since the boundary conditions are satisfied by the velocity, temperature, and concentration.

12.5 RESULTS AND DISCUSSION

12.5.1 Velocity

Influence of various non-dimensional parameters over the velocity profile is demonstrated in Figure 12.2(a)–(c). The results in Figure 12.2(b) show that the velocity of the nanofluid is a decreasing function of porous parameter, w, and suction/injection parameter, Fw; this is due to the fact that with increasing values of w, the porous medium becomes less permeable and thus resists the flow. The increasing values of Fw is related to suction at the surface of the permeable sheet. With increasing wall suction, the fluid molecules are attracted towards the wall and hence the velocity of the nanofluid decreases. The influence of Lorentz force on the velocity profile of normal fluid and $TiO_2 - water$ nanofluid has been depicted in Figure 12.2(a). It can be seen that the velocity of the $TiO_2 - water$ nanofluid is higher than the normal fluid under the action of larger Lorentz force, while in the absence of Lorentz force, the velocity of the normal fluid was higher than the $TiO_2 - water$ nanofluid. The average velocity of the normal fluid is 17.426% higher than the TiO_2 nanofluid in the absence of magnetic field. However, in the presence of higher Lorentz force, the velocity of the nanofluid decreased by 24.703%. It is obvious that the velocity of the fluid under consideration declines for augmented values of magnetic parameter. This is owing to the fact that when M rises, the resistive force caused by the Lorentz force grows, which decreases fluid velocity. A comparative study on the velocity of various nanoparticles in same base fluid has been depicted in Figure 12.2(c). It has been reported that the velocity of the metal oxides of Al and Ti possesses higher velocity than the metals Cu and Ag nanoparticles.

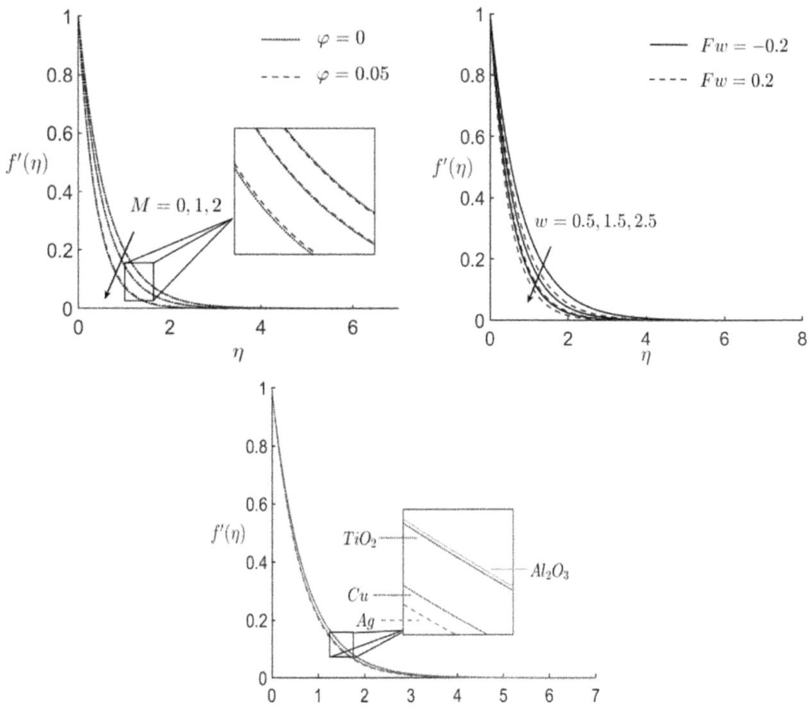

Figure 12.2 Impact of (a) *M* and φ, (b) *Fw* and *w*, and (c) various nanoparticles over $f'(\eta)$.

12.5.2 Temperature

The influence of various parameters on heat diffusion is depicted in Figure 12.3(a) and (b) via line graphs of $\theta(\eta)$. Figure 12.3(a) illustrates the effect of Brownian motion parameter on temperature distribution for $Bi = 0.5$ and $Bi = 1.5$. The Brownian motion parameter's value increases the collision rate of nanoparticles in a nanofluid, converting kinetic energy into heat energy and increasing the fluid's temperature with an increase in the parameter. The Biot number directly influences the convective boundary condition at the surface, enhancing the temperature gradient and subsequently increasing the surface temperature. Figure 12.3(b) portrays the influence of thermophoresis parameter for $w = 10^{-4}$ and $w = 5$. Thermophoresis is the motion of suspended particles in response to the temperature gradient within the system. Thermophoretic force tends to move the fluid molecules towards cold region. Higher thermophoresis parameters increase thermophoretic forces, causing nanoparticle diffusion and thickening of the thermal boundary layer. It is evident that elevated values of *Nt* lead to the penetration of the nanoparticles towards free stream, and thus, the temperature of the nanofluid rises. It is also seen that the rise in temperature by the influence of thermophoresis parameter is accompanied by the porous parameter. Porous parameter $w = 5$ corresponds to the less permeable porous medium

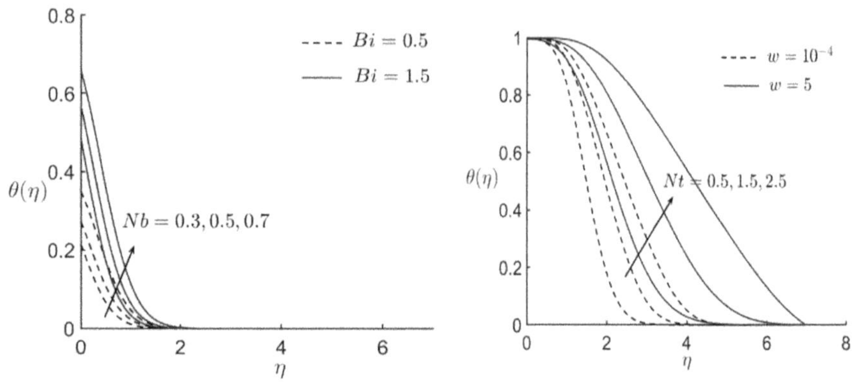

Figure 12.3 Impact of (a) *Nb* and *Bi* and (b) *Nt* and *w*.

which offers restriction to the fluid molecules and thus the molecules are warm. The study aimed to improve heat transfer rate by adding nanoparticles to the base fluid, which in turn enhances the thermal conductivity of the base fluid.

12.5.3 Concentration profiles

The impact of non-dimensional parameters over the nanoparticles diffusion is demonstrated in Figure 12.4(a) and (b). Improved values of *Nt* enhances the concentration while *Nb* decreases the species profile which is shown in Figure 12.4(a). Increasing the *Nt* values boosts the thermophoretic force which in turn moves the nanoparticles from hot to the colder region; as a result, the concentration of nanoparticles enhances. The maximum concentration of nanoparticles has been recorded at $\eta = 2$ for $Nt = 1.5$. For decreasing *Nt* values, the highest concentration region shifts towards the stretched sheet. The concentration profiles of different nanoparticles have been depicted in Figure 12.4(b). Due to low velocity of metal-based nanofluids, the concentration of metal-based nanofluid is higher than the nanofluids of metal oxides.

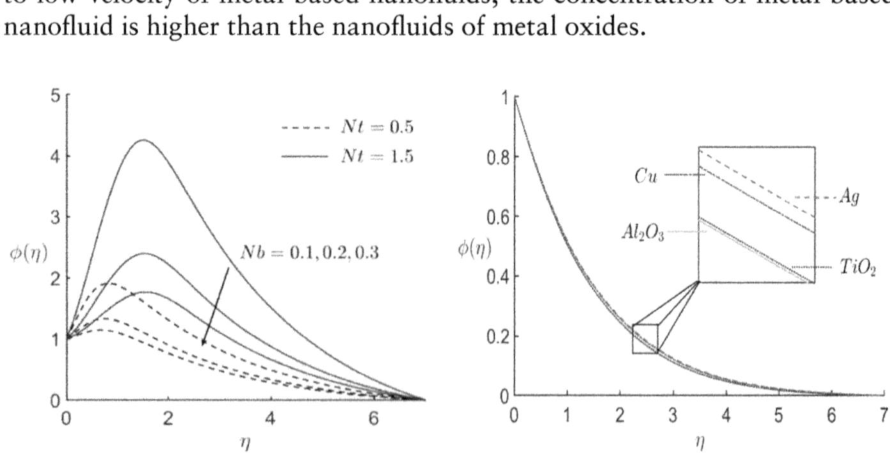

Figure 12.4 Impact of (a) *Nt* and *Nb* and (b) different nanoparticles over $\phi(\eta)$.

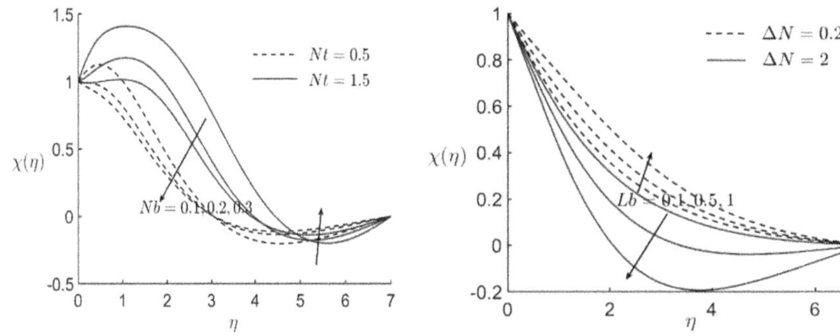

Figure 12.5 Impact of (a) *Nt* and *Nb* and (b) ΔN and *Lb* over $\chi(\eta)$.

12.5.4 Microorganisms

Figure 12.5(a) depicts the impact of *Nb* and *Nt* on microorganism density. The density of microorganisms near the surface declines for augmented values of *Nb* near the surface, while density of microorganisms enhances at free stream for higher *Nb* values. The nature of microorganism's profiles resembles that of nanoparticle concentration for the impact of *Nt*. The impact of *Lb* over the microorganism's profile for variation of ΔN is depicted in Figure 12.5 (b). The augmented values of bioconvection Lewis number decrease microorganism for $\Delta N = 0.2$. However, the trend gets reversed for $\Delta N = 2$.

12.5.5 Engineering key quantities

Influence of thermophoretic force on the Nusselt number for $\varphi = 0$ and $\varphi = 0.05$ under the action of Lorentz force is depicted in Figure 12.6(a). The heat transmission is enhanced by 3.16% by the addition of nanoparticles in water; however, thermophoretic force reduces the surface thermal transfer rate. For lower values of magnetic parameter, the thermal transfer rate is enhanced, while increasing magnetic parameter above $M = 2$ reduces the surface thermal transfer rate.

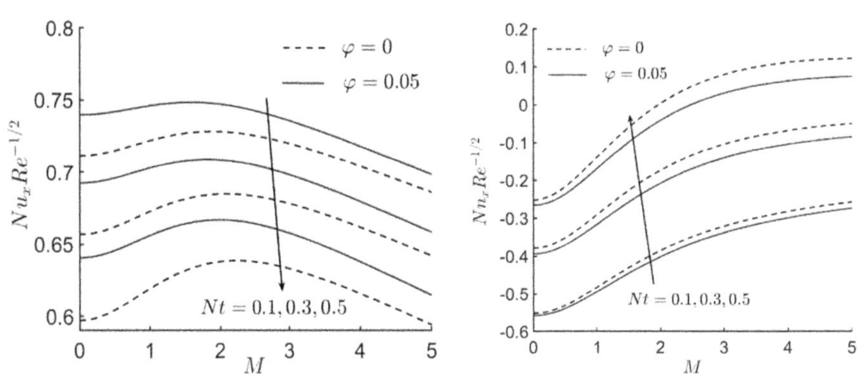

Figure 12.6 Influence of (a) φ, *Nt*, and *M*over Nusselt number and (b) φ, *Nt*, and *M* over density number of microorganisms.

Figure 12.6(b) sketches the influence of Nt, φ, and Lorentz force on density number of microorganisms. The augmented values of Nt tend to move the microorganisms towards higher temperature region; as a result, increasing thermophoretic force enhances the gradient of microorganism density at the surface. The density number of microorganism is an increasing function of magnetic parameter, while solid volume fraction has a contrary effect on it.

12.6 CONCLUSIONS

A mathematical model has been developed for bioconvective flow of $TiO_2 - water$ nanofluid in a porous medium under the influence of Lorentz force containing motile microorganisms. A finite difference scheme via MATLAB BVP4C routine has been implemented to obtain the solution of the problem under consideration. The numerical stability or the convergence of the solution has been validated by conducting the grid point stability test, showing excellent agreement for $h = 0.001$ and $h = 0.0001$. The validity of the current model has also been compared with the existing literatures and found to be well validated. From the above analysis, whole analysis can be concluded as below:

1. In the absence of Lorentz force, the velocity of $TiO_2 - water$ nanofluid is less than that of water; however, when Lorentz force is applied, the velocity of $TiO_2 - water$ is greater than that of water.
2. Biot number and Brownian motion parameter enhance the temperature profile. Also, thermophoretic force accompanied by the porous parameter boosts the temperature rise of the nanofluid.
3. A higher thermophoretic parameter promotes nanoparticle accumulation in cooler areas, while a lower thermophoretic parameter leads to a concentration shift towards the heated surface. Also, a higher Brownian motion parameter results in a diminished concentration profile of the species.
4. Augmented Brownian motion parameter reduces microorganism density near surfaces but increases it in free streams, and this effect is similar to how lower Bioconvection Lewis number impacts microorganism density, but with a reversal at high Bioconvection Lewis number.
5. Adding nanoparticles enhances heat transfer in water, but thermophoretic force and high magnetic fields (> 2) reduce surface thermal transfer rate.
6. The density number of microorganism is an increasing function of magnetic parameter, while solid volume fraction has a contrary effect on it.
7. Metal-oxide-based nanofluids move faster than the metal-based nanofluids.
8. The temperature of metal-oxide-based nanofluid is less than the temperature of metal-based nanofluid.
9. Since, metal-oxide-based nanofluid moves faster than the metal-based nanofluid, the concentration profile for metal-based nanofluid is higher than the metal-oxide-based nanofluid.

Current investigation suggests, for enhancing surface thermal transfer rate, one should consider $TiO_2 - water$ nanofluid as it provides best surface thermal transfer rate. In future, this study can be extended to study the effect of other phenomenon like ohmic heating and viscous dissipation heating due to thermal radiation.

NOMENCLATURE

Bi	Biot number
C_p	Specific heat
F	Non-uniform inertial coefficient of porous medium
f	Non-dimensional velocity field
Fr	Forchheimer parameter
Fw	Suction/injection parameter
k^*	Mean absorption coefficient
k_{fb}	Permeability of porous medium
Lb	Lewis number
M	Magnetic field parameter
Nb	Brownian motion parameter
Nt	Thermophoresis parameter
Pr	Prandtl number
Re	Reynolds number
Sc	Schmidt number
T	Temperature
(u,v)	Velocity in x and y direction
v_w	Mass transfer velocity
(x,y)	Cartesian coordinates

Greek Symbols

φ	Volume fraction
η	Similarity variable
θ	Non-dimensional temperature
ϕ	Non-dimensional concentration
κ	Thermal conductivity
ν	Kinematic viscosity
μ	Dynamic viscosity
ρ	Density
β	Coefficient of thermal expansion
σ	Electrical conductivity
σ^*	Stefan-Boltzmann constant
w	Porous parameter
λ_i	Constants

Subscripts

Ag	Silver
bf	base fluid
nf	Nanofluid
w	wall
∞	Ambient condition

Superscripts

* Asterisk
′ Derivative with respect to η

REFERENCES

Ahmad, B., Ahmad, M.O., Ali, L., Ali, B., Hussein, A.K., Shah, N.A., and Chung, J.D. (2022). Significance of the Coriolis force on the dynamics of Carreau–Yasuda rotating nanofluid subject to Darcy–Forchheimer and gyrotactic microorganisms. Mathematics, 10(16), 1–15.

Bejan, A. (1979). A study of entropy generation in fundamental convective heat transfer. Journal of Heat Transfer, 101(4), 718–725.

Bejan, A., and Kestin, J. (1983). Entropy generation through heat and fluid flow. Journal of Applied Mechanics, 50(2), 475.

Bhattacharyya, A., Sharma, R., Hussain, S.M., Chamkha, A.J., and Mamatha, E. (2022). A numerical and statistical approach to capture the flow characteristics of Maxwell hybrid nanofluid containing copper and graphene nanoparticles. Chinese Journal of Physics, 77, 1278–1290.

Chakrabarti, A., and Gupta, A.S. (1979). Hydromagnetic flow and heat transfer over a stretching sheet. Quarterly of Applied Mathematics, 37(1), 73–78.

Choi, E.J., Ahn, Y., and Yong, D.H. (2007). Size dependence of magnetic properties of co-ferrite nanoparticles. Journal of the Korean Physical Society, 50(2), 460–463.

Choi, S.U., and Eastman, J.A. (1995). Enhancing thermal conductivity of fluids with nanoparticles. Argonne National Lab. (ANL).

Cui, J., Razzaq, R., Farooq, U., Khan, W.A., Farooq, F.B., and Muhammad, T. (2022). Impact of non-similar modelling for forced convection analysis of nanofluid flow over stretching sheet with chemical reaction and heat generation. Alexandria Engineering Journal, 61(6), 4253–4261.

Devi, S.A., and Devi, S.S. (2016). Numerical investigation of hydromagnetic hybrid water nanofluid flow over a permeable stretching sheet with suction. International Journal of Nonlinear Sciences and Numerical Simulation, 17(5), 249–257.

Eastman, J.A., Choi, S.U., Li, S., Yu, W., and Thompson, L.J. (2001). Anomalously increased effective thermal conductivities of ethylene glycol-based nanofluids containing copper nanoparticles. Applied Physics Letters, 78(6), 718–720.

Gautam, A.K., Verma, A.K., Bhattacharyya, K., Mukhopadhyay, S., and Chamkha, A.J. (2021). Impacts of activation energy and binary chemical reaction on MHD flow of Williamson nanofluid in Darcy–Forchheimer porous medium: A case of expanding sheet of variable thickness. Waves in Random and Complex Media, 34(4), 3528–3549.

Ghasemi, S.E., and Hatami, M. (2021). Solar radiation effects on MHD stagnation point flow and heat transfer of a nanofluid over a stretching sheet. Case Studies in Thermal Engineering, 25, 1–12.

Gowda, R.P., Kumar, R.N., Prasannakumara, B.C., Nagaraja, B., and Gireesha, B.J. (2021). Exploring magnetic dipole contribution on ferromagnetic nanofluid flow over a stretching sheet: An application of Stefan blowing. Journal of Molecular Liquids, 335, 1–8.

Hai, N.H., Phu, N.D., Luong, N.H., Chau, N., Chinh, H.D., Hoang, L.H., and Leslie-Pelecky, D.L. (2008). Mechanism for sustainable magnetic nanoparticles under ambient conditions. Journal of the Korean Physical Society, 52(5), 1327–1331.

Hayat, T., Khan, W.A., Abbas, S.Z., Nadeem, S., and Ahmad, S. (2020a). Impact of induced magnetic field on second-grade nanofluid flow past a convectively heated stretching sheet. Applied Nanoscience, 10(8), 3001–3009.

Hayat, T., Qayyum, S., Alsaedi, A., and Ahmad, B. (2020b). Entropy generation minimization: Darcy-Forchheimer nanofluid flow due to curved stretching sheet with partial slip. International Communications in Heat and Mass Transfer, 111, 1–8.

Hosseinzadeh, K., Asadi, A., Mogharrebi, A.R., Khalesi, J., Mousavisani, S., and Ganji, D.D. (2019). Entropy generation analysis of (CH2OH) 2 containing CNTs nanofluid flow under effect of MHD and thermal radiation. Case Studies in Thermal Engineering, 14, 1–13.

Iijima, S. (1991). Helical microtubules of graphitic carbon. Nature, 354, 56–58.

Joshi, N., Upreti, H., and Pandey, A.K. (2022). MHD Darcy-Forchheimer hybrid nanofluid flow via a porous stretching sheet with suction/blowing and viscous dissipation. International Journal for Computational Methods in Engineering Science and Mechanics, 23, 1–9.

Khan, M.I., Hayat, T., Khan, M.I, Waqas, M., and Alsaedi, A. (2019). Numerical simulation of hydromagnetic mixed convective radiative slip flow with variable fluid properties: A mathematical model for entropy generation. Journal of Physics and Chemistry of Solids, 125, 153–64.

Khan, S.A., Khan, M.I., Hayat, T., and Alsaedi, A. (2020). Darcy-Forchheimer hybrid (MoS2, SiO2) nanofluid flow with entropy generation. Computer Methods and Programs in Biomedicine, 185, 1–20.

Khan, W.A., Khan, Z.H., and Rahi, M. (2014). Fluid flow and heat transfer of carbon nanotubes along a flat plate with Navier slip boundary. Applied Nanoscience, 4(5), 633–641.

Lee, S., Choi, S.S., Li, S.A., and Eastman, J.A. (1999). Measuring thermal conductivity of fluids containing oxide nanoparticles. Journal of Heat Transfer, 121(2), 280–289.

Liu, W., Shahsavar, A., Barzinjy, A.A., Al-Rashed, A.A., and Afrand, M. (2019). Natural convection and entropy generation of a nanofluid in two connected inclined triangular enclosures under magnetic field effects. International Communications in Heat and Mass Transfer, 108, 1–13.

Pandey, A.K., and Kumar, M. (2017). Natural convection and thermal radiation influence on nanofluid flow over a stretching cylinder in a porous medium with viscous dissipation. Alexandria Engineering Journal, 56(1), 55–62.

Rawat, S.K., Upreti, H., and Kumar, M. (2021). Numerical study of activation energy and thermal radiation effects on Oldroyd-B nanofluid flow using the Cattaneo–Christov double diffusion model over a convectively heated stretching sheet. Heat Transfer, 50(6), 5304–5331.

Rehman, S.U., Fatima, N., Ali, B., Imran, M., Ali, L., Shah, N.A., and Chung, J.D. (2022). The Casson dusty nanofluid: Significance of Darcy–Forchheimer law, magnetic field, and non-Fourier heat flux model subject to stretch surface. Mathematics, 10(16), 1–14.

Shaiq, S., Maraj, E.N., and Iqbal, Z. (2019). Remarkable role of on transportation of hybrid nanoparticles influenced by thermal deposition and internal heat generation. Journal of Physics and Chemistry of Solids, 126, 294–303.

Sreedevi, P., Sudarsana, P.R., and Chamkha, A.J. (2020). Heat and mass transfer analysis of unsteady hybrid nanofluid flow over a stretching sheet with thermal radiation. SN Applied Sciences, 2(7), 1–5.

Upreti, H., Pandey, A.K., Kumar, M., and Makinde, O.D. (2020). Ohmic heating and non-uniform heat source/sink roles on 3D Darcy–Forchheimer flow of CNTs nanofluids over a stretching surface. Arabian Journal for Science and Engineering, 45(9), 7705–7717.

Wong, K.V., and De, O.L. (2010). Applications of nanofluids: Current and future. Advances in Mechanical Engineering, 2, 1–11.

Yoo, D.H., Hong, K.S., Hong, T.E., Eastman, J.A., and Yang, H.S. (2007). Thermal conductivity of /water nanofluids. Journal of the Korean Physical Society, 51(1), 84–87.

Chapter 13

Nanofluidic applications in sustainable food technology and agriculture

Ashwani Kumar, Mukesh Kumar Awasthi, and Nitesh Dutt

13.1 INTRODUCTION

Nanofluids are a class of engineered fluids that have captured significant interest due to their remarkable thermal properties. One exciting application of nanotechnology lies in the development of nanofluids. These engineered fluids, infused with nanoparticles (1–100 nm), exhibit exceptional thermal, electrical, and optical properties surpassing those of traditional fluids. These fluids are infused with nanoparticles, typically metallic or oxide-based, that significantly enhance the thermal conductivity and heat transfer efficiency compared to traditional fluids [1–2]. This improvement arises from the unique properties of the nanoparticles themselves. By incorporating these nanoparticles into a base fluid, such as water or oil, nanofluids become promising candidates for a wide range of applications, from electronics cooling in computers to medical imaging techniques. Understanding the specific properties of different nanofluid compositions is crucial for optimizing their performance in these diverse applications [3–5].

This chapter explores into the emerging realm of nanofluidic applications within the food technology and agriculture sectors. It highlights the potential of nanofluids to revolutionize processes, enhance product quality, and improve agricultural practices. Das et al. [6] explored the thermal properties of nanofluids relevant to food processing applications. The food industry relies heavily on efficient and precise heating and cooling processes to ensure food safety and quality. McClements [7] has studied the background on encapsulation techniques and their potential application with nanofluids in food. Traditional processes like pasteurization and sterilization often require high temperatures, potentially impacting food quality. Nanofluids, with their superior thermal conductivity, can achieve the same results at lower temperatures, preserving the organoleptic properties (taste, texture, and aroma) of food products. Encapsulation of vitamins, minerals, or other bioactive compounds within nanoparticles allows for targeted delivery within food products. This approach ensures controlled release and improved bioavailability of these vital nutrients for consumers [8–9].

Food spoilage is a significant global challenge, leading to immense economic losses and food insecurity. Traditional methods for preserving food often

DOI: 10.1201/9781003527251-13

involve harsh treatments like high-temperature pasteurization, which can compromise the quality and nutritional value of products. Fortunately, nanotechnology offers promising solutions through the development of nanofluid-based applications. As highlighted by Kahrilas et al. [9], nanofluid coatings on food packaging represent a potential future development in the fight against foodborne pathogens. These coatings could actively inhibit microbial growth on food surfaces, thereby extending shelf life and reducing spoilage. Building upon this concept, Aziz et al. [10] explored deeper into the broader applications of nanotechnology in food production, including the exciting potential of nanofluids in agriculture. Nanoparticles can be harnessed to encapsulate pesticides, herbicides, or fertilizers, creating a targeted delivery system for these chemicals.

Nanotechnology offers exciting possibilities for improving food quality and agricultural practices. Research by Chen et al. [11] explores the use of nanoparticles in agriculture, highlighting their potential for targeted delivery of essential nutrients and agrochemicals. This targeted approach minimizes waste and environmental impact while potentially improving the efficacy of these substances. Additionally, Singh et al. [12] explored into the broader applications of nanotechnology in agriculture, including studies on how nanoparticles can stimulate plant growth and improve crop yields. Building upon these advancements, Liu et al. [13] explored the concept of controlled release technology in agriculture. This technology allows for the gradual release of beneficial compounds over time. By encapsulating vitamins, minerals, or other bioactive compounds within nanoparticles, targeted delivery within food products becomes possible. This approach, as highlighted by the referenced studies, ensures controlled release and improved bioavailability of these vital nutrients for consumers. The combination of targeted delivery with controlled release technology holds immense promise for revolutionizing food fortification and agricultural practices.

13.2 SUSTAINABILITY IN FOOD TECHNOLOGY AND AGRICULTURE

The global population is projected to reach nearly 10 billion by 2050, placing immense pressure on our food systems. To meet this growing demand, food technology and agriculture need to embrace sustainable practices that ensure food security while minimizing environmental impact. Sustainability is a complex concept encompassing social, economic, and environmental aspects. In the context of food technology and agriculture, this translates to three pillars, i.e. environmental sustainability, economic sustainability, and social sustainability.

13.2.1 Environmental sustainability

Ensuring sustainable food production requires minimizing the environmental footprint associated with agricultural practices. This multifaceted challenge encompasses minimizing deforestation, water scarcity, pollution, and greenhouse gas emissions. Agricultural expansion is a major driver of deforestation, leading to habitat loss and reduced biodiversity [14]. Sustainable practices promote efficient land-use strategies, minimizing the need for further deforestation and protecting vital ecosystems. Tilman et al. [14] analyzed the environmental consequences of human alterations to ecosystems, emphasizing the detrimental impact of land-use change for agriculture on biodiversity. Agriculture is a major consumer of freshwater resources. Traditional practices can contribute to water scarcity and pollution through excessive water use and agricultural runoff. Sustainable technologies prioritize water conservation by implementing techniques like drip irrigation. Additionally, minimizing the use of chemical fertilizers and pesticides reduces water pollution from agricultural runoff, protecting water quality for future generations. Mekonnen and Hoekstra [15] assessed global water scarcity and highlighted the need for efficient water management practices in agriculture to ensure water security for future generations. Food production contributes significantly to greenhouse gas emissions, primarily through methane emissions from livestock and nitrous oxide emissions from fertilizers. Sustainable practices promote the adoption of renewable energy sources within agricultural systems to reduce reliance on fossil fuels. Additionally, exploring alternative fertilizer formulations and implementing carbon sequestration strategies can help mitigate greenhouse gas emissions from agriculture (Figure 13.1). Vermeulen et al. [16] discussed the challenges of climate change for food security and emphasized the role of sustainable agricultural practices in mitigating greenhouse gas emissions.

13.2.2 Economic sustainability

Economic sustainability in food systems requires a two-pronged approach, ensuring both producers (farmers) and consumers have fair access and economic security. A persistent challenge for small-scale farmers is receiving low prices for their produce. Sustainable practices actively support fair-trade initiatives, which offer fairer pricing structures. This economic empowerment

Figure 13.1 Key considerations for sustainability in food technology and agriculture.

fosters a more viable and competitive agricultural sector, contributing directly to food security [17]. Ingram [17] delved deeper into this complex relationship between poverty, hunger, and food security, emphasizing the critical role of economically empowered small-scale farmers in achieving sustainable food systems. Beyond the ethical implications, food waste represents a significant economic and environmental burden. Sustainable technologies aim to optimize food production chains, minimize spoilage throughout the supply chain (from farm to fork), and reduce overall waste generation [18]. Gustavsson et al. [18] analyzed the global food waste scandal, highlighting the economic and environmental costs associated with wasted food (Figure 13.1). Their research emphasizes the need for sustainable solutions across the entire food supply chain. Innovation plays a crucial role in developing new sustainable technologies and improving the efficiency of food production. Sustainable practices encourage increased investment in research and development (R&D) within the food and agriculture sectors. This investment fosters the creation of novel technologies that can address existing challenges like resource scarcity and improve overall production efficiency [19]. Spielman et al. [19] analyzed the critical role of agricultural R&D in achieving food security. Their research emphasizes the need for increased investment in research for sustainable food production technologies.

13.2.3 Social sustainability

The social pillar of sustainability in food technology and agriculture emphasizes social equity and food accessibility for all populations. Sustainable practices aim to address critical issues like food insecurity and malnutrition, fostering a world where everyone has access to the nutritious food they need to thrive [20]. Godfray et al. [20] highlighted the urgent need for a global transformation towards sustainable food systems. These systems must prioritize food security, ensuring sufficient, safe, and nutritious food availability for all people. Furthermore, they should address issues of malnutrition, promoting access to a balanced diet and combating deficiencies in essential nutrients. Sustainable food production must prioritize the well-being of those who work within the agricultural sector (Figure 13.1). This includes ensuring safe and fair working conditions for agricultural workers throughout the entire supply chain [21]. The International Labour Organization [21] emphasizes the prevalence of exploitative labor practices in agriculture and calls for promoting decent work and social justice within the sector. This translates to fair wages, safe working environments, and protection of workers' rights. Supporting local food production goes beyond environmental benefits. It also empowers communities, fostering social cohesion and food sovereignty [22]. Patel [22] advocated for a shift toward local food systems, highlighting their advantages for social equity. Local production allows communities to have a greater say in their food choices, promotes cultural preservation, and strengthens local economies.

Additionally, it fosters a sense of food sovereignty, enabling communities to have more control over their food security.

Sustainability in food technology and agriculture demands a holistic approach, balancing environmental, economic, and social considerations. By optimizing resource utilization, promoting fair trade practices, and prioritizing social equity, we can create a sustainable food system that can nourish a growing population while protecting our planet's resources for future generations.

13.3 NANOFLUIDS RELEVANT TO FOOD TECHNOLOGY AND AGRICULTURE

Nanofluids are engineered fluids containing nanoparticles (1–100 nm) suspended in a base fluid (water, oil, or ethylene glycol). These suspensions offer unique properties highly relevant to food technology and agriculture. Figure 13.2 and Table 13.1 show key applications in food processing and agriculture. Traditional methods often require high temperatures for tasks like pasteurization or sterilization, which can negatively impact food quality [4, 8]. Nanofluids containing heat-transfer-enhancing nanoparticles can improve

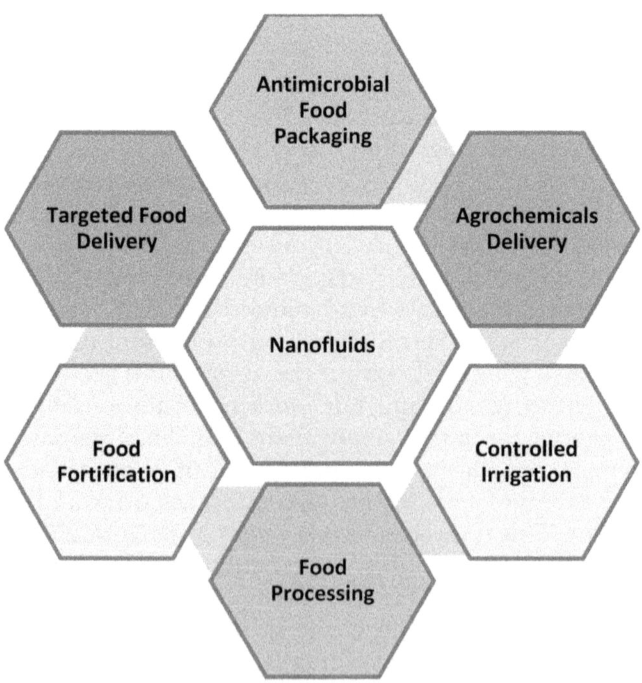

Figure 13.2 Different applications of nanofluids in food technology and agriculture.

Table 13.1 Nanofluid for sustainable food technology and agriculture industry

Application area	*Nanofluid composition*	*Target benefit*	*Potential advantages*	*References*
Food processing: Heat transfer	Metallic nanoparticles (e.g., Cu and Al) dispersed in water or oil	Enhanced thermal conductivity	Lower processing temperatures for pasteurization, sterilization, and blanching with improved product quality (organoleptic properties)	[4, 8]
Food fortification and delivery	Biocompatible polymer nanoparticles encapsulating vitamins, minerals, or bioactive compounds	Controlled release and improved bioavailability of nutrients	Targeted delivery within food products Enhanced absorption by consumers	[7, 10]
Antimicrobial food packaging	Functionalized nanoparticles incorporated into packaging coatings	Inhibit microbial growth on food surfaces	Extended shelf life of perishable food items Reduced food spoilage	[9]
Precision delivery of agrochemicals	Pesticides, herbicides, or fertilizers encapsulated in biodegradable nanoparticles	Targeted delivery of agrochemicals	Minimized waste and environmental impact Improved efficacy against pests, diseases, or weeds Reduced use of chemicals	[11]
Plant growth enhancement	Specific nanoparticles (e.g., cerium oxide)	Potential stimulation of plant growth and crop yield	Improved nutrient uptake by plants Enhanced stress tolerance	[12]
Controlled irrigation	Water-repellent nanoparticles in irrigation system coatings	Reduced water evaporation	Improved water management Targeted delivery of water to plant roots Water conservation in agriculture	[12]

thermal conductivity. This allows for more uniform and efficient heat transfer at lower temperatures, preserving the sensory qualities and nutritional value of food products [7, 10]. Additionally, nanofluids can be used as antimicrobial agents. By incorporating nanoparticles with inherent antimicrobial properties (like silver or zinc oxide), these fluids can extend shelf life and reduce food-borne illness risks [9].

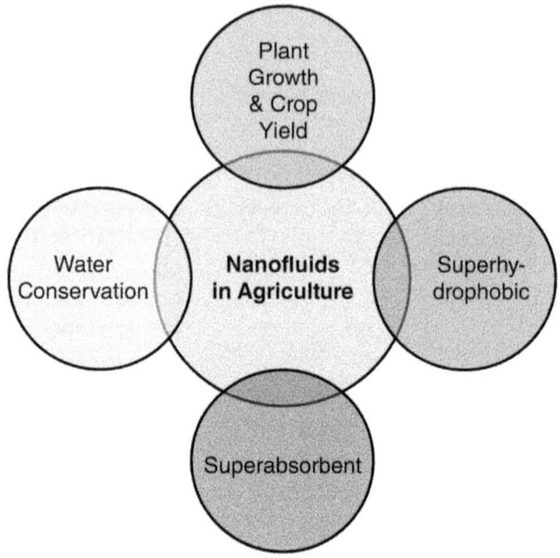

Figure 13.3 Nanofluids application in agriculture.

Figure 13.3 shows that in agriculture, nanofluids hold promise for enhancing irrigation efficiency. By incorporating superhydrophobic and superabsorbent nanoparticles, nanofluids can improve water retention in soil. This reduces water waste and ensures better delivery of water and nutrients to plant roots. Furthermore, specific nanoparticles can be used as targeted pesticides or fertilizers. These targeted nanoparticles can improve efficacy while minimizing environmental impact by reducing overall use of chemicals [11–12].

13.3.1 Challenges of traditional food processing

Maintaining precise temperature control is crucial throughout the food production chain, from pasteurization and sterilization to storage and transportation. However, traditional heating and cooling methods often present significant challenges that negatively impact food quality. One major concern is the detrimental effect of high temperatures on sensitive food components. Pasteurization and sterilization processes, typically relying on boiling water or steam, can degrade essential vitamins, minerals, and flavor compounds. This not only reduces the nutritional value of food but also diminishes its sensory appeal to consumers [23]. Additionally, high-temperature cooking can lead to protein denaturation and textural changes, negatively affecting the overall eating experience [24].

Another challenge associated with traditional methods is the lack of temperature uniformity. Large-scale heating processes can result in uneven temperature distribution within food products. Hot spots can develop, leading to localized overprocessing and further compromising food quality. Furthermore, traditional cooling methods such as air or water chilling can be slow and inefficient, potentially leading to spoilage or the growth of pathogenic microorganisms [6]. These limitations necessitate the exploration of novel heating and cooling technologies that offer greater control and precision. Minimizing thermal damage and ensuring uniform temperature distribution are key factors in preserving food quality and safety while meeting consumer demands for fresh, nutritious products.

13.3.2 Nanofluids for enhanced heat transfer in food processing

Conventional food processing techniques often rely on high temperatures for processes like pasteurization, sterilization, and blanching. While these methods ensure food safety, they can negatively impact the quality of the final product. Nanofluids offer a promising solution for improving heat transfer efficiency and enabling lower processing temperatures. The addition of nanoparticles significantly enhances the thermal conductivity of the base fluid. This allows for more efficient and uniform heat transfer throughout the food product during processing. Peng et al. [25] have shown that nanofluids containing metal oxides like aluminum oxide (Al_2O_3) or silicon dioxide (SiO_2) can significantly improve heat transfer compared to water alone. By enabling lower processing temperatures, nanofluids can minimize thermal damage to food, leading to improved product quality.

The use of nanofluids in food processing holds immense potential for preserving the organoleptic properties (sensory qualities) of food. Lower processing temperatures can help retain the natural flavors, textures, and colors of food products. Additionally, by reducing processing time, nanofluids can minimize the formation of undesirable compounds that can affect taste and aroma. This translates to a higher quality final product with a more desirable consumer experience [6, 25].

13.4 TARGETED DELIVERY OF FOOD COMPONENTS

The human body requires a complex interplay of various nutrients for optimal health. However, challenges exist in ensuring sufficient delivery and absorption of these components, particularly vitamins, minerals, and bioactive compounds. Targeted delivery of food components emerges as a promising solution, utilizing encapsulation within nanoparticles for enhanced functionality. Nanoparticles, microscopic carriers crafted from food-grade materials like proteins or polysaccharides, offer a unique platform for encapsulating these sensitive nutrients. The encapsulation process shields the components from harsh environmental conditions like light, heat, or stomach acidity, preserving their bioactivity until they reach their target site within the body.

13.4.1 Encapsulation

One of the significant advantages of this targeted delivery system lies in its ability to control the release of encapsulated nutrients. By manipulating the size, surface properties, and materials used in nanoparticle construction, scientists can design systems that release their cargo at specific locations within the gastrointestinal tract. Bilal et al. [26] have investigated the encapsulation process and stated that it protects the bioactive compounds from harsh environments, promoting their stability and functionality throughout the food chain. This controlled release allows for a more targeted approach, delivering nutrients precisely where they are needed for optimal absorption. Additionally, encapsulation can improve the bioavailability of nutrients. Bioavailability refers to the proportion of a nutrient that is absorbed and used by the body. Many essential nutrients suffer from poor bioavailability due to factors like limited solubility or degradation during digestion. Encapsulation within nanoparticles can protect these nutrients, enhancing their solubility and facilitating their passage through the digestive system, ultimately leading to improved overall nutritional value for consumers.

13.4.2 Fortifying food products

Targeted delivery of food components holds immense potential for fortifying existing food products and delivering functional ingredients. Food fortification involves strategically adding essential nutrients to address dietary deficiencies in specific populations. Encapsulating vitamins, minerals, or other micronutrients within nanoparticles allows for their incorporation into various food products without compromising taste, texture, or visual appeal. This approach ensures a more targeted and efficient delivery of these vital components, leading to a more nutritious food supply.

Beyond fortification, targeted delivery systems can also be employed to deliver functional food ingredients, which are bioactive compounds that offer specific health benefits beyond basic nutrition. Razavi et al. [27] studied targeted food delivery and highlight that certain minerals might be more effectively absorbed in the small intestine, while probiotics might require delivery to the colon to exert their health benefits. Encapsulation can protect these delicate functional ingredients and ensure their targeted delivery to specific organs or tissues, maximizing their health-promoting effects. Nanoparticles such as liposomes, micelles, and biopolymer-based carriers are used for targeted food delivery. Large-scale production, cost-effectiveness, and ensuring regulatory compliance for food applications are the current challenges for nanoparticles applications [28].

13.5 ANTIMICROBIAL FOOD PACKAGING

Food spoilage poses a significant threat to global food security. Microbial growth is a primary culprit, leading to foodborne illnesses, economic losses, and wasted resources. Antimicrobial food packaging emerges as a promising technology to combat these challenges. By incorporating antimicrobial agents or surfaces, this packaging actively inhibits or kills microorganisms, extending shelf life and reducing spoilage. Food spoilage is the deterioration of food quality caused by microbial growth, enzymatic activity, and chemical reactions. Microbes, including bacteria, fungi, and yeasts, are the leading cause. Their growth leads to undesirable changes in texture, odor, flavor, and appearance, rendering food unfit for consumption. This not only results in significant economic losses but also contributes to food insecurity, especially in resource-limited regions. The World Health Organization estimates that foodborne illnesses affect millions globally each year [29]. Furthermore, food spoilage contributes to food waste, a major sustainability concern.

13.5.1 Nanofluid-based antimicrobial coatings on food packaging

Li et al. [30] investigated that nanotechnology offers innovative solutions for antimicrobial food packaging. Nanofluids, suspensions of nanoparticles in a liquid carrier, are incorporated into coatings applied to packaging materials. These nanoparticles possess unique properties that enhance antimicrobial activity. Silver nanoparticles are widely studied for their broad-spectrum antimicrobial efficacy. They can disrupt microbial membranes, inhibit enzyme activity, and interfere with DNA replication. Other nanomaterials like zinc oxide and titanium dioxide nanoparticles also exhibit promising antimicrobial properties.

Antimicrobial coatings play a crucial role in extending the shelf life of perishable foods. These coatings can be incorporated into the packaging material itself or applied as a separate layer. The mode of action of these coatings varies [31]. Release-based coatings allow the controlled release of antimicrobial agents into the food or headspace, inhibiting microbial growth. Absorption-based coatings remove essential nutrients or create an unfavorable environment for microbial growth. Immobilization-based coatings physically trap microorganisms on the surface, preventing their attachment to the food. These coatings significantly impact the shelf life of various perishable foods. Studies have shown that incorporating silver nanoparticles into coatings for meat packaging can extend shelf life by several days by inhibiting bacterial growth [32]. Similarly, coatings containing essential oils have been shown to extend the shelf life of fruits and vegetables by reducing fungal spoilage

Antimicrobial food packaging holds immense potential for addressing food spoilage and enhancing food security. By leveraging nanotechnology and advanced coatings, this technology can create a sustainable solution for preserving food quality, reducing waste, and ensuring safe food for consumers. However, further research is necessary to optimize the efficacy and safety of these materials, considering potential interactions with food components and potential environmental concerns.

13.6 SUSTAINABLE NANO-BASED AGRICULTURE

Traditional application methods for pesticides, herbicides, and fertilizers often lead to significant environmental concerns. Kah et al. [33] have investigated that spraying chemicals can cause pesticides to drift to unintended areas, contaminating soil, water bodies, and non-target plants. This can harm beneficial insects, disrupt ecosystems, and potentially pose health risks to humans and animals. In further study, Sarkar et al. [34] have investigated that excess fertilizers can leach through the soil profile, contaminating groundwater. Similarly, heavy rainfall can cause applied chemicals to runoff into surface water bodies, leading to eutrophication (excessive nutrient levels) and algal blooms. Many conventional application methods result in uneven distribution of chemicals. This can lead to under-application in some areas, allowing pests and weeds to persist, and over-application in others, wasting resources and potentially harming the environment.

13.6.1 Targeted delivery of agrochemicals

Nanotechnology offers a promising solution to these challenges through the development of nanofluid-based encapsulation for targeted delivery of agrochemicals. Nanofluids are suspensions of nanoparticles (particles with at least one dimension in the range of 1–100 nm) in a liquid carrier. By encapsulating agrochemicals within these nanoparticles, the nanoparticles can be designed to release their cargo slowly over time, ensuring a constant and targeted dose to the target pest, weed, or plant. Nanoparticles can improve the penetration of chemicals into plant tissues or target specific insect or weed structures, leading to a more potent effect with lower overall application rates. By minimizing off-target effects, leaching, and runoff, nanofluid-based delivery systems can significantly reduce the environmental footprint of agrochemicals [34–35]. The combined benefits of reduced waste minimized environmental impact and potentially improved efficacy which can contribute significantly to sustainable agricultural practices. Lower application rates translate to lower production costs for farmers and a reduced risk of environmental contamination. Additionally, improved efficacy can lead to better crop yields and reduced crop losses, further enhancing agricultural sustainability.

R&D efforts are ongoing to explore the potential of nanoparticles for applications beyond targeted delivery of conventional agrochemicals. Nanoparticles themselves can be engineered to stimulate plant growth and crop yield. For instance, nanoparticles containing essential nutrients or plant hormones can be delivered directly to plant roots or leaves, promoting growth and enhancing stress tolerance.

13.6.2 Nanofluid-based coatings for irrigation systems

Nanotechnology can also be integrated with irrigation systems to improve water management in agriculture (Figure 13.3). Nanofluid-based coatings can be applied to irrigation canals or pipes to reduce water evaporation during transport. Additionally, these coatings can be designed to release water slowly and directly to plant roots, minimizing runoff and water waste. Water scarcity is a growing challenge in agriculture. By reducing evaporation and ensuring targeted delivery of water to plant roots, nanofluid-based coatings can significantly contribute to water conservation efforts. This not only benefits the environment but also allows farmers to optimize water use and improve their overall water productivity [36].

13.7 CHALLENGES AND CONSIDERATIONS: NANOPARTICLE TOXICITY AND ENVIRONMENTAL IMPACT

The burgeoning field of nanofluidics offers tremendous potential for advancements in the food and agriculture sectors. However, alongside these advancements lie significant challenges concerning nanoparticle toxicity and environmental impact. Nanoparticles, due to their unique properties, can exhibit unforeseen biological interactions. Their miniscule size allows them to infiltrate biological systems, potentially causing damage to cells and tissues. Research suggests that nanoparticles can induce oxidative stress, inflammation, and genotoxicity [37]. The specific toxicity of a nanoparticle depends on various factors like size, shape, surface chemistry, and agglomeration state [38]. Understanding these factors and their influence on biological systems is crucial for responsible development in the food and agriculture domains.

The environmental impact of nanofluids also necessitates careful consideration. The release of nanoparticles into the environment during production, use, or disposal can pose a threat to ecosystems. Nanoparticles can contaminate soil and water bodies, potentially disrupting ecological processes and harming organisms [39–43]. Furthermore, the long-term fate and behavior of nanoparticles in the environment remain poorly understood, making it difficult to assess their full impact.

13.7.1 Responsible development of nanofluidic applications

To harness the potential of nanofluidics while mitigating potential risks, a focus on responsible development is paramount. This necessitates a multipronged approach encompassing research, risk assessment, and life cycle analysis (LCA). In-depth research is required to elucidate the interaction of nanoparticles with food, agricultural products, and biological systems [44–48]. This includes studies on nanoparticle uptake, distribution, metabolism, and excretion to assess potential health risks associated with food consumption or agricultural applications. A comprehensive risk assessment framework should be established to evaluate the potential hazards associated with the use of nanofluids in food and agriculture [49–53]. This framework should consider factors like nanoparticle properties, exposure routes, and potential ecological effects. LCA is a crucial tool for evaluating the environmental impact of nanofluids throughout their life cycle, from production and use to disposal. LCA can help identify potential environmental hot spots and guide the development of more sustainable practices [54–58].

The responsible development of nanofluidic applications necessitates robust regulatory frameworks. Currently, there is a lack of harmonized global regulations governing the use of nanomaterials in food and agriculture. This regulatory gap poses challenges for ensuring the safety and sustainability of nanofluidic applications in different areas from food technology to agriculture to building to earth air heat exchangers [59–62]. Developing effective regulatory frameworks requires collaboration between scientists, policymakers, and stakeholders from the food and agriculture industries. These frameworks should establish clear guidelines for the characterization, safety testing, and labeling of nanofluids intended for use in these sectors. Additionally, regulations should address the environmental impact of nanofluids throughout their life cycle, promoting sustainable production and disposal practices [63–67].

13.8 CONCLUSION

Nanofluidics, the manipulation of fluids at the nanoscale, offers exciting possibilities for transforming food and agriculture. In this domain, nanofluidics can be used for targeted delivery of nutrients, pesticides, or beneficial microbes to plants and animals. Imagine fertilizers precisely delivered to plant roots or medications directly targeting insects harming crops. Additionally, nanofluidics can be employed for developing intelligent food packaging that monitors freshness or detects contaminants. However, for responsible development, addressing potential risks associated with nanofluidics is essential. Nanoparticles, the building blocks of nanofluidics, can have unique properties that might raise concerns about their toxicity to living organisms and their impact on the environment. Thorough research is requiring understanding these potential risks and developing safe nanomaterials. Furthermore, LCA, which evaluates the environmental footprint of nanofluidics throughout its life cycle, and robust regulations are crucial to ensure the sustainable implementation of this technology. By prioritizing these aspects, we can unlock the immense potential of nanofluidics for revolutionizing food and agriculture while safeguarding human health and the environment.

REFERENCES

1. Eastman, J. A., Choi, S. U., Li, S., Grenfell, W. E., & Guo, J. (2001). Anomalously increased effective thermal conductivity of nanotube-based fluids. *Physical Fluids*, 12(7), 1787–1793.
2. Xuan, Y., & Li, Q. (2000). Heat transfer characteristics of nanofluids. *Journal of Heat Transfer*, 122(2), 480–487.

3. Kakac, S., & Pramuanjaroenkij, A. Heat Transfer Enhancement Using Nanofluids. Springer Science & Business Media: Berlin/Heidelberg, Germany, 2009.
4. Das, S. K., & Choi, S. U. (2006). Heat transfer in nanofluids. *Heat Transfer Engineering*, 27(5), 3–19.
5. Dreaden, T. O., Alkilany, A. M., Huang, X., Murphy, C. J., & El-Sayed, M. A. (2012). Delivery of therapeutic agents using dendritic nanocarriers. *Chemical Society Reviews*, 41(7), 2740–2779.
6. Parkin, K. L., & Fairchild, I. A. Microbial Spoilage of Chilled Foods; Springer: Heidelberg, 2007, 1–14.
7. McClements, D. J. (2015). Encapsulation technologies for delivery of bioactive food components. Food Technology and Biotechnology, 53(1), 1–25.
8. Soysal, Y., Moradi, M., Mousavi, S. M., & Khodagholi, F. (2016). Potential applications of nanofluids in the food industry: A review. *Journal of Food Science and Technology*, 53(1), 427–444.
9. Kahrilas, G. A., Tzianno, E. A., & Ziadi, C. E. (2014). Antimicrobial food packaging: A review of recent developments. *Food Chemistry*, 166, 182–189.
10. Aziz, M., Finkelstein, L., & Hussain, M. A. (2018). Nanotechnologies for agriculture and food production: Techniques and potential applications. *Critical Reviews in Food Science and Nutrition*, 58(16), 2761–2781.
11. Chen, H., & Yuan, X. (2016). Nanoparticles for agricultural biotechnology. *Journal of Nanoscience and Nanotechnology*, 16(1), 79–93.
12. Singh, S., Singh, M., Prasad, S., & Shankar, D. (2017). Applications of nanotechnology in agriculture and allied sciences: A review. *Journal of Nanoscience and Nanotechnology*, 17(2), 1825–1848.
13. Liu, R., Li, M., Chen, L., Sun, Y., & Xo, Z. (2010). Applications of controlled release technology in the agriculture industry: A review. *Journal of Controlled Release*, 142(3), 283–294.
14. Tilman, D., Balzer, C., Hill, J., & Befort, B. L. (2001). Global estimates of productive nitrogen and phosphorus in agroecosystems. *PNAS*, 98(22), 12265–12270.
15. Mekonnen, M. M., & Hoekstra, A. Y. (2011). National water footprint accounts – The green, blue and grey water footprint of production and consumption. *Water Resources Management*, 24(11), 1799–1834.
16. Vermeulen, S., Campbell, B. M., & Ingram, J. S. I. (2007). Climate change and food security. *Global Food Security*, 1(1), 7–13.
17. Ingram, J. S. (2011). The role of capital in food security. *Proceedings of the National Academy of Sciences*, 108(50), 19969–19974.
18. Gustavsson, J., Cederberg, C., Sonesson, U., van Otterdijk, R., & Meybeck, A. (2011). Global food losses and food waste. Food and Agriculture Organization of the United Nations. https://www.fao.org/4/mb060e/mb060e00.htm
19. Spielman, D. J., Byerlee, D., & Jayne, T. S. (2011). Does agricultural research and development in Africa pay? *Journal of Development Economics*, 96(2), 220–235.
20. Godfray, H. C. J., Beddington, J. R., Crute, I. R., Haddad, L., Lawrence, D., Muir, J. F., & Toulmin, C. (2010). Food security: The challenge of feeding 9 billion people. *Science*, 327(5967), 812–818.
21. International Labour Organization. (2016). World Employment and Social Outlook - Agriculture: Transforming the workforce for a sustainable future. https://www.ilo.org/sites/default/files/wcmsp5/groups/public/@dgreports/@dcomm/@publ/documents/publication/wcms_795453.pdf
22. Patel, R. C. (2012). Food sovereignty: power, gender, and the right to food. *PLoS Medicine*, 9(6), e1001223. https://doi.org/10.1371/journal.pmed.1001223. Epub 2012 Jun 26. PMID: 22745604; PMCID: PMC3383747.

23. Rahman, M. S., Al-Maskri, A. Y., Khan, M. I., Shahbazullah, S., & Alam, F. (2007). Thermal degradation kinetics of vitamin C in orange juice: A combined effect of temperature and storage time. *Journal of Food Engineering*, 79(3), 996–1002.
24. Fellows, P. Food Processing Technology: Principles and Practices (2nd ed.). Woodhead Publishing: New Delhi, India, 2000.
25. Peng, X., Li, Y., & Xuan, Y. (2010). Enhanced thermal conductivity of water-based Al2O3 nanofluids by ultrasonic vibration. *International Journal of Heat and Mass Transfer*, 53(17–18), 3606–3612.
26. Bilal, M., Iqbal, W., Niazi, M. A., & Li, W. (2021). Recent advancements in the development of food-grade delivery systems for enhancing nutraceutical bioavailability. *Food & Nutrition Research*, 65(1), 8322.
27. Razavi, S. M., Ahmad, N., &Jafari, S. M. (2021). Application of natural biopolymers for encapsulation of food nutraceuticals: A review. *Trends in Food Science & Technology*, 111, 844–859.
28. Wang, X., Shi, G., Fan, S., Ma, J., Yan, Y., Wang, M., ... Zhang, Y. (2023). Targeted delivery of food functional ingredients in precise nutrition: design strategy and application of nutritional intervention. *Critical Reviews in Food Science and Nutrition*, 1–24. https://doi.org/10.1080/10408398.2023.2193275.
29. World Health Organization. (2020). Foodborne diseases. https://www.who.int/health-topics/foodborne-diseases (Accessed 25 April 2024).
30. Li, Q., Wang, C., & Luo, Y. (2013). Applications and toxicologic effects of graphene oxide and graphene-based nanomaterials. *Nanoscale*, 5(12), 5511–5523.
31. Jorge, B.-V., Editor. *Antimicrobial Food Packaging*. Academic Press, 2016, https://doi.org/10.1016/C2013-0-18941-6.
32. Appendini, P., Hotchkiss, J. H., & Sanchez-Plata, M. (2009). Re-emergence of bacteriophage therapy: A reply to eminent bacteriologists. *Clinical Microbiology and Infection*, 15(3), 338–341.
33. Kah, M., & Hofmann, T. (2014). Nanomaterials and nanotechnology for the delivery of agrochemicals: Strategies towards sustainable agriculture. *Journal of Nanobiotechnology*, 12(1), 6.
34. Sarkar, S., Singh, S., Singh, R. P., & Dubey, S. K. (2018). Agricultural runoff and its management practices: A review. *Environmental Chemistry Letters*, 16(4), 1153–1178.
35. Xiaopeng, H., Li, Z., He, X., Xing, B., & Liu, S. (2014). Targeted delivery of pesticides using nanoparticles. *Journal of Agricultural and Food Chemistry*, 62(24), 5878–5889.
36. Chaudhary, R., Jain, N., Tomar, R., Gupta, P., & Baranwal, A. (2020). Nanoparticle-based delivery systems for the improvement of herbicide efficacy. *Journal of Experimental Nanoscience*, 15(1), 1–22.
37. Nel, A. E., Xia, T., Meng, H., & Wang, Z. (2006). Toxicologic potential of materials at the nano level. *Science*, 311(5761), 622–627.
38. Jaideep, P. T., & Kumar, A. CFD Numerical Investigation of Thermal Performance of Diamond Shape Micro Rectangular Heat Exchanger. In Heat Transfer Enhancement Techniques: Thermal Performance, Optimization and Applications; A. Kumar, N. Dutt, M. K. Awasthi, Editors; John Wiley & Sons, Inc., 111 River Street, Hoboken, NJ 07030, USA and Scrivener Publishing LLC, 100 Cummings Center, Suite 541J, Beverly, MA 01915, USA, 2024; Chapter 07, Pages 159–192.
39. Oberdorster, G., Oberdorster, E., &Oberdorster, J. (2005). Nanotoxicology: An emerging discipline evolving from studies of ultrafine particles. *Environmental Health Perspectives*, 113(7), 823–831.
40. Heidari, M., Thangavel, S., Ghafri, K. A., & Kumar, A. Future Trends and Emerging Research in Nano Fluids for Aerospace applications. In Nanofluids Technology for Thermal Sciences and Engineering: Research, Development, and Applications; M. K. Awasthi, N. Dutt, A. Kumar, Editors; CRC Press: Boca Raton, FL, USA, 2024; Chapter 15, Pages 272–291. https://doi.org/10.1201/9781003494454-15.

41. Handy, R. D., Owen, R., & Valsami-Jones, E. (2008). The ecotoxicology of nanoparticles and nanomaterials: Current status. *Ecotoxicology*, 17(4), 312–320.
42. Kumar, A., Singla, Y., & Namboodri, T. Globalization and International Issues in Sustainable Manufacturing. In Sustainability in Smart Manufacturing: Trends, Scope, and Challenges; S. Shah, H. Nautiyal, G. Gugliani, A. Kumar, T. Namboodri, Y.K. Singla Editors; CRC Press: Boca Raton, FL, USA, 2024; Chapter 01; pp. 1–18, http://dx.doi.org/10.1201/9781003467496-1.
43. Awasthi, M. K., Kumar, A., & Dutt, N. Modeling Rayleigh-Taylor Instability in Nanofluid Layers. In Computational Fluid Flow and Heat Transfer: Advances, Design, Control and Applications; M. K. Awasthi, A. Kumar, N. Dutt, and S. Singh Editors; CRC Press: Boca Raton, FL, USA, 2024; Chapter 13, https://doi.org/10.1201/9781003465171-13.
44. Awasthi, M. K., Dutt, N., Kumar, A., & Kumar, S. (2024). Electrohydrodynamic capillary instability of Rivlin–Ericksen viscoelastic fluid film with mass and heat transfer. Heat Transfer, 53(1), 115–133. https://doi.org/10.1002/htj.22944.
45. Awasthi, M. K., Shukla, A. K., Kumar, A., Yadav, D., & Dutt, N. (2024). Viscous correction to the potential flow analysis of Rayleigh-Taylor instability in a Rivlin-Ericksen viscoelastic fluid layer with heat and mass transfer. Heat Transfer. https://onlinelibrary.wiley.com/doi/10.1002/htj.23076.
46. Yadav, D., Awasthi, M. K., Kumar, A., & Dutt, N. (2024). Swirling capillary instability of Rivlin-Ericksen liquid with heat transfer and axial electric field. *Physics*, 6(2), 828–844. https://doi.org/10.3390/physics6020051.
47. Verma, V., Thangavel, S., Dutt, N., Kumar, A., Weerasinghe, R., Editors, *Highly Efficient Thermal Renewable Energy Systems: Design, Optimization and Applications*, Pages 1–341, 2024; CRC Press: Boca Raton, FL, USA, https://doi.org/10.1201/9781003472629.
48. Awasthi, M. K., Dutt, N., & Kumar, A. Editors, "Nanofluids Technology for Thermal Sciences and Engineering: Research, Development, and Applications" CRC Press, Boca Raton, FL, USA, 2024, ISBN 9781032799117. https://doi.org/10.1201/9781003494454.
49. Awasthi, M. K., Kumar, A., & Dutt, N. Recent Innovation, Technological Development and Futuristic Applications of Nanofluids. In Nanofluids Technology for Thermal Sciences and Engineering: Research, Development, and Applications; M. K. Awasthi, N. Dutt, A. Kumar, Editors; CRC Press: Boca Raton, FL, USA, 2024; Chapter 01, pages 1–25. https://doi.org/10.1201/9781003494454-1
50. Kumar, A., Thangavel, S., Awasthi, M. K., Singh, V. P., & Dutt, N.. Safety and Environmental Considerations in Nano Fluid-Based Systems. In Nanofluids Technology for Thermal Sciences and Engineering: Research, Development, and Applications; M. K. Awasthi, N. Dutt, A. Kumar, Editors; CRC Press: Boca Raton, FL, USA, 2024; Chapter 06, pages 99–122. https://doi.org/10.1201/9781003494454-6
51. Verma, V., Thangavel, S., Dutt, N., Kumar, A., & Weerasinghe, R. Editors. Highly Efficient Thermal Renewable Energy Systems: Design, Design, Optimization and Applications; CRC Press: Boca Raton, FL, USA, 2024; pp. 1–341, https://doi.org/10.1201/9781003472629.
52. Awasthi, M. K., Kumar, A., Dutt, N., & Singh, S. Editors, Computational Fluid Flow and Heat Transfer: Advances, Design, Control and Applications; CRC Press: Boca Raton, FL, USA, 2024, https://doi.org/10.1201/9781003465171.
53. Awasthi, M. K., Dutt, N., Kumar, A., & Hedau, A. Introduction to Mathematical and Computational Methods. In Computational Fluid Flow and Heat Transfer: Advances, Design, Control and Applications; M. K. Awasthi, A. Kumar, N. Dutt, and S. Singh Editors; CRC Press: Boca Raton, FL, USA, 2024; Chapter 01, https://doi.org/10.1201/9781003465171-1

54. Kumar, A., Dutt, N., & Awasthi, M. K., Editor, *Heat Transfer Enhancement Techniques: Thermal Performance, Optimization and Applications*, John Wiley & Sons, Inc., 111 River Street, Hoboken, NJ 07030, USA and Scrivener Publishing LLC, 100 Cummings Center, Suite 541J, Beverly, MA 01915, USA, 2024.
55. Kumar, A., Awasthi, M. K., Dutt, N., & Singh, V. P.. Recent Innovation in Heat Transfer Enhancement Techniques. In *Heat Transfer Enhancement Techniques: Thermal Performance, Optimization and Applications*; A. Kumar, N. Dutt, M. K. Awasthi, Editors; ; John Wiley & Sons, Inc., 111 River Street, Hoboken, NJ 07030, USA and Scrivener Publishing LLC, 100 Cummings Center, Suite 541J, Beverly, MA 01915, USA, 2024; Chapter 01, pages 1–38.
56. Dutt, N., Hedau, A., Kumar, A., Awasthi, M. K., Hedau, S., & Meena, C. S. (2024). Thermo-hydraulic performance investigation of solar air heater duct having staggered D-shaped ribs: Numerical approach. *Heat Transfer*. https://onlinelibrary.wiley.com/doi/10.1002/htj.22998.
57. Dewangan, A. K., Moinuddin, S. Q., Cheepu, M., Sajjan, S. K., Kumar, A. Thermal Energy Storage: Opportunities, Challenges and Future Scope. In Thermal Energy Systems: Design, Computational Techniques and Applications; Kumar, A.; Singh, V.P.; Meena, C.S.; Dutt, N.; Editors; CRC Press: Boca Raton, FL, USA, 2023; Chapter 2; pp. 17–28. http://dx.doi.org/10.1201/9781003395768-2.
58. Ahmadizadeh, M., Heidari, M., Thangavel, S., Naamani, E. A., Khashehchi, M., Verma, V., & Kumar, A.. Technological Advancements in Sustainable and Renewable Solar Energy Systems. In Highly Efficient Thermal Renewable Energy Systems: Design, Design, Optimization and Applications; V. Verma, S. Thangavel, N. Dutt, A. Kumar, R. Weerasinghe Editors; CRC Press: Boca Raton, FL, USA, 2024; Chapter 02; pp. 23–39, https://doi.org/10.1201/9781003472629-2.
59. Dutt, N., Hedau, A. J., Kumar, A., Kumar Awasthi, M., Singh, V. P., & Dwivedi, G. (2023). Thermo-hydraulic performance of solar air heater having discrete D-shaped ribs as artificial roughness. Environmental Science and Pollution Research. https://doi.org/10.1007/s11356-023-28247-9.
60. Reddy, P. N., Verma, V., Kumar, A., & Awasthi, M. (2023). CFD Simulation and Thermal Performance Optimization of Flow in a Channel with Multiple Baffles. Journal of Heat and Mass Transfer Research, 10(2), 257–268. https://doi.org/10.22075/JHMTR.2023.31108.1458.
61. Singh, B., Kaur, P., Yadav, A. K., Awasthi, M. K., & Kumar, A. Design and Analysis of Solar Tracking System for PV Thermal Performance Enhancement. In Heat Transfer Enhancement Techniques: Thermal Performance, Optimization and Applications; A. Kumar, N. Dutt, M. K. Awasthi, Editors; John Wiley and Sons: USA, 2024; Chapter 11, pages 251–268.
62. Verma, V., Thangavel, S., Dutt, N., Kumar, A., & Weerasinghe, R. Introduction to Thermal Energy Storage: Solar, Geothermal and Hydrogen Energy. In Highly Efficient Thermal Renewable Energy Systems: Design, Design, Optimization and Applications; V. Verma, S. Thangavel, N. Dutt, A. Kumar, R. Weerasinghe Editors; CRC Press: Boca Raton, FL, USA, 2024; Chapter 01; pp. 1–22, https://doi.org/10.1201/9781003472629-1.
63. Chitt, M., Thangavel, S., Verma, V., & Kumar, A. Green Hydrogen Productions: Methods, Designs and Smart Applications. In Highly Efficient Thermal Renewable Energy Systems: Design, Design, Optimization and Applications; V. Verma, S. Thangavel, N. Dutt, A. Kumar, R. Weerasinghe Editors; CRC Press: Boca Raton, FL, USA, 2024; Chapter 16; pp. 261–276, https://doi.org/10.1201/9781003472629-16.
64. Singh, V. P., Kumar, A., & Awasthi, M. K.. Augmentation of Solar, Geothermal and Earth-Air Heat Exchanger in Sustainable Buildings. In Heat Transfer Enhancement Techniques: Thermal Performance, Optimization and Applications;

A. Kumar, N. Dutt, M. K. Awasthi, Editors; ; John Wiley & Sons, Inc., 111 River Street, Hoboken, NJ 07030, USA and Scrivener Publishing LLC, 100 Cummings Center, Suite 541J, Beverly, MA 01915, USA, Chapter 06, pages 119–158.

65. Chitt, M., Thangavel, S., Kumar, A., Singh, V. P., Meena, C. S., Building Performance Prediction and Optimization. In *Sustainable Technologies for Energy Efficient Buildings*; C. S. Meena, A. Kumar, V. P. Singh, A. Ghosh, Editors; chapter 11, pages 241–257; CRC Press: Boca Raton FL, https://doi.org/10.1201/9781003496656-11.
66. Khashehchi, M., Thangavel, S., Rahmanivahid, P., Heidari, M., Moazzeni, T., Verma, V., & Kumar, A. Solar Desalination Techniques: Challenges and Opportunities. In Highly Efficient Thermal Renewable Energy Systems: Design, Design, Optimization and Applications; V. Verma, S. Thangavel, N. Dutt, A. Kumar, R. Weerasinghe Editors; CRC Press: Boca Raton, FL, USA, 2024; Chapter 19; pp. 305–329, https://doi.org/10.1201/9781003472629-19.
67. Dutt, N., Hedau, A., Kumar, A., Awasthi, M. K., & Singh, V. P. (2024). Effect of Duct Length Variation on Solar Air Heater Performance for Smooth and D-Shaped Roughened Absorber Plate. Heat Transfer, 1–29. https://onlinelibrary.wiley.com/doi/10.1002/htj.23115.

Index

For Product Safety Concerns and Information please contact our EU representative GPSR@taylorandfrancis.com
Taylor & Francis Verlag GmbH, Kaufingerstraße 24, 80331 München, Germany

www.ingramcontent.com/pod-product-compliance
Ingram Content Group UK Ltd.
Pitfield, Milton Keynes, MK11 3LW, UK
UKHW020011200926
475702UK00014B/1824

* 9 7 8 1 0 3 2 8 6 3 7 3 3 *